Urban Forests, Trees, and Greenspace

Urban forests, trees, and greenspace are critical in contemporary planning and development of the city. Their study is not only a question of the growth and conservation of greenspaces, but also has social, cultural, and psychological dimensions. This book brings a perspective of political ecology to the complexities of urban trees and forests through three themes: human agency in urban forests and greenspace; arboreal and greenspace agency in the urban landscape; and actions and interventions in the urban forest.

Contributors include leading authorities from North America and Europe from a range of disciplines, including forestry, ecology, geography, landscape design, municipal planning, environmental policy, and environmental history.

L. Anders Sandberg is Professor and former Associate Dean in the Faculty of Environmental Studies at York University, Toronto, Canada. His two most recent books are *The Oak Ridges Moraine Battles: Development, Sprawl and Nature Conservation in the Toronto Region* (2013) and *Urban Explorations: Environmental Histories of the Toronto Region* (2013).

Adrina Bardekjian is a PhD candidate at the Faculty of Environmental Studies at York University, Toronto, Canada, where she studies under-represented narratives and strategic visioning for urban forestry praxis. She is an urban forestry researcher, writer, and educator, and works with a number of organizations on diverse projects and initiatives.

Sadia Butt is a PhD candidate at the Faculty of Forestry at the University of Toronto, Canada. She has worked in urban forestry for the last 15 years as a practitioner, researcher, and volunteer in raising urban forest awareness through environmental education.

"This book is the first to use the lens of political ecology to understand urban forests and finally provide us with a much needed politicized view of these spaces, based on case studies from around the globe."

– *Guy Baeten, Professor of Human Geography, Lund University, Sweden*

"This wonderful volume incorporates new theoretical ground in the field of political ecology. Diverse chapters bring nuanced and balanced attention to the social and ecological dynamism that shapes the myriad ways through which urban forests constitute and are constituted by everyday lived experiences, politics, and institutional dynamics of today's cities."

– *Patrick Hurley, Associate Professor and Chair, Environmental Studies, Ursinus College, Collegeville, PA, USA*

"A collection that makes urban political ecology accessible to a wide readership without losing the sophistication needed to handle the complexity and contradictions of the field. Its compelling narratives stretch from unruly trees and people in diverse cities around the world, to original ways of thinking about nature and the city."

– *Alan Mabin, Research Fellow, Capital Cities project, University of Pretoria, South Africa*

Urban Forests, Trees, and Greenspace

A Political Ecology Perspective

Edited by
L. Anders Sandberg,
Adrina Bardekjian,
and Sadia Butt

First published 2015
by Routledge
2 Park Square, Milton Park, Abingdon, Oxfordshire OX14 4RN

and
by Routledge
711 Third Avenue, New York, NY 10017

First issued in paperback 2016

Routledge is an imprint of the Taylor & Francis Group, an informa business

British Library Cataloguing-in-Publication Data
A catalogue record for this book is available from the British Library

Library of Congress Cataloging-in-Publication Data
Urban forests, trees, and greenspace : a political ecology perspective / edited by L. Anders Sandberg, Adrina Bardekjian, and Sadia Butt.
pages cm
Includes bibliographical references and index.
1. Urban forestry. 2. Urban landscape architecture. 3. City planning. I. Sandberg, L. Anders, 1953– editor of compilation. II. Bardekjian, Adrina, editor of compilation. III. Butt, Sadia, editor of compilation.
SB436.U735 2014
635.9′77–dc23
2014003126

ISBN 13: 978-1-138-28257-5 (pbk)
ISBN 13: 978-0-415-71410-5 (hbk)

Typeset in Baskerville
by Keystroke, Station Road, Codsall, Wolverhampton

Contents

List of Illustrations ix
List of Contributors xiii
Preface xvii

1 **Introduction** 1
L. ANDERS SANDBERG, ADRINA BARDEKJIAN, AND SADIA BUTT

PART 1
Human Agency in Urban Forests and Greenspace 17

2 **Urban Forests are Social Natures: Markets, Race, Class, and Gender in Relation to (Un)Just Urban Environments** 19
HAROLD PERKINS

3 **From Government to Governance: Contribution to the Political Ecology of Urban Forestry** 35
CECIL C. KONIJNENDIJK VAN DEN BOSCH

4 **A Genealogy of Urban Forest Discourse in Flanders** 47
ANN VAN HERZELE

5 **Institutions, Law, and the Political Ecology of Urban Forests: A Comparative Approach** 61
BLAKE HUDSON

6 **Manufacturing Green Consensus: Urban Greenspace Governance in Singapore** 77
NATALIE MARIE GULSRUD AND CAN-SENG OOI

7 The Places of Trees in Honduras: Contributions of Public Spaces and Smallholders 93
J. O. JOBY BASS

PART 2
Arboreal and Greenspace Agency in the Urban Landscape 109

8 (Urban) Places of Trees: Affective Embodiment, Politics, Identity, and Materiality 111
OWAIN JONES

9 Order and Disorder in the Urban Forest: A Foucauldian–Latourian Perspective 132
IRUS BRAVERMAN

10 Four Arboricultures of the Tokyo Metropolis: High and Low, West and East, From Edo to 2020 147
JAY BOLTHOUSE

11 The Unruly Tree: Stories from the Archives 162
JOANNA DEAN

12 Seeking Citizenship: The Norway Maple (*Acer platanoides*) in Canada 176
BRENDON M. H. LARSON

13 Queering the Urban Forest: Invasions, Mutualisms, and Eco-Political Creativity with the Tree of Heaven (*Ailanthus altissima*) 191
DARREN PATRICK

14 The Thin End of the Green Wedge: Berlin's Planned and Unplanned Urban Landscapes 207
CYNTHIA IMOGEN HAMMOND

PART 3
Actions and Interventions in the Urban Forest 225

15 "A Few Trees" in Gezi Park: Resisting the Spatial Politics of Neoliberalism in Turkey 227
BENGI AKBULUT

16 Constructing New York City's Urban Forest: The Politics and Governance of the MillionTreesNYC Campaign 242
LINDSAY K. CAMPBELL

17 Reimagining Ecology in the City of Cape Town: Contemporary Urban Ecological Research and the Role of the African Centre for Cities 261
PIPPIN ANDERSON

18 Cultivating Citizen Stewards: Lessons from Formal and Non-Formal Educators 277
GREGORY SMITH

19 Learning and Acting through Participatory Landscape Planning: The Case of the Bräkne River Valley, Sweden 292
HELENA MELLQVIST AND ROLAND GUSTAVSSON

20 Art, Enchantment, and the Urban Forest: A Step, a Stitch, a Sense of Self 307
KATHLEEN VAUGHAN

Index 322

List of Illustrations

Figures

2.1 Tall and proud. Early springtime view of vigorous and manicured street trees in Milwaukee's affluent East Side neighborhood, USA. 23
2.2 Under the wire. Early springtime view of small and mangled street trees in Milwaukee's impoverished Harambee neighborhood. 24
2.3 Arboricide. Wintertime view of a recently planted tree struck down by vandals in Milwaukee's impoverished Lynden Hill neighborhood. 27
3.1 Visualization of the four dimensions of the Policy Arrangement Approach. 38
3.2 The urban forest as 'powerscape': Boboli Gardens in Florence, Italy. 39
3.3 Management of Central Park in New York City, USA, represents a 'governance with government' approach, with the city's park department and the private Central Park Conservancy sharing responsibility. 41
3.4 Stockholm's Urban National Park represents a complex governance arrangement. 41
4.1 "The close-to-nature forestry" concept—influential in discourse on forest expansion—uses natural processes in primeval forests as a prime source of inspiration. 50
4.2 The forest image was changed from one massive entity to an open concept of interacting land uses. 55
4.3 Farmers' protest against the Park Forest project: "Where may I graze now?" 57
5.1 Forests currently on the outskirts of urban areas will be the urban forests of tomorrow, and should be managed with that potentiality in mind. 63
5.2 "Local" forest in south Alabama, USA. 67
5.3 Forested watershed in the southern part of the US state of Alabama. 68
6.1 Supertree Grove by day. Gardens by the Bay, Singapore. 78
6.2 Tourists in front of green wall and waterfall. 85
6.3 Citizens engaged in community gardening, Singapore. 86

7.1 Santa Lucia, Francisco Morazan, former mining center north-east of Tegucigalpa, Honduras. 96
7.2 Llano Yarula, La Paz, Bolivia. 98
7.3 El Chichicaste, El Paraiso, Honduras. 99
8.1 The giant beech tree with missing half of canopy. 113
8.2 Newly planted tree in Munich, Germany. Swathed in fabric and supported with multiple cables. 123
8.3 The verticality of many tree species makes them an ideal spatial medium for city streets (Munich). 124
8.4 The Party in the Square within the spaces formed by the trees. 127
9.1 The natural grid, Buffalo, NY, USA. 136
9.2 Trees, grates, and other street furniture, Cambridge, MA, USA. 137
9.3 Dig-Safe, Ithaca, NY, USA. 141
10.1 Map of central Tokyo, Japan, depicting sites mentioned in the text. 148
10.2 Four images of Tokyo's High City arboricultures. 150
10.3 Four images of Tokyo's Low City arboricultures. 152
10.4 Four images of the arboricultures of Tokyo's western periphery. 155
10.5 Four images of the arboricultures of Tokyo's eastern periphery. 158
11.1 Child planting a tree in an Ottawa suburb to celebrate Canada's Centennial, 1967, Ottawa, ON. 168
11.2 Ottawa's Central Park looking east, 1920s. 170
12.1 An old-growth sugar maple (*Acer saccharum*) forest in southern Ontario, Canada. 176
12.2 Canada's $20 polymer banknote, which was released on November 7, 2012. 177
12.3 The contrasting leaves and samaras of sugar maple and Norway maple. 182
12.4 Forty-year old Norway maple (*Acer platanoides*) trees along a residential street in Guelph, Ontario, Canada. 184
13.1 *Ailanthus*. 30th Street and 11th Ave, Manhattan, NY, USA, 2002 (removed 2002). 197
13.2 *Ailanthus altissima* on Detroit Tree of Heaven Woodshop's tree farm as pictured for an exhibition at the SMART Museum of Art (Chicago, IL, USA). 200
13.3 *Tehnică pentru o viaţă* (Technique for a life time). 202
14.1 Design for green roof, Canadian Embassy. 17 Leipziger Platz, Berlin, Germany (completed 2005). 208
14.2 Canadian Embassy (middle-right) at Leipziger Platz, Berlin, with the towers of Potsdamer Platz to the left. 215
14.3 The green roof of the Canadian embassy in Berlin, 2010. 219
15.1 Protestors in Gezi Park, Istanbul, Turkey, on the third day of protests. 233
15.2 Barricades on a main artery leading to Taksim Square, Istanbul. 235
15.3 Gezi Park Library, Istanbul. 236
16.1 Launching the MillionTreesNYC campaign as a public–private partnership. 249

16.2 Tree stewardship in East New York, Brooklyn, NY, USA. 251
16.3 Volunteer stewards at a MillionTreesNYC reforestation planting. 254
17.1 The flora of Cape Town, South Africa, is predominantly Fynbos, a low and scrubby vegetation type. 265
17.2 The Green Point Urban Park, constructed as a soccer world cup legacy project, presents a significant urban greenspace in post-apartheid Cape Town. 269
17.3 Cape Town is renowned for its exceptional plant diversity, and views around biodiversity vary among the populace. 271
18.1 Urban Ecology Center, Riverside Park, Milwaukee, WI, USA. Urban Ecology Center, 2013. 280
18.2 Student-built and designed Peace Park, Egleston Square, Boston, MA, USA. 284
18.3 Student-written and illustrated land use history of Graham Oaks Nature Park, Wilsonville, OR, USA. 287
19.1 Old stone bridge south of Bräkne-Hoby, Sweden, the main village along the valley. 297
19.2 Landscape change at Örseryd, Sweden. 298
19.3 A combined flour and saw mill at Björstorp, Sweden, 2 km north of the coast line. 298
20.1 *Nel mezzo del cammin: Bois Summit/Summit Woods* (2013), digital and hand embroidery on textile assemblage, by Kathleen Vaughan. 308
20.2 *Nel mezzo del cammin: Bois Summit/Summit Woods* (2013), detail, by Kathleen Vaughan. 311
20.3 *Nel mezzo del cammin: Bois Summit/Summit Woods* (2013), detail, by Kathleen Vaughan. 315

Table

16.1 DPR's PlaNYC funding. 247

Contributors

Bengi Akbulut is an Istanbul-based political economist and activist whose joint and independent work has appeared in the *Cambridge Journal of Economics* and *Development and Change* among others. She holds a PhD from the University of Massachusetts, Amherst, USA, and her research focuses broadly on political economy of development, including issues of political ecology, agrarian and environmental change, state-society relationships, non-capitalist economies, social and environmental movements, and gender and household work.

Pippin Anderson is Lecturer with the Department of Environmental and Geographical Science, and runs the Urban Ecology CityLab at the African Centre for Cities. She is based at the University of Cape Town in South Africa and has a research interest in the ecology of peopled landscapes.

Adrina Bardekjian is a PhD candidate at the Faculty of Environmental Studies at York University, Toronto, Canada, where she studies under-represented narratives and strategic visioning for urban forestry praxis. She is an urban forestry researcher, writer, and educator, and works with a number of organizations on diverse projects and initiatives.

J. O. Joby Bass is Associate Professor of Geography at the University of Southern Mississippi in Hattiesburg, USA. A cultural geographer with interests in human and human–environment relationships and the landscapes they create, he has focused much of his research on the use of repeat photography in Latin America.

Jay Bolthouse is a PhD candidate at the Department of Natural Environmental Studies at the University of Tokyo, Japan. His work focuses on the constantly transforming relationship between trees and cities from the perspectives of historical geography and political ecology.

Irus Braverman is Professor of Law and Adjunct Professor of Geography at SUNY Buffalo, USA.

Sadia Butt is a PhD candidate at the Faculty of Forestry at the University of Toronto, Canada. She has worked in urban forestry for the last 15 years as a practitioner, researcher, and volunteer in raising urban forest awareness through environmental education.

Lindsay K. Campbell is a research social scientist with the USDA Forest Service Northern Research Station, who has worked for 12 years out of the New York City Urban Field Station. She holds a PhD in geography from Rutgers University, USA, and her research focuses on urban stewardship, sustainability, and environmental policymaking.

Joanna Dean is Associate Professor of History at Carleton University, Ottawa, Canada. She teaches Canadian environmental history and conducts research into the history of urban trees.

Natalie Marie Gulsrud is a PhD candidate in the Department of Parks and Urban Landscapes at the Danish Center for Forest, Landscape, and Planning at the University of Copenhagen, Denmark. She has worked for over 5 years in the public health sector and her main research interest relates to the crossover between urban sustainability policy, green city marketing, and place making.

Roland Gustavsson is Professor at the Department for Landscape Architecture, Planning and Management at the Swedish University of Agricultural Sciences, Alnarp, Sweden. In his book *The New Landscape* (1994) he tries to bridge rural and urban cultures, taking the reader from a deepened understanding of landscape and landscape management to a series of solutions to deal with change.

Cynthia Imogen Hammond is Associate Professor and Chair of the Department of Art History, Concordia University, Montreal, Canada, where she teaches interdisciplinary, collaborative approaches to the study of urban landscapes and architectural history. She is the author of the book *Architects, Angels, Activists and the City of Bath, 1765–1965* (2012).

Blake Hudson holds a joint appointment as Associate Professor at the Law Center and the School of the Coast and Environment at Lousiana State University, USA, where he teaches natural resources and environmental law. He recently authored the book *Constitutions and the Commons: The Impact of Federalist Structures on Local, National and Global Resource Governance* (2014).

Owain Jones is Professor of Environmental Humanities at the School of Humanities and Cultural Industries at Bath Spa University, UK. Funded by a number of academic research projects he has studied and written upon various aspects of nature–society relations, including trees and forests; water, tides, and communities; and intersecting processes of memory, affect, and agency.

Cecil C. Konijnendijk van den Bosch heads the landscape department at the Swedish University of Agricultural Sciences, Alnarp, Sweden, and is also a part-time professor of greenspace management at the University of Copenhagen, Denmark. He has authored books such as *The Forest and the City—The Cultural Landscape of Urban Woodland* (2008) and is editor-in-chief of the journal *Urban Forestry & Urban Greening*.

Brendon M.H. Larson is Associate Professor in the Faculty of Environment at the University of Waterloo, Canada. He speaks and writes widely on perceptions of conservation issues, and recently wrote *Metaphors for Environmental Sustainability: Redefining our Relationship with Nature* (2011).

Helena Mellqvist is a PhD student at the Department for Landscape Architecture, Planning and Management at the Swedish University of Agricultural Sciences, Alnarp, Sweden. During the past 10 years, she has worked as a landscape architect developing a great interest and knowledge in fostering dialogue between people with different backgrounds and perspectives using the landscape as shared basis.

Can-Seng Ooi is the Director of the Center for Leisure and Culture Services Research and Professor at the Copenhagen Business School, Denmark. He has been conducting sociological research on various aspects of Singapore society for more than two decades.

Darren Patrick is a PhD Candidate and Ontario Trillium Scholar in the Faculty of Environmental Studies at York University, Toronto, Canada, whose research brings queer studies/theory and urban studies/theory together to tell complex ecological stories. He has recently published in *Social and Cultural Geography* and has served as the Academic Reviews Editor for a special issue of *Undercurrents: Journal of Critical Environmental Studies.*

Harold Perkins is Associate Professor in the Department of Geography at Ohio University, USA. He is an urban political ecologist who studies the governance of forests, parks, and waterways.

L. Anders Sandberg is a Professor and former Associate Dean in the Faculty of Environmental Studies at York University, Toronto, Canada. His two most recent books are *The Oak Ridges Moraine Battles: Development, Sprawl and Nature Conservation in the Toronto Region* (2013) and *Urban Explorations: Environmental Histories of the Toronto Region* (2013).

Gregory Smith is a Professor in the Graduate School of Education and Counseling at Lewis & Clark College in Portland, Oregon, USA. He writes and speaks about integrating place and community learning into the curriculum as a means for advancing education for sustainability.

Ann Van Herzele works as Senior Researcher at the Nature & Society group, Research Institute for Nature and Forest (INBO), Brussels, Belgium. She has long-standing experience both in practice and research in the field of communication and participation in environmental planning and management.

Kathleen Vaughan is an artist and Associate Professor of Art Education at Concordia University in Montreal, Canada. Her work about place, home, and belonging has an increasingly environmental orientation.

Preface

This book started as a conference idea three years ago and culminated in the conference "Urban Forests and Political Ecologies: Celebrating Transdisciplinarity," (UFPE), held on April 18–20, 2013, at the University of Toronto. The conference was a huge success where academics, professionals, environmentalists and artists mingled to present and discuss their work on urban forests, trees and greenspace.

We would like to thank TD Friends for the Environment Foundation as the main sponsor of the conference. Its public outreach mandate was met wonderfully by the diverse attendees in the practicing, professional and artistic communities.

We would also like to thank the three conference partners: the Faculty of Forestry at the University of Toronto, the Faculty of Environmental Studies at York University and the Humber Arboretum and Centre for Urban Ecology. In particular, we would like to acknowledge Amalia Veneziano and the conference organizing committee for supporting the initial idea for the UFPE conference, the volunteers and our art curator, Catherine Campbell, for their dedication to this unique event.

Sponsors from the University of Toronto included: the Faculty of Forestry, the Forestry Alumni Class of 1954 and the Department of Fine Art. Sponsors from York University included: the Faculty of Environmental Studies, the Vice-President Academic and Provost Rhonda Lenton, Vice-President Research and Innovation Robert Haché, Associate Vice-President Research Lisa Phillips, Canada Research Chair in Sustainability and Culture, the CITY Institute, York University Libraries and Osgoode Hall Law School. Other sponsors included: the International Union of Forest Research Organizations, the International Society of Arboriculture Ontario Chapter, the Network in Canadian History and the Environment, the Gosling Research Institute for Plant Preservation at the University of Guelph, DAVEY Resource Group, Tree Canada, the Ontario Urban Forest Council, Trees Ontario, the Toronto Cancer Prevention Coalition, the Town of Oakville and Dr. Robert Bateman.

We also want to thank the many participants of the conference who stimulated the conversation around urban forests. Some of the participants at the conference have developed their talks into chapters for this book. But there are also many other contributors who we invited subsequent to the conference in order to broaden the geographical scope and theoretical depth of the book.

In addition to our own reviews of the contributors' chapters, we have also relied on external reviewers and some of the contributors for critical assessments. We would like to thank the following for their assistance: Jay Bolthouse, Matthew Gandy, Hilary Inwood, Owain Jones, Ute Lehrer, Terry G. McGee, Harold Perkins, Cate Sandilands, Lisa Wallace and Gerda Wekerle. We would also like to acknowledge all the other participants at the conference who presented papers and artwork and contributed to the discussions, especially Roger Keil, Professor at the Faculty of Environmental Studies at York University and at the time Director of the CITY Institute, who provided an extensive end-of-the-conference commentary on the two-day proceedings.

Tim Hardwick and Ashley Wright at Earthscan showed tremendous confidence and support for our efforts and the anonymous reviewers of the initial book proposal provided sober advice and great encouragement. The collaboration of all the above individuals for the conference, as well as the interest and support by Earthscan, have made this book possible.

L. Anders Sandberg would like to express a thank you to his wife Maria Legerstee for her support and inspiration. He would like thank his co-editors for their enthusiasm and efforts to make both the conference and this book a reality.

Adrina Bardekjian would like to thank her partner, Chant Asaduryan, for his continual encouragement and creative insights. She would also like to express a sincere thanks to her co-editors, the contributing authors and conference participants for their inspiration and dedication to this work.

Sadia Butt is grateful for the opportunity to have worked with her co-editors, conference presenters and contributing authors. Their work continues to be both a source of knowledge and inspiration for future paths.

1 Introduction

L. Anders Sandberg, Adrina Bardekjian, and Sadia Butt

> The world is networked, and always has been, but it is NOT flat, not socially and not ecologically. Neither is it a simple pyramid of predictable power with cybernetic power from above; we have always also had self-organization from below, . . . The challenge is to mesh social, ecological and technological domains in theories and models of rooted networks, relational webs and self-organized assemblages, all laced with power, and linked to territories across scale.
>
> (Rocheleau and Roth, 2007: 436)

The oft-quoted statement by noted forester Jack Westoby that "forestry is not about trees, it is about people" is especially apt when considering urban forests, trees, and greenspace. Green areas are a function and product of the decisions and passions of people who manage, speak for, and often live close to them, including developers, forest experts, politicians, residents (be they homeowners, renters, squatters or the homeless), and environmentalists.

Urban forests and greenspace are clearly embraced by many cities. Witness Central Park in New York City, Tiergarten in Berlin, the Meiji Shrine in Tokyo, Table Mountain in Cape Town, and the Gardens by the Bay in Singapore as evidence of the presence of urban parks in some of the largest cities of the world. These urban forests and green areas are recreational havens for local residents, attractions to tourists, even magnets for development and capitalist growth, and city icons that are well known across the world.

But urban forests, trees, and greenspace are not always embraced in all places, at all times, in similar ways, by all people. Green areas are often compromised in the quest for capitalist growth. Sometimes the destruction of such spaces is deliberate, a function of city builders and planners making conscious decisions to sacrifice trees to provide room for urban housing subdivisions, infrastructure, and industrial and commercial buildings. In other cases, greenspaces are occupied by rural to urban migrants who occupy land "illegally", and use trees for building material and fuel wood in a desperate attempt to create and build a new life in the city.

Urban forests and trees can also elicit passions and conflicts among individuals. There are tensions between different visions of what urban forests, trees, and greenspace should look like. A shade tree for one person may constitute a blocked view for another. A gnarly or old and crumbling tree may be a source of admiration and beauty for some but a safety hazard to others. Urban forests and trees are also often unruly and unpredictable agents that resist the act of being managed. Trees contract diseases, attract insects, fall down during storms, and shed leaves, debris, and fruit that upset urban residents. People may prefer trees for aesthetic reasons, others may prefer utility trees that create shade or bear fruit, while still others do not like trees at all because of the work related to the clean-up of falling fruits, seeds, and leaves or because trees shade their vegetable gardens. Professional foresters and arborists have different preferences too. Some like native trees because of their connections to local ecologies and natural and cultural histories, others favour exotics because of their ease of growth and management.

Urban forestry or the science, culture, and art of managing trees—these practices suggest—is a politics that changes in space, scale, and time. In this book, we use a political ecology perspective to address some of the political issues in urban forest and greenspace management. Political ecology we define as the study of the dynamics of networks surrounding the historical and socio-economic context and political decisions, controversies, actions, and material and ecological processes that occur in and around greenspace and trees in the city (Rocheleau, 2008). One of the central premises of political ecology is that the concept of power exerts considerable control over the shape and distribution of urban forests. Such power comes in many shapes. Elected politicians do play a role but, equally importantly, political ecologists typically consider the social and economic power of wealthy and propertied institutions, groups, and individuals, as well as the discursive power of experts as relevant in setting the policy agenda and framing the debates around urban forests. Such frames can silence and marginalize the voices and interests of other groups, both humans and non-humans. But there are also resistances, contingencies, and challenges that emanate from marginalized human and non-human actors and it is in the interactions of these networks that the urban forest and greenspaces are formed.

In the following, we elaborate further our conception of the political ecology perspective on urban forests, trees, and greenspace. But before we do so, we present a critique of some of the current dominant approaches to the topic so as to situate our own approach and what we seek to contribute to the field.

Dominant Themes in Urban Forestry

Currently, almost 50 per cent of the world's population inhabits 3 per cent of the earth's land cover that is urbanized, making urban living an important subject for discussion and research (Singh et al., 2010). Parallel to this situation, there has been a steady growth in urban forestry and ecology research (Carreiro and Zipperer, 2011).

Urban forestry as a field of practice and research was well established by the 1980s in both North America and Europe (Konijnendijk et al., 2006). But research in developing countries, which have an average 44 per cent of their populations in urban areas, is on the rise. Urban forestry studies in countries such as India, the Philippines, Bolivia, and Brazil have been investigating diverse management issues, identifying many similar factors that impact urban forest management and conservation throughout the world (Andersson, 2003; Singh et al., 2010). In the US, urban and community forestry as a field and concept is further advanced due to the inclusion of a formalized urban forestry program at the federal and state levels. However, the top-down structure is not prevalent in practice in other nations. Research that has been accumulated under Tree US Aid projects has instigated exploration in this area that seeks to define an urban approach to community forestry, which has largely been studied through a livelihood approach.

Yet it is seldom recognized that such areas are political, that is, the product of human-based decisions that are not written in stone but that are open to interpretation and challenge. Instead, the urban forest is largely the unquestioned preserve and reserve of urban tree experts and professionals. The apolitical nature of urban forest management is expressed in several ways.

First, urban forest and greenspace are often seen as unidimensional, as providers of ecological, economic and health services for city populations (Salzman, 2006; Poudyal et al., 2012). From this perspective, urban trees constitute a part of the terrestrial lungs of the world that sequester carbon emissions in an era of climate change. They also offer an urban aesthetic that boosts property values of homes and whole neighbourhoods; cool residents by providing shade in the summer and wind breaks and warmth in the winter; stabilize soils and absorb water during storms; improve air and water quality; furnish urban wildlife with habitat and sustenance (Adams, 2005); serve as attenuators of urban noise, heat island mitigators, and biodiversity vessels; improve human health and well-being; and constitute a recreational haven for city dwellers in an otherwise artificial and often alienating urban environment. It is seldom, however, that trees and forests are considered as having non-monetary value or values unto themselves. And though the ecological services above are increasingly considered from a distributional perspective, that is, from the viewpoint of who may benefit more or less from these services (Van Herzele et al., 2005), urban trees are still presented as unproblematic public goods rather than as private goods that are bought, sold, and distributed unevenly across urban space.

A second problem that is related to dominant urban forest practice and research is the focus on developing, rather than questioning, the techniques and technologies used to document and evaluate the ecological services provided by trees and greenspace. This includes the use of Geographic Information System data to calculate how much urban area can be reforested, the carbon sink potential of the urban forest, and the many other ecological and further services provided by urban trees. One example is the United States Department of Agriculture (USDA) Forest Service's i-Tree software program, a computer program that estimates the economic value of the urban forest and is marketed globally as a management tool for urban forest managers. There are also modelling techniques that record the benefits of the urban forest climatologically, especially as an asset to municipal infrastructure (McPherson et al., 1997; Nowak et al., 2002; Millward and Sabir,

2010). Such universal technologies and techniques risk homogenizing urban forest ecologies and functions thereby neglecting the uniqueness of local forests and trees and the local values, knowledge, and management practices that are associated with them.

A third problem with urban forest practices and research is that they are based on pluralist assumptions of governance that are largely uncritical of present power structures. Pluralism assumes that all stakeholders in current governance structures have, or can have, an equal voice. The objective for pluralists is, therefore, to seek to incorporate more voices into existing governing structures rather than to challenge and seek to conceive of different governance models. In the US, for example, research on urban forest management has focused on citizen participation after the advent of a national initiative to financially support municipalities to conserve urban forests. Such research has focused largely on profiles of citizens who participate in the federally funded Tree US Aid, defining the characteristics that increase participation and the willingness to pay for and support tree protection ordinances. The main premise of these studies is the recognition that effective management cannot be made solely with the few resources available to municipalities, but that public participation is economically sensible.

Pluralist interpretations also assert that an interdisciplinary approach is integral to informing the policy and governance to the on-ground management of urban forests (Rowntree, 1998; Elmendorf and Luloff, 2001; Kenney, 2003; Carreiro et al., 2008). In this instance, management and governance structures move away from an urban forestry for aesthetics alone to a more ecosystem-based and inclusive management strategy (Carreiro and Zipperer, 2011). Some scholars have explored multicultural aspects of urban forests in order to develop a more inclusive urban forest management (Fraser and Kenney, 2000).

A fourth aspect of the apolitical nature of urban forest practice is the existence of professional organizations that claim expertise and exclusive membership. The dominant strands of urban forestry research and practice are supported and maintained by various networks and conferences that share and develop strategies for sustainable urban forestry management. Such professional organizations resemble what Haas (1992) refers to as epistemic communities that are exclusive in membership and that adhere to similar sets of thought and practices surrounding urban forestry. Such thoughts and practices are not absolute but politically influenced as well as debated within the epistemic community (Forsyth, 2003). William Saunders and Bernhard Fernow wrote seminal accounts of urban forest management over a century ago (Saunders and Macoun, 1899; Fernow, 1910). Since then, there has been an exponential growth in studies on urban forests, trees, and greenspace management that is natural science-based and applied in nature (Carreiro et al., 2008). In 1997, the European Forum on Urban Forestry was established to promote discussion between urban forestry professionals, scientists, and policy-makers. The Forum is associated with the International Union of Forest Research Organizations' urban forestry group, as well as with several European and Nordic networks for urban forestry. In Canada, where the editors of this book reside, the Canadian Urban Forest Conference was established in 1993. It is led by the Canadian Urban Forest Network, for which Tree Canada, the only national urban forest organization, is the secretariat. The Network is guided by a national steering committee whose mandate is to implement a Canadian Urban Forest Strategy (2013–2018). There is also a large body of international scholars and professionals that study, teach, and practice urban forestry. Professional and academic journals that address urban forestry issues include the *Journal of Urban Forestry and Arboriculture* and *Urban Forestry and Urban Greening*.

Most of the works of the above scholars tend to be apolitical and work more with governance design, empirical data, and applied solutions. These endeavours are certainly important but when designing policies it is also useful to take into account the proper historical and place-based contexts within which policy is implemented. In addition, it is important to address directly issues of environmental justice or how to educate the public on urban forest issues. In these ways, this book provides a comprehensive overview of multiple themes running throughout urban forestry.

Urban Forestry and Political Ecology

In this book we use a political ecology perspective to structure the chapters of the various contributors. We use the term perspective to denote that there are many variations of political ecology and that we are not committed to any one in particular (Forsyth, 2003; Robbins, 2004; Keil, 2003, 2005; Zimmer, 2010). The concept of a network, however, may be a particularly useful conceptual tool in political ecology. It is sometimes equated with Actor Network Theory, a theory associated with scholars such as John Law (2009) and Bruno Latour (2005). Actor Network Theory has been criticized for being theoretically weak and methodologically arbitrary because it does not identify any prime mover(s) in social and ecological change. But we still feel it is a powerful tool in appreciating the indeterminacy and contingency of change in the urban forest and the infinite number of (f)actors in urban social and ecological change. The power of a network approach lies in its ability to link disciplines and to "illuminate the ever complex connections between local and transnational social-ecological change and to understand complexity" (Birkenholtz, 2012: 303). But it also lies in looking for and evaluating the differential strengths of an infinite number of human and non-human actors and how they interact and change over time as articulated, for example, by Diane Rocheleau and Robin Roth in the quotation that begins this chapter.

In this book, the contributors identify their own specific set of actors and prime movers that vary from chapter to chapter. Some contributors work explicitly within a political ecology perspective. Others do not, though we have asked them to engage with the approach in some way. In the following chapters, we seek to bring out the unique approach of each contributor and their situatedness within the book as a whole. We identify three themes. In the first section, "Human Agency in Urban Forests and Greenspace", the contributors develop theoretical and conceptual tools to examine how human actors affect urban forests, trees, and greenspace. These include the ways in which class, race, and gender affect the access and decisions made over urban forests, as well as the ways in which political economy, neoliberalism, institutional and legal structures and frames, and world city competition influence, and are implicated in, urban forest ecologies.

In the second section, "Arboreal and Greenspace Agency in the Urban Landscape", contributors consider in more detail the role of the non-human, the arboreal, and the green, in affecting and being affected by the human world in different ways. In some instances, the effects are benign—trees and greenspace representing a positive influence on people's sense of well-being, as expressed in sentiments that extend beyond the rational mind. On other occasions trees and urban forests may act unruly and unpredictably, behaving in manners that impact negatively on humans, like they do when growing too tall, extending their roots into sewer pipes and house foundations, and falling down or shedding branches on cars and power lines during storms.

In the final section, "Actions and Interventions in the Urban Forest", the contributors write from the perspective of participants in shaping urban greenspace and trees as activists, embedded researchers, educators, planners, ecologists, and artists. Though academic in training, and therefore being reflexive in their approach, the contributors in this section also point to the passions and tactics behind taking action in everyday situations and the practical problems associated with such pursuits.

In the end, we hope that the sum of the whole of the book amounts to more than the individual chapters and that readers will be encouraged to read, compare, enjoy, and be inspired by all the chapters and what they bring to urban forestry and political ecology discourses.

Human Agency in Urban Forests and Greenspace

A political ecology approach acknowledges that the urban forest is a political and social construct, a creation of people's preferences and choices, and the function of the visions of some humans prevailing over others in a process of institutional deliberation and contestation. From this perspective, the urban forest is defined by specific ideas and interests that both include and favour, exclude and marginalize, and typically benefit some individuals or groups of people more than others. Political ecologists consider urban forests as human constructs that are not the exclusive preserve of urban forest professionals but a variety of interest groups, including residents, environmentalists, social scientists, and artists. In this context, urban forests and greenspace are arbitrary constructs that are as much human as biological categories.

Political ecology also recognizes that human discourses and technologies, such as the i-Tree software program identified above, are political instruments that can create path dependencies that frame management options in specific ways while ignoring others. In these instances, human relationships matter in the urban forest, especially the concept of political, discursive, and economic power.

Neoliberal policies are associated with a series of processes, the most prominent being the retrenchment of state interventions in the economy, the cut back of state funding for social and environmental programs, increased reliance on volunteer labour, and the growth of the market as a solution to societal problems, and the commodification of previous public goods. Such measures accentuate capitalism as the prime mover in social, political, and economic affairs, and more particularly its global reach in affecting the extent, character, and access to urban forests and greenspace.

In this book, Harold Perkins shows that urban forests are social natures that shape, and are shaped by, markets, race, class, and gender. In spite of the benefits provided by forests, especially in an era of rapid climate change, Perkins notes that they are not equally accessible to everyone. In the urban setting, trees are in fact primarily privately purchased, planted, owned, and maintained. The ability to pay thus largely determines the distribution of trees in the capitalist city. In the United States, wealthy and white neighbourhoods are typically more treed than poor Black and Hispanic neighbourhoods. But the pattern is not ubiquitous. Black communities sometimes live in abundantly treed inner-city neighbourhoods that were once occupied by white residents. And sometimes non-white residents prefer non-treed areas because they provide plenty of sun to grow vegetables and flowers. Perkins also explores the implications of the growth of voluntary organizations and volunteer labour in forest planting and conservation programs in the city. Such programs are part of the neoliberal agenda and tend to be highly classed and racialized, with an over-representation of activity in white and wealthy communities. Perkins is by no means opposed to such schemes but laments their unequal distribution and their tendency to displace similar programs operated by the state.

Cecil C. Konijnendijk van den Bosch then devises a dual framework for building and assessing effective policy in the urban forest sector. He sees the current policy climate as a move from "governance by government" to "governance with (or even without) government", a state of affairs where government roles decrease in favour of multi-stakeholder deliberations, and a situation that contains both possibilities and limitations. One aspect of the framework, based on the work by Van Tatenhove et al. (2000) considers policy actors and their coalitions, the rules of the game, the division of power between actors, and policy discourses in building innovative new policy, such as seeing the urban forest as a commons that provides essential ecological services to society as a whole. But van den Bosch then considers the neoliberalist policy agenda and its tendency to limit the value of the urban forest to monetary values and neglect its intrinsic, social, and cultural values.

Ann Van Herzele explores how a forest-centred discourse at first initiated by a small group of foresters gained credibility in Flanders, a part of Belgium that has traditionally not been focused on woodland creation, and definitely not near cities. She shows the power of discursive strategies employed in steering and supporting a drive toward establishing and expanding urban forests. Foresters used their profession, changing notions of expertise, and an idealized forest image to carve out an urban forest discourse that took on the shape of normalcy and unchallenge-ability. However, changes were noted when it came to the realization of projects. The city-centred discourse of the planning professionals has had an influence, and local politicians, residents, and farmers have had some success in protesting against the intrusion of trees, forests, and recreational users on farmlands. Van Herzele shows the power of discourse and its variable impact on different spatial jurisdictions.

Blake Hudson examines how institutional and legal discourse can have differential impacts on urban forest policy in the US and Canada. Focusing in particular on the peri-urban forest—the forest that is now threatened by urban sprawl—he highlights, in the face of feeble federal regulatory inputs, a strong to weak use of state and provincial regulation depending on local histories, cultures, and property regimes. In the southern United States, for example, where private property ownership of forests is widespread, there is very little support for prescriptive regulatory approaches to forest conservation. In the US and Canadian Pacific Northwest, by contrast, where public ownership of forests is widespread and environmentalist sentiments are strong, legally prescriptive regulatory efforts at the state/provincial level, especially in the United States, has the potential to protect urban forests if the government should choose to value the public and environmental service functions of forests higher than the rights of private forest or woodlot owners to indiscriminately cut their trees or sell their lands to developers.

Urban forests and greenspace are also scaled in relation to global capitalism and its current neoliberal conjuncture (Heynen, 2003; Heynen et al., 2006). This is reflected in the embrace of urban forests and greenspace by urban elites and city planners in most parts of the world. Forests and greenspace may here constitute potential attractors for footloose capital investments and professional elites who roam the world for the most preferential treatment and comfortable living conditions. However, the concept of city parks subscribed to by elites throughout the world is not always to the benefit of the whole population (Byrne and Wolch, 2009; Byrne et al., 2009; Byrne, 2012). In Third World countries, the pursuit of environmental agendas by local elites, including the planting of trees and the establishment of parks, may serve to both control and displace the poor and marginalized (Baviskar, 2010; Ghertner, 2010).

Natalie Marie Gulsrud and Can-Seng Ooi investigate how urban forests and ecology can be used to attract capital and tourism to the city-state, Singapore. In the process, the government has used greenspace to both discipline local residents and local forest ecologies. Singaporeans have thus been commanded and mandated to build and maintain often imported flora to beautify the city. In its most recent manifestation, Singapore's green strategy is all about erasing the local ecology and replacing it with an artificial green brand. The city-state has since labelled itself a Garden City and has earned accolades as Asia's greenest city.

J. O. Joby Bass tackles the long-standing myth that countries of the South are losing their green areas to deforestation. He shows instead that in Honduras urban and peri-urban areas are being re-forested, likely as a result of the impact of forest conservation discourse from the global North on local political and economic elites and the livelihood possibilities that forest patches constitute for the urban and peri-urban poor. He thus notes, using repeat photography, that urban plazas have gone from treeless to treed places, occupied by civic forests that provide a symbol for a new era that values trees as central aspects of the Honduran economy and culture.

He also notes the growing presence of trees surrounding urban areas—settlement forests, which provide a source of income and food for their owners. Such forests provide fuel or fence posts in peri-urban areas and fruit in urban areas.

Arboreal and Greenspace Agency in the Urban Landscape

A second element of political ecology pertains to a consideration of non-human nature, in our case trees and forests themselves, as agents or actors. This does not suggest an environmental determinism but an interaction between the human and non-human where the latter is taken seriously. In these processes, the boundaries between humans and non-human nature become arbitrary. Many scholars have struggled to even erase the boundary between nature and culture, using terms such as cyborg, nature-cultures, trans-corporeality and cosmopolitics to describe their interaction (Haraway, 1991; Hinchcliffe et al., 2005; Alaimo, 2010; Stengers, 2010, 2011).

Political ecology also acknowledges the importance of nature's unruliness, the roles played by urban trees themselves and the non-human animals that live in, depend on, and pass through them. Such actors often influence the urban environment in quirky, unpredictable, and unintended ways (Haila and Dyke, 2006). Trees, just like humans, can be seen to labour in the urban forest.[1]

In his chapter, Owain Jones proposes that trees can be significant actors in the city, affecting humans in profound ways through multiple registers beyond the rational and self-reflexive mind, such as touch, feel, hearing, smell, and emotions. In some instances, Jones suggests, trees and forests can provide humans with emotional support and comfort that are essential for mental and physical health. Trees, for Jones, possess cultures and communicate amongst themselves as well as with humans, though it is difficult to record and appreciate such exchanges. Jones thus advocates for non-representational theories that seek to capture the affective spectrum that mediates the human and non-human relations. Affect, in short, he proposes, can sometimes constitute an important actor in the human and non-human networks that make up urban forests.

Irus Braverman provides a different perspective from that of Jones. For her, urban trees are more contestants and rivals than allies and comforters to human beings. She investigates the agency of trees through the perspectives of the law and the efforts of bureaucrats and experts to manage city trees depending on their position above, at, or below ground. To Braverman's experts, trees are objects that call for order and discipline. At the same time, trees fight back and act disorderly. Bureaucrats and experts respond in different ways that depend on their abilities to tackle the unique properties of the trees. Above ground, experts exert tight control, because they can readily space trees when planted as well as devise means to prune their canopies. At ground level, control is negotiated as experts struggle to provide human navigability around trees while trees themselves require moisture and buckle pavements in the process. Below ground, by contrast, there is complete anarchy as roots grow and spread indiscriminately and are never planned for but only dealt with when causing trouble. Braverman also acknowledges the usefulness of Actor Network Theory in her analysis, though here trees as actors are portrayed very differently from Jones' account. In the rest of the contributions to this section, the authors engage with the tension and complementarity of Jones' and Braverman's explanations.

Jay Bolthouse calls our attention to the agency and culture of trees by exploring four different arboricultures of Tokyo, the world's largest metropolis. Bolthouse suggests that an urban forest is never homogenous but it is co-constitutive of the cultures and material forces that surround it. In Tokyo, the green islands of forest in the central High City reinforce the link between nature, nation, and a nurturing imperial figure. And on the peripheries, the forests are in part shaped by the historical legacies of protecting the city from floods and winds, elite aspirations to create verdant neighbourhoods, and recent efforts to forest reclaimed lands at the waterfront to soften the heat island effect. Bolthouse also warns against a too-mechanistic application of neoliberal critiques in the Japanese context where, for example, volunteer labour planting trees may have more to do with cultural traditions than filling the gaps of shrinking state funds. He also cautions against a fetishization of trees as desirable by all, because many Tokyo residents resent trees for blocking wind and sun that dry laundry and air out futons. Bolthouse argues for a political ecology that is just as mindful of general politico-metabolic processes as the unique aspects of people, place, and culture.

Joanna Dean describes three persistent historical narratives that have shaped the management of city trees: the narrative of service, the narrative of power, and the narrative of heritage. Then she tells a counter-narrative, drawing on the histories of three tree species, the Manitoba maple (*Acer negundo*), the ornamental crabapple (*Malus sp.*), and the Lombardy poplar (*Populus nigra*) in the city of Ottawa, Canada, to explore the unruliness of urban trees, and their resistance to civic management. Trees, Dean concludes, have agency, perhaps even arboreal sensory functions, which need to be acknowledged and respected by humans.

Brendon M. H. Larson then challenges the view that another urban tree, the now-common Norway maple (*Acer platanoides*) in Canada, is an invasive species that displaces and damages local flora and is undeserving of a place in Canadian urban ecologies. The issue recently gained national prominence when the leaves of the Norway maple appeared on the new Canadian $20 banknote, and many Canadians raised their voices in protest. Larson, by contrast, argues that the Norway maple is long-deserving of Canadian citizenship. Larson blurs the lines of the often taken-for-granted binary of native species being good and exotic ones being bad.

Darren Patrick questions the conventional view of one particular tree, the Tree of Heaven (*Ailanthus altissima*), as an invasive weed species, by considering the species in three different relationships. In one case, the High Line greenspace in New York City, where the Tree of Heaven was once abundant, existing in relative harmony with particular human ecologies (a gay cruising area), developers and city planners replaced the tree with an ordered and disciplined ecology as part of a redevelopment which further gentrified the neighbourhood. In the case of Detroit's Tree of Heaven Woodshop, the creators formed a mutualistic relationship with the tree, acknowledging its beauty, utility, and even critical capacity by finding ways to cure its wood for different ways of wood-making. This practice supported their artistic and political efforts to understand Detroit's post-industrial social and ecological transformation. In a third case, the "Ghetto Palm" project, two Moldavian artists use the now-ubiquitous and cursed, yet resilient, Tree of Heaven to challenge its moniker as a "bad" species. They embrace the concepts of alien species and invasive biology to trace the tree as a global resistor of human plans. Patrick uses the term "queer" both as a noun to denote marginalized humans and non-humans and as a verb to reveal possibilities for re-imagining and challenging that position.

Cynthia Imogen Hammond, finally, posits another thesis about greenspace in the urban context. She takes to task the suggestion that the presence and character of the green roof of the Canadian embassy at Leipziger Platz, Berlin, is a result of its progenitor, the landscape architect Cornelia Hahn Oberlander, a Berlin Holocaust survivor, and her attempt to reinsert herself and her Canadian heritage into her city of birth. Hammond argues instead that Berlin's long history of greenspace planning and the self-seeding vegetation in its war-torn landscapes affected the emergence of the green roof as a thin end of a larger green wedge that seeks to connect the city centre to the green areas and forests beyond the city. Hammond reminds us that the seemingly unique and personal might be a function of much larger and historical (f)actors, some of them non-human in character. She also calls attention to the embassy greenspace which, though providing limited public access, constitutes a unique space for the non-human to act out its own existence and destiny in the urban setting.

So what roles do trees and greenspace play in the fate of human societies? Do they bind or do they divide humans? Jones probably provides the right answer when he writes that "it is not 'either or'—both go on". In some instances, trees combine with political economy energies to create gated ecologies for the wealthy that shut out the poor. On other occasions, the affective transmissions generated between trees and people may *short circuit* economy and politics and set up different ethics and politics of bodies in relation to each other.

Actions and Interventions in the Urban Forest

Political ecology is also about direct and indirect actions, and political interventions in and through the urban landscape. We here pay particular attention to an examination of the urban forest through activism, embedded research, education, planning, and art. Though most of our contributors in this section are academics, a very prominent part of their lives constitutes public engagement with people and the non-human world. Their perspective, therefore, contains an interesting tension between the rational and the affective, the academic and the professional, as well as the theoretical and the practical.

Bengi Akbulut stays the closest to the radical aspects and roots of political ecology. She interprets the public demonstrations against the razing of trees at Gezi Park, Istanbul, in 2013 as a popular resistance and opposition to the neoliberalizing policies in Turkey that seek to introduce private property rights, commodity relationships, international commerce, and gentrification to the country. She also gestures towards the possibilities of the Turkish popular uprising being a wider means by which to build alliances between groups who have hitherto been isolated from each other but are now facing a unified enemy in the neoliberalization of Turkey. The uprising, therefore, bears resemblance to the Occupy movement that earlier occurred in the western industrialized world.

Lindsay K. Campbell, working from the standpoint of an embedded researcher, analyses an urban forestry campaign administered via a public–private partnership in New York City. In her analysis, she is both sympathetic to and critical of the interpretations of scholars like Harold Perkins and Nik Heynen who see the neoliberalizing aspects of the forestry program in a largely negative light. Campbell acknowledges the problems of the elite dimensions of the program, as illustrated by the support of actress Bette Midler, Mayor Bloomberg, and their entourage of wealthy benefactors, and the use of volunteer labour in the face of public cutbacks. At the same time, she challenges casting urban forestry as neoliberalism at work, noting the massive municipal investment in tree planting throughout the city, targeting first the neighbourhoods with the fewest street trees, and the empowerment of individuals in taking part in the program.

Pippin Anderson writes from the perspective of an ecologist who is part of the African Centre for Cities' Urban Ecology CityLab at the University of Cape Town, an organization that aims to reframe urban ecology research as not only a biological but a social and political endeavour. Anderson notes that the Cape region generally lacks trees but is instead host to the unique Fynbos flora, a bushy low-to-the-ground-growing vegetation, which constitutes a global biodiversity hotspot. The uneven distribution of such open greenspace in Cape Town, and the way trees insert themselves in such a landscape, is often a politically and a scientifically contested exercise and sometimes deviates from expected patterns. In one example, Anderson points to a wealthy neighbourhood protesting the replacement of exotic trees with the indigenous Fynbos vegetation because of the latter's provision of a forest and shade for recreational purposes. In another instance, she points to a historically Black neighbourhood on Cape Flats, the apartheid area of Cape Town, where the residents have worked in collaboration with state authorities to bring back the Fynbos vegetation. Cape Town's CityLab constitutes a significant effort to grapple with political and ecological issues in an applied manner in the unique context of the rapidly urbanizing African subcontinent.

Gregory Smith is an advocate for environmental education. He illustrates how urban forest education and place-based participatory pedagogies can encourage a closer and more inclusive student engagement with local urban forests. He shows that there are informal and formal ways of learning that ultimately impact how people build their knowledge about urban forests. The educators Smith writes about are often seen as local heroes, who work diligently to improve the academic achievement of an increasingly impoverished and stressed student population. Few of them have the space or ability to move the structural economic constraints and political obstacles that may be responsible for social and environmental degradation in the first place. But they are nevertheless progressive actors who contribute to the affect between people and trees and local ecologies that have the potential to politicize and build the bridges identified by Bengi Akbulut and Lindsay Campbell.

Helena Mellqvist and Roland Gustavsson write as landscape planners and architects about their concern for the invasion of urban ecologies into some of the marginalized areas of the European countryside. They explore their role in a participatory landscape-planning exercise in the Bräkne River Valley of south-eastern Sweden where the local community ponders what strategies to employ to create livelihoods that both resist and accommodate the external urban forces that envelope them. Mellqvist and Gustavsson insist that the traditional open landscape, the landscape cleared and cultivated by generations of farmers, matters for building a sustainable landscape. They also remind us, like Pippin Anderson, that trees and forests do not necessarily constitute the only forms of greenspace. They prompt us to think about the physical landscape and its scenic properties, and how they speak to different sensual registers that drip of significance for human well-being and that need to be appreciated and conserved.

Kathleen Vaughan is an academic and artist. She is part of a rapidly growing area of practice- and place-based research in the form of art-making. In her account, she walks with her dog through several forested areas in the city of Montreal seamlessly sewing together the artistic and the academic, and the personal and the professional. She feels that her textile-based maps of urban forests can speak to the affective registers of artists and their audience. On her walks, she is aware and takes note of political issues like urban gentrification and public access but the main objective is for her artwork to move the viewer from personal aesthetic pleasure to active engagement and advocacy. The beauty or enchantment of trees and woods and art, she proposes, can be a strong instrument for social, political, and environmental change. Urban trees and forests, then, can affect the development of a sense of self through arts practice and embodied learning, and through these processes public engagement and forest stewardship can be enhanced.

Conclusion

As in any collection of essays, it is impossible to be comprehensive in covering such a broad topic as urban forests, trees, and greenspace. Even a political ecology perspective is insufficient to narrow the scope to claim comprehensiveness. There is certainly more work needed to advance the theory of political ecology itself; our modest goal here is to apply some of its basic principles to urban forests. As such, there is certainly room to extend political ecology analyses to other parts of the world. There is also room, as elsewhere in political ecology, to engage more fully with the natural and ecological sciences. We do hope, though, that the following chapters will begin to broaden the discussion surrounding the political nature of urban forests and greenspace and widen the conversation about the relevance of political ecology in pursuing that goal.

Notes

1 Non-human animals, though not a focus of this book, can similarly be seen to labour in the urban forest. Forest insects, such as the Asian Emerald Ash Borer (*Agrilus planipennis*) and the European Brown Spruce Longhorn Beetle (*Tetropium fuscum*), which have caused extensive "damage" to forests stands in North America, are not seen as much as "pests" but as hapless refugees who are simply trying to eke out a livelihood in a new and foreign environment. Political ecologists, rather than controlling their presence, consider the wider context which brought the insect to North America in the first place, and then seek to respect and work out resolutions that challenge the root causes of their presence (Perkins, 2007). The same perspective can be applied to the white-tailed deer that is typically blamed for over-browsing the forests at the rural–urban fringe. Political ecologists, in this instance, call attention to the few measures that are taken to target a shrinking deer habitat, the smorgasbord of food that is presented by private gardens, and anti-hunting and anti-predator climate that exist in such areas as key actors in understanding the "abundance" of deer and the "over-browsing" of forests.

References

Adams, C., Lindsey, K., and Ash, S. (2006) *Urban Wildlife Management,* Taylor and Francis Group, London: 287–303.

Alaimo, S. (2010) *Bodily Natures: Science, Environment, and the Material Self,* Indiana University Press, Bloomington, IN.

Andersson, K. (2003) 'What Motivates Municipal Governments? Uncovering the Institutional Incentives for Municipal Governance of Forest Resources in Bolivia', *Journal of Environment and Development,* 4(12): 5–27.

Baviskar, A. (2010) 'Urban Exclusions: Public Spaces and the Poor in Delhi', in B. Chaturvedi (ed) *Finding Delhi: Loss and Renewal in a Megacity*, Penguin, New Delhi.

Birkenholtz, T. (2012) 'Network Political Ecology: Method and Theory in Climate Change Vulnerability and Adaptation Research', *Progress in Human Geography*, 36(3): 295–315.

Byrne, J. (2012) 'When Green is White: The Cultural Politics of Race, Nature and Social Exclusion in a Los Angeles Urban National Park', *Geoforum*, 43(4): 595–611.

Byrne, J. and Wolch, J. (2009) 'Nature, Race, and Parks: Past Research and Future Direction for Geographic Research', *Progress in Human Geography*, 33(6): 743–765.

Byrne, J., Wolch, J., and Zhang, J. (2009) 'Planning for Environmental Justice in an Urban National Park', *Journal of Environmental Planning and Management*, 52(3): 365–392.

Carreiro, M., Song, Y., and Wu, J. (eds) (2008) *Ecology, Planning, and Management of Urban Forests: International Perspectives*, Springer, New York City.

Carreiro, M. and Zipperer, W. (2011). 'Co-adapting Societal and Ecological Interactions following Large Disturbances in Urban Park Woodlands'. *Journal of Austral Ecology*, 36: 904–915.

Elmendorf, W. and Luloff, A. (2001) 'Using Qualitative Data Collection Methods When Planning for Community Forests', *Journal of Arboriculture*, 27(3): 139–151.

Fernow, B.E. (1910) *The Care of Trees in Lawn, Street and Park. With a List of Trees and Shrubs for Decorative Use*, Henry Holt and Company, New York City.

Forsyth, T. (2003) *Critical Political Ecology: The Politics of Environmental Science*, Routledge, London.

Fraser, E. and Kenney, W. (2000) 'Cultural Background and Landscape History as Factors Affecting Perceptions of the Urban Forest', *Journal of Arboriculture*, 26(2): 106–113.

Ghertner, D.A. (2010) 'Green Evictions: Clearing Slums, Saving Nature in Delhi' in P. Robbins, R. Peet, and M. Watts (eds) *Global Political Ecology*, Routledge, London.

Haas, P. (1992) 'Epistemic Communities and International Policy Coordination: Introduction', *International Organization*, 46(1): 1–35.

Haila, Y. and Dyke, C. (eds) (2006) *How Nature Speaks: The Dynamics of the Human Ecological Condition*, Duke University Press, Durham, NC.

Haraway, D. (1991) 'A Cyborg Manifesto: Science, Technology, and Socialist Feminism in the Late Twentieth Century', in D. Haraway (ed) *Simians, Cyborgs, and Women: The Reinvention of Nature*, Routledge, New York City.

Heynen, N. (2003) 'The Scalar Production of Injustice Within the Urban Forest', *Antipode*, 35(5): 980–998.

Heynen, N., Kaika, M., and Swyngedouw, E. (eds) (2006) *In the Nature of Cities: Urban Political Ecology and the Politics of Urban Metabolism*, Routledge, Oxford.

Hinchcliffe, S., Kearnes, M., Degan, M., and Whatmore, S. (2005) 'Urban Wild Things: A Cosmopolitical Experiment', *Environment and Planning D: Society and Space*, 23(5): 643–658.

Keil, R. (2003) 'Urban Political Ecology', *Urban Geography*, 24(8): 723–738.

Keil, R. (2005) 'Progress Report—Urban Political Ecology', *Urban Geography*, 26(7): 640–651.

Kenney, W. (2003) 'A Strategy for Canada's Urban Forests', *Forestry Chronicle*, 79(4): 785–789.

Konijnendijk, C., Ricard, R., Kenney, A., and Randrup, T. (2006) 'Defining Urban Forestry—A Comparative Perspective of North America and Europe', *Urban Forestry & Urban Greening*, 4(3–4): 93–103.

Latour, B. (2005) *Reassembling the Social: An Introduction to Actor-Network Theory*, Oxford University Press, Oxford.

Law, J. (2009) 'Actor Network Theory and Material Semiotics', in B. Turner (ed) *The New Blackwell Companion to Social Theory*, Blackwell, London.

McPherson, E., Nowak, D., Heisler, G., Grimmond, S., Souch, C., Grant, R., and Rowntree, R. (1997) 'Quantifying Urban Forest Structure, Function, and Value: The Chicago Urban Forest Climate Project', *Urban Ecosystems*, 1: 49–61.

Millward, A.A. and Sabir, S. (2010) 'Structure of a Forested Urban Park: Implications for Strategic Management', *Journal of Environmental Management*, 91(11): 2215–2224.

Nowak, D., Crane, D., Walton, J., Twardus, D., and Dwyer, J. (2002) 'Understanding and Quantifying Urban Forest Structure, Functions, and Value', in W. Kenney, J. McKay, and P. van Wassaneaer (eds) *Proceedings of the 5th Canadian Urban Forest Conference*, Ontario Urban Forest Council, Markham, Ontario.

Perkins, H. (2007) 'Ecologies of Actor-Networks and (Non)social Labor within the Urban Political Economies of Nature', *Geoforum*, 38(6): 1152–1162.

Poudyal, N., Siry, J., and Bowker, J. (2012) 'Market-based Approaches Toward the Development of Urban Forest Carbon Projects in the United States', in J. Garcia and J. Diez (eds) *Sustainable Forest Management—Current Research*, InTech, www.intechopen.com/books/sustainable-forest-management-current-research/markets-based-approaches-towards-the-development-of-urban-forestry-carbon-projects-in-the-united-sta (accessed 3 June 2013).

Robbins, P. (2004) *Political Ecology: A Critical Introduction*, Blackwell, Malden, MA.

Rocheleau, D. (2008) 'Political Ecology in the Key of Policy: From Chains of Explanation to Webs of Relation', *Geoforum*, 39(2): 716–727.

Rocheleau, D. and Roth R. (2007) 'Rooted Networks, Relational Webs and Powers of Connection: Rethinking Human and Political Ecologies', *Geofourm*, 38(3): 433–437.

Rowntree, R. (1998) 'Urban Forest Ecology: Conceptual Points of Departure', *Journal of Arboriculture*, 24(4): 62–71.

Salzman, J. (2006) 'A Field of Green? The Past and Future of Ecosystem Services', *Journal of Land Use and Environmental Law*, 21: 133–151.

Saunders, W. and Macoun, W. (1899) *Catalogue of Trees and Shrubs*, Bulletin 2. Central Experimental Farm, Ottawa, Ontario.

Singh, V.S., Pandey, D.N., and Chaudhry, P. (2010). *Urban Forests and Open Green Spaces: Lessons for Jaipur, Rajasthan (India)*. Occasional Paper No 1/2010. Rajasthan State Pollution Control Board, www.urbanforestrysouth.org/resources/library/citations/urban-forests-and-open-green-spaces-lessons-for-jaipur-rajasthan-india/ (accessed 3 June 2013).

Stengers, I. (2010) *Cosmopolitics I*, University of Minnesota Press, Minneapolis, MN.

Stengers, I. (2011) *Cosmopolitics II*, University of Minnesota Press, Minneapolis, MN.

Van Herzele, A., De Clercq, E.M., and Wiedemann, T. (2005) 'Strategic Planning for New Forests in the Urban Periphery: Through the Lens of Social Inclusiveness', *Urban Forestry & Urban Greening*, 3(3–4): 177–188.

Van Tatenhove, J.P.M., Arts, B., and Leroy, P. (2000) *Political Modernization and the Environment: The Renewal of Environmental Policy Arrangements*. Kluwer Academic Publishers, Dordrecht.

Zimmer, A. (2010) 'Urban Political Ecology: Theoretical Concepts, Challenges, and Suggested Future Directions', *Erdkunde*, 64(4): 343–353.

Part 1

Human Agency in Urban Forests and Greenspace

2 Urban Forests are Social Natures

Markets, Race, Class, and Gender in Relation to (Un)Just Urban Environments

Harold Perkins

Introduction

Many people do not realize that trees in the city are part of an urban forest. Trees seem so out of place in some heavily built environments that they appear to be verdant interlopers among the concrete and rectilinear shapes surrounding them. But there is a growing awareness among researchers, planners, and even the general public that urban forests exist and serve ecological functions or 'services' to people living in even the most densely built urban corridors (Jim and Chen, 2009). The benefits provided by a healthy, well-managed forest are so extensively documented by academics that it is impractical to list them all here. Some of the best-known benefits, however, include sequestration of greenhouse gases, reduction of heat island effects, reduced power demand, and storm water retention (see Nowak and Dwyer, 2000). Given these benefits, efforts to grow a small number of trees into a forest are welcome strategies across the urban milieu, from the concrete corridors of downtown to the far-reaching cul-de-sacs of sprawling suburbia. It is not surprising then that planners, city foresters, and the general public are all incorporating more trees into their corner of the metropolitan landscape.

Despite this growing awareness of the benefits of urban forests, there is an ever-present risk that we still treat trees as mundane and randomly dispersed objects while ignoring the power-laden processes shaping their distribution across the city. Benevolent public and private planting programs and a general belief that trees just 'grow wherever' mask the fact that in many urban forests the majority of trees are planted by people—all of whom live in specific geographical contexts and circumstances. The fact that so many trees are planted by people is an important consideration for the distribution of this valuable urban resource. Urban trees, and by extension the urban forest they comprise, represent a nexus of social, economic, and political happenings in the city (Heynen et al., 2006). They are literally a living record of past and present relationships between actors who occupy different positions within urban society and wield power differentially according to their position.

Of course this line of thinking is different than the conventional wisdom that urban tree growth is determined by environmental conditions including the availability of light, water, nutrients, and the troublesome presence of soil compaction and disease-causing organisms (Jim, 2000). Certainly biological drivers are important determinants of tree growth in cities. My argument here is that we keep these in mind while further considering how biological drivers of forest structure and distribution operate in the context of human endeavors. In other words, the biological imperative of photosynthesis that supports the growth of tree canopy occurs in the spatial context of race, class, ethnicity, gender, culture, and markets that produce and underpin the larger urban setting. The task is for us to elucidate precisely how it is that people harness the power of photosynthesis in the urban forest canopy to their benefit according to explicitly spatialized and uneven societal processes. Only then can we obtain a more comprehensive picture of all the factors impacting the composition and distribution of the urban forest.

A more complete picture is important because we live in an urban world. The human population increasingly lives in cities and there is little doubt that this trend will accelerate. People are migrating to cities in the developing world and often live and work in difficult environmental conditions once there (Davis, 2007). Climate change associated with rising temperatures and shortages of water promise to make life in cities even more difficult for vulnerable migrants seeking an urban way of life in the future. If trees can do for us even a fraction of what we think they can in urban settings, it will be extremely important going forward that urban forests are better incorporated into these destination cities where a billion or more people are expected to relocate in the coming decades. This of course will be no simple task as (re)forestation efforts will have to be managed across a continuum of spatial scales (Heynen, 2003). Small-scale efforts to forest individual communities will continue to be important goals for achieving localized environmental justice. However, those efforts should also be incorporated and complemented by efforts at regional, national, and international scales where global ecological concerns of biodiversity, resilience, and climate change mitigation can better be addressed. Understanding social, political, and economic drivers of urban forest canopy across varied geographical contexts is, therefore, essential to an appropriate expansion and intensification of urban forestry to the planetary scale. Unfortunately, discrimination in commodity markets based on race/ethnicity, socio-economic standing, and gender prevents many people living in cities from benefiting from urban forests. In what follows I address these constraints and then go on to discuss/critique what is currently being done to expand access to these groups of people. I conclude this chapter by discussing briefly why including people historically marginalized in forestry governance should be central to achieving environmental justice in cities throughout the world.

Social Nature and the Political Ecology of the Urban Forest

We tend to think of trees as growing on their own without the assistance of people. Certainly trees are organisms with metabolic properties just like other plants and animals in the biosphere. In order to survive they photosynthesize inputs from their environment and engage in evapotranspiration by releasing oxygen and water vapor among other chemical compounds. Like other organisms, trees grow, die, and eventually decay and contribute to the nutrient cycle. However, urban political ecologists push us to further consider how these biophysical processes are related to human actions (Castree, 2000). They in fact want us to reconsider urban trees and forests as a produced 'social nature' where human behavior affects biological processes in profound ways (Perkins, 2007). This is not to say that the photosynthesis that trees accomplish is somehow carried out by people. Rather it suggests that we humans harness the process of photosynthesis and tree growth according to our needs and wants.

People intervene in the process of urban forest dynamics in ways that produce a forest that would otherwise look much different if left to its own devices. Proof of our ability to create new forest ecologies according to our social capacities is everywhere present. Most urban trees are explicitly planted by someone (Heynen, 2003). This means then that much of the forest exists by human design. Urban forestry departments plant trees on the parking strip along roadways using hardy species that can withstand heat and soil compaction. People are even adept at altering the genetic makeup of trees to resist diseases endemic to urban forests. "Valley Forge" and "New Harmony" are new elm cultivars designed by scientists to resist the ravages of Dutch elm disease (DED) (US National Arboretum, 1995). As another example of our ecological interventions, forestry departments and property owners alike prune and maintain trees to make them take certain shapes for aesthetic and structural purposes. All of these examples demonstrate the efficacy of our social interventions into forest ecology.

What this means is that people are constantly in the process of co-producing urban ecologies with the trees in their neighborhood, therefore, making human activity fundamentally ecological. David Harvey insists it is ridiculous that our thinking on ecological matters has resolutely excluded human activity from that realm of relational thinking (1996). But expanding our understanding of humans as ecological actors allows us to consider urban forest ecologies as artifacts of human creativity, will, ingenuity, and labor. However, urban political ecologists note that the production of these urban artifacts is the result of, and contributes to, arrangements of power and the flow of resources in the city. Thus the work of co-producing socio-natural environments that comprise cities is never politically neutral; it always serves some particular purpose within society and the resulting ecology is usually uneven (Swyngedouw and Heynen, 2003). The task then for urban political ecologists is to recognize the socio-natural forest for the trees by elucidating the societal factors that influence who works on forests, what composition forests assume (i.e. species proportions), and what shape they take in terms of geographical distribution. All of these outcomes have tremendous importance when we consider trees to be valuable urban resources that are not shared by all urban dwellers.

Racialized, Classist, and Gendered Forests in a Market-Mediated Urban Landscape

Capitalism is the dominant mode of economic organization in most cities in the world today. Fundamental to capitalism is the existence of property; virtually everything in the city belongs to someone or some institution. If a capitalist system is to function successfully, there needs to be a market that mediates the production, exchange, and consumption of property. If we measure wealth as a function of the accumulation of property under capitalism, markets are thus important mediators of the distribution of that wealth. There are fundamental distinctions in the property market that are important to the distribution of property in any capitalist city. There are private forms of property owned by individuals and firms, and there are properties held in common (public). In a capitalist economy, private property is favored over public property that is usually subject to state management. But what does the property relation under capitalism have to do with urban forests?

Urban trees are frequently treated as property in commodity form in a capitalist setting (Perkins et al., 2004). This is in part the case because many trees are produced and sold to consumers by private nurseries and distributors. Individuals purchase their trees with money and then plant them on their own private property. The act of planting trees on private property also makes them commodities as they are then rooted in the exchange value of the land they are planted on as expressed in urban real-estate markets. So what all this means in a capitalist economy is that money is necessary to access the market for trees sold as individual commodities or as part of a larger property. We also know that wealth is not evenly distributed in a capitalist system, so some people and government agencies have more purchasing power to buy trees than others. Thus there is an inherent tendency for trees as commodities to be unevenly distributed in capitalist cities, even on public lands. Studies on trees bear this out (Heynen and Perkins, 2005; Landry and Chakraborty, 2009).

Conventional wisdom tells us that the market for capitalist commodities is ruled by an invisible hand (Smith, [1776] 2009). Markets are supposed to function in ways that are color-blind and politically neutral. However, we now know this position is false. Access to markets for commodities is determined by many important factors including race, class, and gender, just to name a few social distinctions (Pulido, 2000). By extension, access to urban trees (and private property with urban trees already growing on it) is mediated by the complex positions of various groups in society categorized according to these distinctions (Romm, 2002). Access to commodity markets is in part differentiated according to these distinctions because they have a bearing on access to income producing jobs and by extension purchasing power. Take race and/or ethnicity as an example.

Race/Ethnicity and Inequitable Urban Forests

The relationship between race/ethnicity and urban forest inequity is complex. However, most studies conducted on race/ethnicity and the distribution of urban trees in US cities suggest that non-Hispanic whites tend to live in more verdant neighborhoods than African-Americans and Hispanic/Latinos. A study conducted in New Orleans suggests canopy cover in its neighborhoods is lower in proportion to increasing numbers of African-American residents (Talarcheck, 1990). Research carried out several years later in Milwaukee indicates that census tracts predominantly populated with non-Hispanic whites are strongly correlated with higher rates of tree cover while Hispanics were negatively correlated with canopy cover (Heynen et al., 2006). Yet another study conducted in Tampa finds that canopy cover also decreases there as the proportion of African-American residents increases in neighborhoods (Landry and Chakraborty, 2009). There exist some likely explanations for these racial/ethnic discrepancies in canopy cover.

Cities sometimes under-invest in public property in African-American and Hispanic neighborhoods, leading to variations in the number and vitality of trees present on public right-of-ways. Minorities' purchasing power is on average lower compared with non-Hispanic whites because of discrepancies in employment opportunities and compensation. Little disposable income is an economic constraint placing the purchase of trees far below priorities like paying the mortgage/rent/utilities and putting food on the table (Perkins, 2009). Minority populations living in US cities also tend to live in rental housing as their low incomes restrict their access to the housing market. As Perkins et al. (2004) suggest, people who rent their

Figure 2.1 Tall and proud. Early springtime view of vigorous and manicured street trees in Milwaukee's affluent East Side neighborhood, USA.

Source: Harold Perkins, 2006.

Figure 2.2 Under the wire. Early springtime view of small and mangled street trees in Milwaukee's impoverished Harambee neighborhood.

Source: Harold Perkins, 2006.

homes tend not to have opportunities to plant trees (more on this in the following subsection) and also tend to live in relatively un-canopied neighborhoods.

Interestingly, however, one study conducted in Baltimore found evidence that some African-Americans there live in neighborhoods with relatively high rates of vegetation, including trees (Troy et al., 2007). A potential explanation for this, according to the study, is that some African-Americans are now living in formerly white neighborhoods with 'historical legacies' of past tree plantings coupled with increasing numbers of volunteer trees growing on vacant lots. There is also evidence in Milwaukee that some African-Americans live in relatively canopied census tracts (Heynen et al., 2006). Like Baltimore, however, Heynen et al. largely attribute the cause for this correlation in Milwaukee to the fact that African-Americans live in some of the oldest neighborhoods in the city with an historical planting legacy. They go on to note that disinvestment in many of these properties, especially by absentee landlords, has caused volunteer trees to be counted in canopy measures, too. Disinvestment in these older Milwaukee neighborhoods has created a problematic fence-line forest of fast-growing boxelder and silver maples that causes extensive damage to roofs, foundations, and fences (Perkins, 2006). The fence-line forest causes an antagonistic outlook concerning urban trees to emerge among some African-American residents given these problems. The future of these poorly maintained trees and the otherwise valuable canopy they provide is in doubt as people remove them because of the problems they cause.

It is important to note here that there are differences regarding the perceptions of the value of trees among different groups residing in the city and that these perceptions potentially have an impact on canopy distribution as well. For example, I have noted that some residents in African-American neighborhoods in Milwaukee and Detroit prefer not to have trees planted near their homes. Hmong communities in Milwaukee are concerned that trees will shade their sun-loving urban gardens, too. Similar stories about residents not wanting trees come from some neighborhoods in Baltimore (Grove et al., 2006) and New York City (Susman, 2009). Potential property damage by trees and interference with other kinds of land uses are prominent reasons why some residents resist the planting of additional trees. Some of this antagonism also is the likely result of the imposition of forestry management policies on neighborhood residents in conjunction with a lack of communication between foresters and residents living in the central city (Romm, 2002). This is a missed opportunity to communicate with residents about how trees can be an amenity that benefits the people living in central cities by saving them energy costs while providing them with a better quality of life. It is also a missed opportunity to share urban environmental governance with residents that might want other forms of green infrastructure in addition to, or besides, trees (Perkins, 2011). Racial/ethnic considerations of forest inequity are highly related to class, or socio-economic standing.

Class/Socio-economic Standing and its Relation to Urban Forest Inequity

Trees were planted in upper- and middle-class neighborhoods in the 19th and early 20th centuries in many cities in the Northeastern and Midwestern US, but working-class tenement neighborhoods remained mostly un-forested. By the middle of the 20th century, however, many non-Hispanic white working-class families moved into single-family homes with yards and their relationship with trees began to change. Historical accounts suggest that American elms, for example, were planted ubiquitously in newly created white, working-class neighborhoods since their leafy benefits quickly contributed to the higher quality of life that workers achieved (Campanella, 2003). Green cathedrals towering over the streets essentially worked to blur class lines in the US as the working class began to live in verdant neighborhoods that closely resembled middle-class standards in quality and aesthetics. This was particularly the case in newly emerging suburbs in the post-Second World War era. But of course white flight and deindustrialization meant that many of these upwardly mobile, white, working-class families left the city and its earliest suburbs for newer settlements at the farthest fringes of the metropolitan area. Outmigration occurred around the same time that DED was destroying urban elm forests in the US. The relationship between urban forests and class status had once again changed.

The connection between lower socio-economic standing and the absence of urban trees is once again quite strong, giving the distribution of urban forests a significant class-based dimension today. The urban population that remains in post-industrial US cities is more diverse and poorer relative to what it once was. Jobs and white residents left the city taking with them the capital needed to support public and private forestry efforts. Much of the remaining housing stock is now rented out to low income/minority groups. DED of course ended up being a devastating disturbance since many cities' canopies were predominantly comprised of elms (Schreiber and Peacock, 1979). One result of all of these changes is that large tracts of old rental housing in the central city are occupied by some of the poorest residents in a metropolitan area, and the trees lost to DED in these locations were never replanted on private property. This is largely the case in Milwaukee, for example, where census tracts with low socio-economic standing and larger percentages of rental housing feature significantly lower levels of canopy than other neighborhoods in the city (Perkins et al., 2004; Heynen et al., 2006). The few wealthier homeowners who stayed in the city replaced their trees while people living in poorer neighborhoods mostly did not. Evidence suggests a similar distribution of trees according to socio-economic considerations in Tampa (Landry and Chakraborty, 2009).

Housing tenure and socio-economic standing are closely correlated. Middle- and upper-income groups tend to own and stay in their homes for a long time while poorer people tend to rent and move around more frequently (van der List et al., 2002). This has bearing on the prospects for reforestation on both public and private lands in the city. There are discrepancies in the number of trees planted on the right of way in wealthy neighborhoods full of homeowners versus poorer neighborhoods with more rental homes (Heynen and Lindsey, 2003). Even if the city forestry department or a local nonprofit organization does its best to plant trees on the right of way in poor neighborhoods, their future is anything but certain. Trees planted by urban foresters on the right of way in the poorest neighborhoods of Milwaukee and Detroit, for example, have short life expectancies due to neglect and vandalism (personal communication). Homeowners also have a number of incentives to plant trees on private property that renters do not.

Planting trees in a proper location is a form of property maintenance that adds value to a home in the long term (Anderson and Cordell, 1988). Trees can increase exchange (market) values of private properties because they provide property owners with certain kinds of use-values, including enhanced aesthetics and the shading and sheltering structures that reduce costs for cooling and heating. Renters usually have little to no power to make maintenance decisions. Also, they are unlikely to purchase and plant trees given the investment will take 30 years to mature and pay dividends in reduced energy bills. Unlike homeowners, renters must secure permission from the landlord if they want to plant trees. All of these factors suggest that upper- and middle-class homeowners are more likely to plant trees where they live than lower-class renters (Perkins et al., 2004). Scale up to the entire metropolitan area, and it becomes evident that the difference between wealthy homeowners and poor renters will have distributional effects on existing

Figure 2.3 Arboricide. Wintertime view of a recently planted tree struck down by vandals in Milwaukee's impoverished Lynden Hill neighborhood.

Source: Harold Perkins, 2006.

and future urban canopy cover. In addition to race/ethnicity and class, gender roles in society also impact the production of urban forests.

A Masculine Arboricultural Industry and its Gendered Urban Forests

Middle-class, white women were instrumental in the early history of urban forestry in the US. Women's clubs in the late 19th and early 20th centuries embraced the budding conservation movement and with it they pushed for the planting of trees along streets in cities throughout the US (Merchant, 1985; Campanella, 2003). As Merchant goes on to suggest, women's clubs embraced forestry in the city because they believed conservation in the city was tantamount to conserving home, family, womanhood, and whiteness. In other words, city beautification as espoused by women in these clubs was supposed to reform immigrant/working-class residents whom they believed threatened the gendered, racialized, and classist relations predominant in society at the turn of the century. Reflection on women's club involvement in urban forestry in the US thus indicates the 'progressive' women who participated in these arboricultural activities were actually reinforcing, rather than challenging, the important societal norms of their time. Urban forestry was, therefore, able to quickly become a technical and masculine profession which left its women progenitors behind (Merchant, 1985). This is important because professionalized urban forestry organizations (both public and private) at a variety of scales have become the primary means through which most forestry-related decisions are made and carried out in the US.

Data on women's participation in formal urban forestry organizations is sparse today, but it is accepted that men continue to dominate the field (Pinchot Institute, 2006). For example, Kuhns et al. (2002, 2004) found that only 10 percent of urban forestry professionals in the US were women and that many women working in the industry perceive some form of discrimination against them in the workplace. These findings are not surprising because gender roles assigned by society to workers determine what those workers are—and are not—allowed to do for a living (McDowell and Massey, 1984; Harvey, 1998). The same kind of thinking that men should be firefighters, doctors, and lawyers, while women should be nurses, teachers, and secretaries has historically prohibited women from managing and working directly on urban trees in a professional setting. Thus, if we look at the structure of urban forests, particularly the component located on public rights of way, we see a masculine urban forest ecology (Heynen et al., 2007). Its structure and physical form represents decades of intellectual and manual arboricultural work done almost exclusively by white men who prioritize shade trees. We have to ask ourselves if something has been missed by not allowing women to more fully participate in the science and decision-making regarding the urban forest around them. For example, would women choose the same tree species to plant as men traditionally have, or would they perhaps prioritize fruit trees instead? Would they practice the same methods of planting and maintenance? These are questions we do not have good answers for at this point because women have largely been kept out of formalized urban forestry. There are, however, increasing numbers of women working in the professional urban forestry sector (Kuhns et al., 2004; Pinchot Institute, 2006; Heynen et al., 2007). It remains to be seen if they will change the form and structure of the urban forest or carry on the industry's predominantly masculine traditions.

So Race/Ethnicity, Class, and Gender Contribute to Uneven Forests. What is Being Done about it?

Academics know urban forests are unevenly distributed according to many variables, but especially by socio-economic standing of neighborhood residents. There is recognition among formalized urban forestry organizations that urban forests need to be distributed more equitably, too (American Forests, 2013). I am, therefore, skeptical that more studies are needed to prove that the urban poor and minorities live in locations featuring less canopy cover than their wealthier counterparts. It seems intuitive that when houses and trees are treated as a commodity, people who have the ability to pay for them are more likely to purchase homes and trees than those who cannot afford them. We do need, however, to understand what is being done to provide underserved urbanites access to urban trees and forests. As it turns out, nonprofit organizations and community groups are leading the expansion of urban forests into marginalized neighborhoods.

Their efforts are comprised of two prominent strategies to better distribute urban forests beyond the wealthiest and whitest sections of cities. Both strategies have their roots in the belief that the civil sector can solve problems better than expensive, top-down bureaucracies like municipal forestry departments. One strategy is voluntarism with its emphasis on the role of nonprofit organizations in coordinating and carrying out urban service provision; the other is volunteerism where community groups and individuals provide their own services. Both voluntarism and volunteerism for urban forestry have become more prominent in an era of government fiscal austerity during the last 30 years (Perkins, 2009). US cities, for various reasons, including deindustrialization and a loss of tax base, have less money to spend on public works than they did three decades ago. Cash-strapped large cities, especially in the US Northeast and Midwest, scaled back or eliminated altogether their spending on public services like urban forestry. Detroit is currently undergoing bankruptcy proceedings, for example, leaving little money for its police force, fire department, and roadways (Davey and Williams Walsh, 2013). Trees fall far behind these needs on the public works priorities list.

Voluntarism is one way that the civil sector responds to these urban fiscal and service provision emergencies. Voluntarism is a term describing the plethora of professional, yet nonprofit organizations emerging alongside the shrinking government sector that fills the continued need for urban service provision in the wake of municipal budget cuts (Wolch, 1990). A substantial number of nonprofit organizations seek to reforest central cities, including The Greening of Detroit, Baltimore Tree Trust, Greening Milwaukee, and Trees Atlanta. Despite their supposed separation from the state, these organizations largely draw their budgets from grants that they receive from a variety of Federal, state, and local government agencies (Perkins, 2009). They also receive a substantial amount of their operating budgets through donations from the private sector, including philanthropies, corporations, and individuals. These organizations seek to create, influence, and promote urban forest policies. One of their biggest goals, however, is to encourage volunteerism—getting local citizens to volunteer their time and efforts for urban forests.

Volunteerism happens when citizens work on something because they feel they need to tackle a problem on their own (Fyfe and Milligan, 2003). Citizen groups sometimes volunteer to inventory and assess their urban forests; more frequently they plant trees on special occasions like Arbor Day. Increasingly, professionalized nonprofit organizations like the ones previously mentioned coordinate volunteer work on trees as it can be rather technical and funding for urban forestry projects can be hard to generate, especially in inner-city neighborhoods (Perkins, 2011). Whether the volunteerism is coordinated by nonprofit organizations or not, the idea behind volunteerism is that citizens be empowered to inventory, maintain, plant, and surveil their own forests. These kinds of efforts are happening in a diverse range of neighborhoods from rich to poor in cities all over the Western world. The

NeighborWoods program is a good example of one of these nonprofit organizations that encourages volunteerism. A study of NeighborWoods volunteers suggests affluent people who give their time and energy to plant trees do so because they view it as a way to learn how to maintain their property while enhancing its value. Less affluent volunteers were said to do it for the sake of neighborhood revitalization (Makra and Andresen, 1990).

Another important corollary to volunteerism as coordinated by nonprofit organizations is the idea that inner-city residents gain skills necessary for certain kinds of employment through their volunteer experiences. Nonprofit urban forestry organizations with funding from government sources and private donors are encouraging and coordinating African-American and Hispanic/Latino youth to become stewards of their local urban environment. In Milwaukee, for example, 'at-risk' youth are taught how to plant trees in urban nurseries, care for them, and eventually replant them in elderly neighbors' yards as part of a service-learning program. The goal of the program run by Greening Milwaukee and funded by government grants and private donations is to get 'kids off the street corner' and learning skills that will help them to get a job someday. Evidence of success by these programs in getting a diverse range of participants is still mixed at best, however.

People who volunteer to work on trees still tend to be affluent and white (Perkins, 2009; 2011). Minorities are under-represented in many programs for a number of reasons. Residents living in central cities often have serious social and economic issues preventing them from placing trees high on their list of priorities. We also have to keep in mind that there are cultural preferences at work here, too. One study indicates various minority groups possess different aesthetic preferences for greenery among their homes that translates into fewer trees (Grove et al., 2006). Inner-city residents are sometimes suspicious of the actions of nonprofit groups from outside their neighborhoods as well (Battaglia, 2010). It can take months if not years for nonprofit forestry groups to earn the trust of residents who feel abandoned or even abused by their city governments (personal communication).

All of these issues make the relationship between voluntarism and volunteerism in urban forestry problematic. The laudable goals of nonprofit organizations to reforest cities may be falling short of being racially/ethnically and socio-economically inclusive. It is even my suggestion that some efforts of nonprofit groups to use minority volunteers for urban greening amount to market-based forms of social engineering (Perkins, 2009; 2011). Such initiatives also absolve the state from stepping in and comprehensively investing in urban environmental amenities like trees. Regardless of these damning charges, a lack of participation in forestry programs by people living in minority communities is highly problematic in relation to the goal of increasing canopy cover equity across the city. This critique, it should be noted, is not meant to dismiss the importance of community participation in urban forestry. Rather, community participation should be encouraged and buttressed by state agencies with diverse staff who are accountable to the public for their dedicated investment in equitably distributed urban forests and other forms of green infrastructure (Romm, 2002).

In Conclusion: The Environmental Justice Implications of Urban Forests as Uneven Social Nature in 21st-Century Cities

Increasingly scholars recognize that a lack of access to urban vegetation based on the color of people's skin, their class standing, and their gender constitutes an environmental injustice perhaps as serious as the presence of environmental disamenities such as landfills and polluting factories (Heynen et al., 2006). This is in part the case because trees and other forms of greenery have the potential to mitigate heatwave events that kill hundreds of vulnerable urban residents every year and seriously injure many more. Poor and minority communities disproportionately exist in sectors of the city most likely to experience the warmest temperatures during extreme heat events (Jesdale et al., 2013). Trees shade buildings and keep them cooler while the evapotranspiration occurring in the canopy simultaneously cools the atmosphere around them. Heavily canopied neighborhoods can thus be considerably cooler than those that are not forested. Vigorous urban forests, therefore, have the potential to mitigate extreme heat events that otherwise feature increased urban mortality among the city's most vulnerable people.

If climate scientists are correct about the future of our changing planet, we should expect increased incidences of extreme heatwaves in the US and elsewhere, and more heatwave-related mortality in cities as a result of rising temperatures (Diffenbaugh and Ashfaq, 2010). This makes vigorous and evenly distributed urban forests key to mitigating some of the worst effects of climate change on vulnerable urban populations the world over. We have at our disposal a socio-natural toolbox in urban forests for mitigating heat-related illness and death. In fact one of the side benefits of trees is they sequester carbon while cooling the environment. So enhancing the urban forest everywhere possible is one way we can also reduce (if only temporarily) the amount of carbon available in the atmosphere that exacerbates long-term temperature trends and extreme heat events. Of course many of the concerns cited in this chapter about uneven urban forests and environmental injustice are situated in Western contexts where shade trees are viewed favorably and prioritized by middle- and upper-class communities. It is thus important to reiterate that planting shade trees everywhere without communicating and working with residents will not be the solution to environmental injustice. Urban forestry should not preclude the existence of urban gardens where communities depend on them to enhance food security, nor should shade trees be prioritized over other culturally preferred forms of vegetation like fruit/nut trees that also mitigate the effects of the urban heat-island. What we have to do instead is consider how urban forests can be a substantial, yet democratic, part of any effort to create green, sustainable, urban futures. This will require extensive consultation with groups historically at the margins of forestry governance, including racial/ethnic minorities, lower classes, and women. This is certainly a challenge.

Discriminating market relations are exported to almost all growing urban centers in the developing world and the evidence thus far demonstrates cities are highly polarized in terms of wealth and quality of life as a result (Davis, 2007). This means uneven urban forestry and associated injustice implications are likely to continue to be an important problem in a variety of urban contexts around the world. If commodity markets and related forces continue to restrict access to trees and other forms of vegetation to the world's wealthiest neighborhoods, we will fail to engage in a greener and more equitable politics that supports human life everywhere. Governments, planners, academics, and residents therefore need to find ways to overcome the constraints of 20th-century urbanism that have traditionally restricted forest access to privileged people. If successful, greener governance for all people living in 21st-century cities is possible.

References

American Forests (2013) 'Urban forest restoration program in Seattle Washington', www.americanforests.org/our-programs/global-releaf-projects/alcoa-foundation-and-american-forests-global-releaf-partnership-for-trees/2013-projects/urban-forest-restoration-program/ (accessed 08 August 2013).

Anderson, L. and Cordell, H. (1988) 'Influence of trees on residential property values in Athens, Georgia (USA): A survey based on actual sales prices', *Landscape and Urban Planning*, 15: 153–164.

Battaglia, M. (2010) 'A multi-methods approach to determining appropriate locations for tree planting in two of Baltimore's tree-poor neighborhoods', Master's thesis, Ohio University, OH.

Campanella, T. (2003) *Republic of Shade: New England and the American Elm*, Yale University Press, New Haven, CT.

Castree, N. (2000) 'Marxism and the production of nature', *Capital & Class*, 72: 5–37.

Davey, M., and Williams Walsh, M. (2013) 'Billions in debt, Detroit tumbles into insolvency', *The New York Times*, www.nytimes.com/2013/07/19/us/detroit-files-for-bankruptcy.html?pagewanted=all&_r=1& (accessed 9 August 2013).

Davis, M. (2007) *Planet of Slums*, Verso, London.

Diffenbaugh, N., and Ashfaq, M. (2010) 'Intensification of hot extremes in the United States', *Geophysical Research Letters*, 37(15): 1–5.

Fyfe, N. and Milligan, C. (2003) 'Out of the shadows: Exploring contemporary geographies of voluntarism', *Progress in Human Geography*, 27(4): 397–413.

Grove, J., Troy, A., O'Neil-Dunne, J., Burch, W., Cadenasso, M., and Pickett, S. (2006) 'Characterization of households and its implication for the vegetation of urban ecosystems', *Ecosystems*, 9: 578–597.

Harvey, D. (1998) 'The body as an accumulation strategy', *Environment and Planning D: Society and Space*, vol 16, pp. 401–421

Harvey, D. (1996) *Justice, Nature, and the Geography of Difference*, Wiley-Blackwell, London.

Heynen, N. (2003) 'The scalar production of injustice within the urban forest', *Antipode: A Journal of Radical Geography*, 35(5): 980–998.

Heynen, N. and Lindsey, G. (2003) 'Correlates of urban forest canopy cover: Implications for local public works', *Public Works Management and Policy*, 8(1): 33–47.

Heynen, N. and Perkins, H. (2005) 'Scalar dialectics in green: Urban private property and the contradictions of the neoliberalization of nature', *Capitalism, Nature, Socialism*, 16(1): 99–113.

Heynen, N., Perkins, H., and Roy. P (2006) 'The political ecology of uneven urban green space: The impact of political economy on race and ethnicity in producing environmental inequality in Milwaukee', *Urban Affairs Review*, 42(1): 3–25.

Heynen, N., Perkins, H., and Roy, P. (2007) 'Failing to grow "their" own justice? The co-production of racial/gendered labor and Milwaukee's urban forest', *Urban Geography*, 28(8): 732–754.

Jesdale, B., Morello-Frosch, R., and Cushing, L. (2013) 'The racial/ethnic distribution of heat risk-related land cover in relation to residential segregation', *Environmental Health Perspectives*, 121: 811–817.

Jim, C.Y. (2000) 'The urban forestry programme in the heavily built-up milieu of Hong Kong', *Cities*, 17(4): 271–283.

Jim, C.Y. and Chen, W.Y. (2009) 'Ecosystem services and monetary values of urban forests in China', *Cities*, 26: 187–194.

Kuhns, M., Bragg, H., and Blahna, D. (2004) 'Attitudes and Experiences of Women and Minorities in the Urban Forestry/Arboriculture Profession', *Journal of Arboriculture*, 30(1): 11–18.

Kuhns, M., Bragg, H., and Blahna, D. (2002) 'Involvement of Women and Minorities in the Urban Forestry Profession', *Journal of Arboriculture*, 28(1): 27–34.

Landry, S. and Chakraborty, J. (2009) 'Street trees and equity: Evaluating the spatial distribution of an urban amenity', *Environment and Planning A*, 41: 2651–2670.

McDowell, L. and Massey, D. (1984) 'A woman's place?', in D. Massey and J. Allen (eds) *Geography Matters! A Reader*, Cambridge University Press in Association with the Open University, Cambridge.

Makra, E. and Andresen, J. (1990) 'NeighborWoods: Volunteer community forestry in Chicago', *Arboricultural Journal*, 14: 117–127.

Merchant, C. (1985) 'The Women of the Progressive Conservation Crusade: 1900–1915', in K. Bailes (ed) *Environmental History: Critical Issues in Comparative Perspective*, University Press of America, Lantham, MD, pp. 153–175.

Nowak, D. and Dwyer, J. (2000) 'Understanding the benefits and costs of urban ecosystems', in J.E. Kuser (ed) *Handbook of Urban and Community Forestry in the Northeast*, Kluwer Academic/Plenum Publishers, New York City.

Perkins, H. (2006) 'Laboring through neoliberalization: The cultural materialism of urban environmental transformation', Ph.D. dissertation, Department of Geography, University of Wisconsin-Milwaukee, WI.

Perkins, H. (2007) 'Ecologies of actor-networks and (non)social labor within the urban political economies of nature', *Geoforum*, 38: 1152–1162.

Perkins, H. (2009) 'Out from the (green) shadow? Neoliberal hegemony through the market logic of shared urban environmental governance', *Political Geography*, 28: 395–405.

Perkins, H. (2011) 'Gramsci in green: Neoliberal hegemony through urban forestry and the potential for a political ecology of praxis', *Geoforum*, 42: 558–566.

Perkins, H., Heynen, N., and Wilson, J. (2004) 'Inequitable access to urban reforestation: the impact of urban political economy on housing tenure and urban forests', *Cities*, 21(4): 291–299.

Pinchot Institute for Conservation (2006) 'Understanding the role of women in forestry: A general overview and a closer look at female forest landowners in the U.S.', Online research report. www.google.com/url?sa=t&rct=j&q=&esrc=s&source=web&cd=1&ved=0CCoQFjAA&url=http%3A%2F%2Fwww.pinchot.org%2F%3Fmodule%3Dup loads%26func%3Ddownload%26fileId%3D68&ei=4c8pU_O0Oum4yAH7u4D4AQ

&usg=AFQjCNHAaiYr9gdal0RP2KXoUav15m24Ug&bvm=bv.62922401,d.aWc (accessed 19 March 2013).

Pulido, L. (2000) 'Rethinking environmental racism: White privilege and urban development in Southern California', *Annals of the Association of American Geographers*, 90(1): 12–40.

Romm, J. (2002) 'The coincidental order of environmental justice', in K. Mutz, G. Bryner, and D. Kenney (eds) *Justice and Natural Resources: Concepts, Strategies, and Applications*, Island Press, London, pp. 117–137.

Schreiber, L. and Peacock, J. (1979) 'Dutch Elm Disease and its control', Government Printing Office, Washington, DC.

Smith, Adam. ([1776] 2009) *The Wealth of Nations*, Thrifty Books, Blacksburg, VA.

Susman, T. (2009) 'A tree grows (and dies) in Brooklyn', *Los Angeles Times*. http://articles.latimes.com/2009/oct/15/nation/na-trees15 (accessed 9 August 2013).

Swyngedouw, E. and Heynen, N. (2003) 'Urban political ecology, justice, and the politics of scale', *Antipode: A Radical Journal of Geography*, 35: 898–918.

Talarcheck, G. (1990) 'The urban forest of New Orleans: An exploratory analysis of relationships', *Urban Geography*, 11: 65–86.

Troy, A., Grove, M., O'Neil-Dunne, J., Pickett, S., and Cadenasso, M. (2007) 'Predicting opportunities for greening and patterns of vegetation on private urban lands', *Journal of Environmental Management*, 40: 394–412.

United States National Arboretum, Agricultural Research Service (1995) 'Ulmus Americana cultivars "Valley Forge" and "New Harmony"', US National Arboretum, Cultivar Release, Floral and Research Nursery Plants Research Unit, p.1.

van der List, A., Gorter, C., Nikamp, P., and Rietveld, P. (2002) 'Residential mobility and local housing-market differences', *Environment and Planning A*, 34: 1147–1164.

Wolch, J. (1990) *The Shadow State: Government and Voluntary Sector in Transition*, The Foundation Center, New York City.

3 From Government to Governance

Contribution to the Political Ecology of Urban Forestry

Cecil C. Konijnendijk van den Bosch

Urban Forestry and Governance

Both in an environmental management and forestry context, the concept of governance has emerged during recent years in recognition of the increasing (alleged) complexity and variety of decision-making (Rayner et al., 2010). In brief, governance has been defined as any efforts to coordinate human actions towards goals. Kjær (2004: 10) defined governance as "the setting of rules, the application of rules, and the enforcement of rules." In governance, actors are searching for control, steering, and accountability. Various authors have recognized the complexity of, for example, wicked forest problems and forestry contexts (Hull, 2011), ruling out simple or standardized governance solutions and models.

Van Tatenhove et al. (2000) introduced the concept of 'political modernisation' within the field of environmental policy-making, stressing the shifting relations between state, market, and civic society. Building on this work, Arts and Leroy (2006) described how roles and responsibilities in environmental policy-making have changed and been renewed during the past decades. Environmental governance is no longer purely state-dominated, but it also involves civic society, as well as the market, with a range of associated interactions, new institutions, and new practices. Political modernisation can then be seen as a move away from 'governance by government' to 'governance with government' or even 'governance without government' (see Kleinschmit et al., 2009 for an application within forestry).

The use of, for example, the word 'modernisation' shows an implicit positive view of the emergence of governance with or without governance in environmental policy-making. Part of the argumentation behind this normative perspective relates to an enhanced focus on the involvement of citizens and other stakeholders in decision-making. Many scholars have, for example, stressed the need to involve local residents in decisions on their own living environment (see Van Herzele et al., 2005 and Janse and Konijnendijk, 2007 for an overview). From a political ecological perspective, however, the shift from 'governance by government' to 'governance with/without government' has been regarded as ambiguous (e.g. Swyngedouw, 2005; Perkins, 2009, 2013). On the one hand, certain groups in society have been able to obtain a greater role in environmental decision-making. On the other hand, idealized models of normative horizontal, non-exclusive and participatory stakeholder-based governance often ignore the neoliberal political-economic context in which these are embedded (Swyngedouw, 2005). Here it can be seen as problematic that the state withdraws from decisions on environmental matters, as it can be questioned whether, for example, public interests and environmental and social values are sufficiently considered in multi-actor networks and under market-oriented conditions.

When having a closer look at environmental governance, awareness of these political ecological analyses is important. Sound analysis of environmental governance needs to consider issues of power, conflicts, and who has access to decision-making (Robbins, 2004). Moreover, political ecological study can identify what the main political narratives ('storylines') in environmental governance are, and how (and why) key concepts are defined by actors and stakeholders.

Issues of power, access, and differing narratives and definitions so fundamental to political ecology are also central to the governance of woodland, trees, and associated vegetation, one of the specific fields within environmental governance. But in an urban context, perhaps surprisingly, the explicit use of a governance perspective in the field of urban forestry is from a more recent date (Lawrence et al., 2013) in spite of the implicit inclusion of a governance perspective in many definitions of what urban forestry is and stands for. In an earlier article (Konijnendijk, 2003), for example, I provided a more normative framing of the field of urban forestry based on the literature and recent European experiences. Urban forestry was defined as being: (1) integrative, (2) socially inclusive, and (3) strategic, while (4) also embracing its urban mandate. Integration refers to looking 'horizontally' beyond sectoral and resource boundaries (for example, from street tree to peri-urban woodland) as well as to 'vertical' integration of public authorities and other actors at different governance levels. Urban forestry's socially inclusive nature relates to equity issues and the wider involvement of stakeholders and urban residents. In spite of the traditionally strong role of (especially local) authorities, urban forestry developed as a collaborative approach involving the public sector, the private sector, and civic society. As discussed earlier, however, a political ecological perspective requires a more critical take on this normative view of urban forestry and related governance, as 'governance with/without governance' should not necessarily always be seen as largely desirable and positive. This contribution will therefore include this more critical view in line with political ecology.

The same can be said regarding the different narratives that have dominated urban forestry, for example, when comparing the development of the field in North America as compared to Europe (Konijnendijk et al., 2006). In Europe, based on the continent's strong heritage of 'city forestry', urban woodland often became the focus when urban forestry was introduced as a concept. In North America, much more attention has traditionally been given to street trees. However, from its outset urban forestry has envisaged integrating different elements of the urban green structure that have often been the domain of different professions and public authorities, that is, trees along roads and on squares, parks and urban woodland. Moreover, urban forestry considers both public and private trees. Although the value of taking such an integrative perspective has been stressed, it has also become clear that decision-making processes are complex, for example, due to the large amount of actors and stakeholders involved (Lawrence et al., 2013). Moreover, urban forestry is typically dealing with so-called 'wicked problems', where no easy solutions are likely to be found. It will be interesting to consider the neoliberal roots and focus of governance transformation in urban forestry, as for example highlighted by Perkins (2009), as reflected in a withdrawal of government involvement and a greater emphasis on the marketization of urban forest benefits.

This chapter aims to analyse current urban forest governance through the lens of both the political arrangement approach (Van Tatenhove et al., 2000, see below) and political ecology. It describes and analyses some of the trends in urban forestry governance worldwide and looks at political ecologically relevant aspects of governance, such as actors and their respective roles, resources, power, and discourses.

A Framework for Analyzing Governance

As described above, governance is seen here as efforts—typically at the more strategic level—to direct human action toward common goals and, more formally, as the setting, application, and enforcement of generally agreed-to rules. When political modernisation is 'translated' into a shift to multi-actor governance, and when the importance of so-called policy arrangements in governance is stressed, the same authors' policy arrangement approach can be applied as an analytical framework for studying governance. Policy arrangements have been defined as temporary stabilization of the substance and organization of the policy domain.

The policy arrangement approach provides a structured approach to analyzing and understanding policy arrangements as the way in which a certain policy domain—such as urban forestry—is shaped in terms of organization and substance. Policy arrangements can change according to four dimensions, namely: (1) actors and their coalitions involved in the policy domain; (2) rules of the game; (3) division of power and other resources between the actors; and (4) current policy discourses. Actors refer to those individuals and organizations who are involved in governance processes and who can form coalitions with one another to strengthen their impact on decision-making. Power is used in this process to promote one's specific interests. 'Rules of the game' refer to institutions, and to the regulations, legislation, and procedures, relevant to a certain policy domain. Discourses have been defined as "(. . .) an ensemble of ideas, concepts and categories through which meaning is given to social and physical phenomena, and which is produced and reproduced through an identifiable set of practices" (Hajer and Versteeg, 2005: 176). These dimensions also touch upon some of the themes that are central to political ecology, such as the division of power and the development and articulation of different policy discourses. The four dimensions and their inter-linkages are visualized in Figure 3.1.

A common way to apply the policy arrangement approach as an analytical framework is to focus on one of the four dimensions and then analyse the other dimensions from this focus dimension. For example, a specific discourse can be selected to analyse its implications for and linkages with actors, rules of the game,

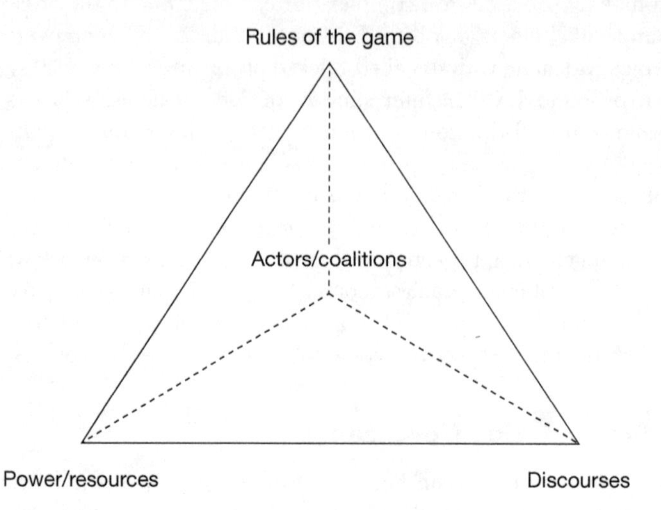

Figure 3.1 Visualization of the four dimensions of the Policy Arrangement Approach.
Source: Liefferink, 2006.

and the division of power and resources. Shifting power relations can be the starting point for considering changes in, for example, actor roles and coalitions, rules of the game, and the adaptation or emergence of discourses. The policy arrangement approach is used in the next sections to analyse governance and governance change in urban forestry.

Actor Perspective

In many countries, urban forestry has seen shifts in the roles and involvement of different actors. This concerns shifts between government, market, and civic society, but also within these different groups. In Europe, for example, the management of woodland in and near towns was initially often in the hands of local communities, for example, in the form of commons (Konijnendijk, 2008). Olwig (1996) described how town meetings were held to regulate the use of these commons. The commons were used for grazing, but also as a venue for community events and festivals. The commons thus did not only have an important economic role, they also contributed to strengthening community bond and identity.

However, large parts of urban forestry history were highly elitist, with kings, queens, other nobility, and later the bourgeoisie conserving forest and tree resources for their own purposes, and not in the least for hunting (Konijnendijk, 2008). Thus urban forests were real 'powerscapes' which were meant to evoke awe and enhance prestige (Figure 3.2).

Power, both over nature and over other human beings, was often reflected in the design of the landscape, and frequently fences or walls were used to close off these lands. Legislation to secure the use rights of the owners was often very

Figure 3.2 The urban forest as 'powerscape': Boboli Gardens in Florence, Italy.
Source: Cecil C. Konijnendijk van den Bosch.

stringent and was enforced by harsh punishments. Although the benefits of these elitist urban forests were only available to a few, strong legislation and enforcement meant that these greenspaces were maintained largely unchanged over the course of time and urban development (Konijnendijk, 2008).

The 'birth' of modern urban forestry, however, was connected to a much stronger role of local government and a search for serving public interests. North American municipalities, for example, were looking for more rational and strategic ways of managing their street-tree populations, especially at a time when diseases such as Dutch elm disease posed major challenges to sustainable management. In Europe, urban forestry could start building on centuries of history of municipal woodland ownership.

As mentioned, many narratives and (normative) definitions of urban forestry have stressed its partnership and participatory dimensions. After the elitist era of urban forestry, the role of civic society in general, and local residents in particular, has gradually increased. In many cases, public involvement has related to more strategic decision-making, but there is also increasing attention to involving local residents in the actual management and maintenance of urban forests and greenspaces. This is the case in Denmark, for example, where municipalities have expressed their interest in setting up new modes of greenspace management partly based on volunteering. However, practical implementation to date has lagged behind ambitions. Denmark has a long tradition of volunteering in the social sector and amateur sports, but people seem more reluctant when local greenspaces are concerned. Volunteering in urban forest management is more widespread in other parts of the world, for instance the UK and North America (Frøik Molin and Konijnendijk, in prep.). An interesting development in this respect is the increase in community gardening projects across the globe. These projects are often bottom-up, initiated by communities of local residents, and involving people from different segments of society who create new coalitions that sometimes also involve government agencies, at least as facilitators. Here the idea of the common is re-discovered, often on a very local scale.

It is important to take a critical stance on the emergence of multi-actor networks in governance as well, as they mostly have emerged in a neoliberal context, with a withdrawal of government intervention. This raises concerns over how far urban forests are still considered 'public goods' and to what extent their multiple benefits are secured and provided to all (Perkins, 2009).

Rules of the Game Perspective

The traditional commons were managed by local communities according to sets of formal and (very often) informal rules. With current developments in urban forestry leaning toward governance with or even without government, new rules of the game are required. Recent history has seen the emergence of new hybrid models of urban forestry involving public and private actors, as well as civic society. A well-known example is Central Park in New York City (Figure 3.3), which is now successfully managed by a partnership of the city's parks department and a local trust, the Central Park Conservancy.

In Europe, national urban parks have been set up in Stockholm, Sweden, and a series of Finnish cities (Konijnendijk, 2008; Ernstson and Sörlin, 2009; Figure 3.4). These national urban parks are typically managed by a partnership of public actors involving different municipalities, regional and national public bodies—but also environmental and other NGOs have key positions in management.

New rules of the game are also required for community gardening, where local authorities often play a rather minor role, but, for example, have to 'lend out' urban space and allow the growing of crops on urban land. When municipal authorities object to community-driven gardens, guerrilla gardening movements might arise, where 'illegal' greenspaces and plantings are established in an ultimate, anarchistic example of governance without government.

New rules of the game need to take the drawbacks and challenges associated with new network governance constellations into account. As discussed by Swyngedouw (2005) and Perkins (2009), it is often no longer clear who the new networks represent, what legitimacy they have, and how they can enforce their strategies. As also discussed by Jones and Cloke (2002), there is a risk that governance arrangements become elitist in nature and exclude other actors. Concerns about this have been raised in various urban forest governance contexts (Janse and Konijnendijk, 2007).

Figure 3.3 Management of Central Park in New York City, USA, represents a 'governance with government' approach, with the city's park department and the private Central Park Conservancy sharing responsibility.

Source: L. Anders Sandberg.

Figure 3.4 Stockholm's National Urban Park represents a complex governance arrangement.

Source: Cecil C. Konijnendijk van den Bosch, 2008.

Power Perspective

The division and allocation of power between different urban forestry actors has also shifted over time. In the commons set-up, local communities shared power, while elitist urban forests saw a strong and sometimes violent enforcement of power by the privileged few. Different layers and sectors of government have also been engaged in power struggles over urban forests, for example where national interests and policies have clashed with local ones.

In modern urban forestry, power is often held by a wider range of actors, although public bodies typically still have most to say. The example of the Stockholm National Urban Park shows that the division of power poses challenges as responsibilities can sometimes be unclear and decision-making can be slow, as also highlighted by Swyngedouw (2005) for environmental decision-making at large. Even in cases of urban forest governance without government, where local communities have taken over urban forest and greenspace management, power relations can be complex. Often parts of the local community can feel excluded, as a small group of local residents puts a lot of time into management and 'claims' the area, sometimes being unaware of other interests (Jones and Cloke, 2002).

Government no longer has a power monopoly in a (public) urban forestry context, which can be considered problematic as the power relations replacing governmental power are often unclear. This also makes enforcement problematic and can mean that urban forests are not properly conserved and lose the battle against other urban interests. Proponents of more participatory urban forests, however, argue that the involvement of multiple stakeholders will actually strengthen the role of urban forestry on the local political agenda, as there is a broader stakeholder base advocating urban forestry issues.

Discourse Perspective

Discourses related to democratization and environmental governance have resulted in greater attention to involvement of local communities in the management of urban forests and greenspace. Local Agenda 21 was an important milestone in this development. Authors have stressed the importance of involving local residents in urban forestry (Van Herzele et al., 2005; Janse and Konijnendijk, 2007) for a series of reasons, including enhancing ownership, legitimacy, and improving the quality of decision-making. However, it has also been made clear that meaningful public involvement is not always easy to achieve and often will require a step-wise process of moving from governance by government to sharing power and responsibility. As this chapter has discussed, the call for more public involvement is closely connected to the current dominance of neoliberalism and its focus on a retreating government and on marketization of public services.

On a wider scale, the neoliberal discourse has led to an increasing focus on the economic benefits of natural resources, due to the prevalence of economic reasoning in politics. The emergence of the ecosystem service concept is a good example. Ecosystem services are the benefits that humans derive from the multitude of resources and processes provided by ecosystems. Natural landscapes are no longer appreciated 'as they are', but increasingly in terms of the (quantifiable) benefits they provide to society. In urban forestry, there has been an increasing focus on benefits and their monetary valuation in particular since the 1990s. In the United States, the work of the US Forest Research Northeastern Research Station in Chicago and other cities has focused on regulating ecosystem services provided by urban trees, such as cooling, air pollution reduction, and stormwater regulation (McPherson et al., 1994). In Europe, an example of the focus on the monetary valuation of urban forest benefits was the hedonic pricing studies carried out by Liisa Tyrväinen in Finland (2001). The work of the US Forest Service has developed further over the years, resulting in a suite of software programmes under the heading 'i-Tree', where entering information about the local tree population and tree cover makes it possible to estimate the economic benefits of the urban forest in terms of regulating service provision (i-Tree, 2013). i-Tree is currently implemented across the globe. In spite of its obvious values in a neoliberal political climate, it represents one specific, economic political narrative and does not give a lot of attention to the cultural ecosystem services of urban forests. Apart from developing a suite of ways to value urban forests and the services that the forests provide, it will be important to find better ways to involve local communities in the use and implementation of i-Tree. In North America, this has already happened, as non-governmental organizations (NGOs) and local communities have been able to argue their cases based on i-Tree data.

However, i-Tree is very much part of its neoliberal context and the trend to quantify greenspace benefits in monetary terms. The ecosystem service approach does include a block of cultural services, but these have been very difficult to operationalize. The same holds true for i-Tree, where important social and psychological values of urban forests are not included. Moreover, the biodiversity values of urban forests are not covered either. Thus, i-Tree is caught between the need to speak the economic language of neoliberalism and the public service and multiple-use role of urban forests.

Perspective

With urban forests increasingly being planned, developed, and managed through governance with (and sometimes even without) government, it could be argued that the management of woodland and trees in towns is coming full circle. During medieval times, local communities often had intricate rules in place for the management and use of local woodland resources. However, gradually the elites took over and claimed urban forests for their own benefits.

Scholars such as Elinor Ostrom (Ostrom et al., 1994) have argued that the co-management of natural resources, such as forests, could provide an important and promising model for the future, where natural resources come under increasing pressure. Urban forests can play an important role in showing the way, building on a rich heritage of local community involvement. But obviously times have changed, and the 'new commons' will require governance arrangements and rules of their own. Much can be learnt from the many examples across the world.

These cases also show that urban forestry governance should always be context-dependent, and governance arrangements need to be adapted to local conditions and local actors.

The above concerns primarily the publicly owned part of the urban forest. However, the large majority of urban trees are located on private property. As argued by Kenney (personal communication), urban forests can thus also be regarded partly as 'inverted commons' where their privately owned component is concerned. Private trees still provide a range of ecosystem services to the public, such as mitigating climate change, stormwater runoff regulation, and—in many cases—aesthetic experiences.

As this chapter shows, the emergence of urban forests as 'new commons' should also be critically reviewed from a political ecological perspective. The new commons emerge in a neoliberal context where focus is on a retreating government, public spending cuts, and marketization of greenspace benefits.

When urban forests develop into the new commons, the role of public actors (and local authorities in particular) needs to be re-addressed. Local urban foresters, arborists, and greenspace managers still have a lot to contribute in terms of knowledge and expertise. However, from being the main, and sometimes even sole, decision-makers their roles may shift into that of facilitators and advisors, working together with local communities in the management of trees and woodland. They also play a crucial role in providing frames and legitimacy for new network governance arrangements.

In some cases, urban forestry might initially still require governance by government, as local actors are unable to solve conflicts over urban forest resources. It is also the task of the government to secure valuable urban forest resources where the commercial claims on public land are too strong. Here local communities and interest groups can act as watchdogs, making sure that the neoliberal discourse does not automatically give a lower priority to greenspaces. Instruments such as i-Tree can help to argue for the case of urban forestry, but need to be supplemented by research and information on the importance of, for example, the cultural ecosystem services provided by urban forests. A recent systematic review of the benefits of urban parks showed that we know too little about the impact of greenspaces to build social cohesion and contributing to local place identity (Konijnendijk et al., 2013). These aspects are central to the commons concept and thus need further political ecological study that can assist in gaining a greater understanding of the important aspects of power in, and access to, the urban forest.

References

Arts, B. and Leroy, P. (2006) *Institutional Dynamics in Environmental Governance*, Springer, Dordrecht.

Ernstson, H. and Sörlin, S. (2009) 'Weaving protective stories: connective practices to articulate holistic values in Stockholm National Urban Park', *Environment and Planning A*, 41(6): 1460–1479.

Frøik Molin, J. and Konijnendijk, C.C. (2014) 'Between big ideas and daily realities—A governance perspective on volunteering in Danish municipal green space maintenance', *Urban Forestry & Urban Greening*, 13(3), DOI: http://dx.doi.org/10.1016/j.ufug.2014.03.006.

Hajer, M. and Versteeg, W. (2005) 'A decade of discourse analysis of environmental politics: Achievements, challenges, perspectives', *Journal of Environmental Policy & Planning*, 7(3): 175–184.

Hull, R.B. (2011) 'Forestry's conundrum: High value, low relevance', *Journal of Forestry*, 109(1): 50–56.

i-Tree (2013) http://www.itreetools.org/ (accessed March 19, 2013).

Janse, G. and Konijnendijk, C.C. (2007) 'Communication between science, policy and citizens in public participation in urban forestry—Experiences from the NeighbourWoods project', *Urban Forestry & Urban Greening*, 6(1): 23–40.

Jones, O. and Cloke, P. (2002) *Tree Cultures—The Place of Trees and Trees in their Place*, Berg, Oxford and New York City.

Kjær, A.M. (2004) *Governance*, Polity Press, Cambridge and Malden, MA.

Kleinschmit, D., Böcher, M. and Giesen, L. (2009) 'Discourse and expertise in forest and environmental governance—An overview', *Forest Policy & Economics*, 11(5–6): 309–312.

Konijnendijk, C.C. (2003) 'A decade of urban forestry in Europe', *Forest Policy and Economics*, 5(3): 173–186.

Konijnendijk, C.C. (2008) *The Forest and the City: The Cultural Landscape of Urban Woodland*, Springer, Berlin.

Konijnendijk, C.C., Annerstedt, M., Maruthaveeran, S. and Nielsen, A.B. (2013) *Benefits of Urban Parks—Systematic Review of the Evidence*, Report for International Federation of Parks and Recreation Administration (Ifpra), University of Copenhagen and Swedish University of Agricultural Sciences, Copenhagen and Alnarp, www.ifpra.org/images/Newsletters/IfpraBenefitsOfUrbanParks.pdf (accessed March 20, 2014).

Konijnendijk, C.C., Ricard, R.M., Kenney, A. and Randrup, T.B. (2006) 'Defining urban forestry—A comparative perspective of North America and Europe', *Urban Forestry & Urban Greening*, 4(3–4): 93–103.

Lawrence, A., De Vreese, R., Johnston, M., Sanesi, G. and Konijnendijk van den Bosch, C.C. (2013) 'Urban forest governance: Towards a framework for comparing approaches', *Urban Forestry & Urban Greening*, 12(4): 464–473.

Liefferink, D. (2006) 'The dynamics of policy arrangements: Turning round the tetrahedron', in: B. Arts and P. Leroy (eds) *Institutional Dynamics in Environmental Governance*, Springer, Dordrecht, pp. 45–68.

McPherson, E.G., Nowak, D.J. and Rowntree, R. (eds) (1994) *Chicago's Urban Forest Ecosystem: Results of the Chicago Urban Forest Climate Project*, General Technical Report NE-186, Northeastern Forest Experiment Station, USDA Forest Service, Radnor, PA.

Olwig, K.R. (1996) 'Reinventing common nature: Yosemite and Mount Rushmore—A meandering tale of a double nature', in: W. Cronon (ed) *Uncommon Ground: Rethinking the Human Place in Nature*, W.W. Norton & Company, New York City and London, pp. 379–408.

Ostrom, E., Gardner, R. and Walker, J. (1994) *Rules, Games, and Common Pool Resources*. University of Michigan Press, East Lansing, MI.

Perkins, H.A. (2009) 'Out from the (Green) shadow? Neoliberal hegemony through the market logic of shared urban environmental governance', *Political Geography*, 28: 395–405.

Perkins, H.A. (2013) 'Consent to neoliberal hegemony through coercive urban environmental governance', *International Journal of Urban and Regional Research*, 37(1): 311–327.

Rayner, J., Buck, A. and Katila, P. (eds) (2010) *Embracing Complexity: Meeting the Challenges of International Forest Governance*, IUFRO World Series vol 28, IUFRO, Vienna.

Robins, P. (2004) *Political Ecology: A Critical Introduction*, Wiley-Blackwell, New York City.

Swyngedouw, E. (2005) Governance innovation and the citizen: The janus face of governance-beyond-the-state, *Urban Studies*, 42(11): 1991–2006.

Tyrväinen, L. (2001) 'Economic valuation of urban forest benefits in Finland', *Journal of Environmental Management*, 62: 75–92.

Van Herzele, A., Collins, K. and Tyrväinen, L. (2005) 'Involving people in urban forestry—A discussion of participatory practices throughout Europe', in: C.C. Konijnendijk, K. Nilsson, T.B. Randrup and J. Schipperijn (eds) *Urban Forests and Trees—A Reference Book*, Springer, Berlin, pp. 207–228.

Van Tatenhove, J.P.M., Arts, B. and Leroy, P. (2000) *Political Modernization and the Environment: The Renewal of Environmental Policy Arrangements*, Kluwer Academic Publishers, Dordrecht.

4 A Genealogy of Urban Forest Discourse in Flanders

Ann Van Herzele

Introduction

This chapter is about the "career" of an emerging discourse within spatial planning in Flanders (an autonomous region in northern Belgium), specifically the need to create an urban forest near to every city and town. The claim is remarkable given the absence of any tradition in Flanders of woodland creation near urban areas. Urban forests have since long been popular recreational assets elsewhere in Europe, most notably in cities of Eastern and Central Europe (Konijnendijk, 2003). In Flanders, by contrast, the concept was almost unknown until the 1990s. Nevertheless, since then, the idea has come to motivate an ever-widening circle of public sector actors and environmental groups. As a result, urban forest projects are shooting up across Flanders.

Starting from this initial observation and informed by Foucauldian genealogy, this chapter aims to explicate how urban forest discourse gained prominence in current land-use debates. According to Foucault (1972: 54), discourses are practices that systematically form the objects of which they speak. Thus, discourse is not just about ideas and concepts, as represented in speech and text, discourse is also actively "practiced". The purpose of discourse analysis is to reveal how the words we use to conceptualize and communicate end up producing the very things or objects of which we speak (Graham, 2011). A genealogical approach to discourse analysis traces the development and constitutive or political effects of discourses over time while also showing the possibilities that are excluded. Strategies that permit control over discourse have a major role to play in triggering these effects. These activities include, for example, using appealing storylines to capture policy attention and mobilize policy action (Van Herzele, 2006). But equally important, controlling and channelizing the discourse also implies discursive strategies of exclusion, for instance, discrediting alternative interpretations (Winkel, 2012).

In what follows, it is first shown how a particular discourse of forest expansion—and a variant of it focusing specifically on urban forest creation—has its roots in a long-lasting political struggle for attention and power by a small forestry sector group. For this group, the stakes were related to the position of forests within spatial politics as well as the identity of the forester as a professional. Next, the case of a park forest project near Ghent, Belgium, illustrates what can happen when a forest-centred discourse comes to interfere with local understandings and politics and, more broadly, a city-centred discourse of planning professionals. The data for the case study were obtained for the period 1997 to 2007 from various documentary sources, including official policy texts, commissioned reports, opinion pieces in magazines, press releases, public reactions in meetings and on the internet, 100 recorded interviews in the field, and also confidential material, such as minutes from meetings, and the exchange of letters between politicians, officials, and others (Van Herzele, 2006; Buizer and Van Herzele, 2012). The chapter ends with some recent developments (up to 2013) and a brief reflection on the changes in discourse and strategy that have occurred since the research was first conducted.

First Struggles for Political Attention

Whereas policy discourse on the creation of urban forests became evident in the 1990s, it is rooted in the wider political debate on forestry dating back to the early 1970s. At that time, forest policy was a nationally regulated matter. Those concerned with the forest observed that forests in Flanders were increasingly under the threat of urbanization and its effects. They accused the state of not doing enough to protect the forest and in some instances even taking an active role in forest-destructive projects (as an initiator or at least a favourable party in political bargaining with developers). Another issue of complaint was the apparent imbalance in political attention. Resources for forestry were mainly directed to the Walloon part of Belgium, where almost all of the state-owned forests were located. The national forestry administration was even spoken of as "a French-speaking bastion".

In 1970, a handful of progressive foresters—dismayed with the unfavourable forest situation in Flanders—founded the "Flemish Forestry Association" (FFA). Its aims were presented as three key messages: forest preservation, forest expansion, and multi-functionality. In this context, a first storyline for forest expansion was launched (Van Miegroet, 1971). The story started with the history of deforestation in Flanders since 1846 and contrasted it with the afforestation policy in Wallonia over the same period. And, as is typical for policy stories, it then made the move from diagnosis to prescription for action: it was claimed that Flanders should be reforested to the level it was in 1846.

Meanwhile, FFA members took every opportunity to alert the politicians to their cause. They widely promoted the forest with campaigns, such as tree-plantings and "the week of the forest", which were all covered extensively by the media. Much was achieved during the 1970s and 1980s. The strong demand for regionalization of forestry matters gradually turned into reality: a Flemish forest administration was established, the need to open up the state forests for recreation was accepted, and increased budgets were set aside for forest acquisition and recreational equipment. The concept of multi-functional forestry was secured as a basic principle in the new Forest Decree of 1990. It can be concluded that a small circle of foresters had actually succeeded in attracting political attention and initiating some important changes in established institutional rules.

The term "small" is significant here. Successes were based largely on the charisma of individuals and their relations with those in key political and administrative decision-making positions, as well as in the media. This was a weak foundation for a more broadly and substantially based support. The FFA was unable to raise its number of members above a few hundred and its image remained one of a "tribe of experts".

In the same period, ecologists (and other nature defenders) began to set themselves up as leaders of the nature conservation movement. The need to protect endangered species and natural habitats received much political attention. Institutionalization followed rapidly with a (framework) law on nature conservation (1973), and later on, the establishment of a separate administration for nature (1980) and the institute of nature conservation (1985). But the division of forest and nature institutions also fuelled controversy. Specifically, it positioned foresters and ecologists at opposite ends to one another in an arena of competition. Also, their discourses on what "nature" is about and what it should look like appeared to be largely incompatible. The nature concept of conservationists—strongly influenced by Dutch examples—was primarily directed at open types of vegetation and less to woodland. Their views were perceived as hostile, both to the forests and the foresters. Moreover, the fact that their message quickly came to attract great political attention became a new source of frustration to foresters. The design of the "Main Green Structure Plan" for Flanders (early 1990s) is one striking example. Both the perceived lack of appreciation of forestry and nature conservationists' claims for authority led to great indignation on the part of foresters. But what has constrained forest discourse from gaining a wider appeal? In the next section closer attention is paid to the concepts and ideas central to the professional forest discourse.

Close-to-Nature Forestry and Multi-Functionality

At Ghent University a particular "close-to-nature forestry" model was promoted, as traditionally applied in Switzerland, Germany, and Slovenia. The basic idea is to use the natural processes in primeval forests as a prime source of inspiration. Accomplishing "ecological stability" is the ultimate goal (Figure 4.1). This model favoured an idealized image, representing the forest as a vast area under permanent tree cover, with closed canopy and trees of various ages, in a mature (or "climax") stage of development, closed-off from external disturbances, and importantly, under the professional control of the forester.

For a long time, the tree-oriented and managerial vision suppressed alternative visions introduced by scientists outside the forestry sector, most notably the creation of habitats (e.g. gradients in forest edges and gaps) supporting species of conservation concern. Forestry scientists continued to marginalize such biodiversity-oriented approaches for being "entirely artificial" and unable to meet the wider societal demands, such as forest recreation and wood consumption (Lust, 1980). And what is more, "the forester" was presented as the only professional qualified for the forest-management task. For example: "Under no circumstances, forest management may be consigned to persons who have a one-sided approach to the forest, who are not familiar with the forestry methods and who do not know

Figure 4.1 "The close-to-nature forestry" concept—influential in discourse on forest expansion—uses natural processes in primeval forests as a prime source of inspiration.

Source: Ann Van Herzele.

the international forestry world" (Van Miegroet in FFA's newsletter, July 1989). By positioning their own professionalism at the centre, the circle of those taking part in the discourse became extremely limited. The exclusionary effect was truly felt by ecologists, for instance (personal communication, 2004): "In those times, in order to be allowed to put in a word about the forest, one had to be an agricultural engineer in forestry studies and preferably a graduate of Ghent University too."

Another concept in forestry science—which later became strategically crucial—is multi-functionality. Whereas the idea was basically inspired by the concept of "multiple-use silviculture" developed in the US, it was particularly the narrowly production-oriented forest policy after the World Wars that had fuelled a counter-discourse as a consequence, giving greater prominence to environmental and social aspects of forestry. The broadening view was also in line with the increasing demand for outdoor recreation. Until that time, the majority of the forests (including those publicly owned) were not accessible to the general public. However, the emerging ideas of "social forestry" led to ambivalent feelings among state foresters. On the one hand, the philosophy of care and respect—also related to the close-to-nature forest concept, representing the forest as a "living entity" and "complex ecosystem"—appeared difficult to reconcile with recreation. Forest recreation was spoken of as something "unavoidable" and efforts should go to "guiding away visitors" and "damage limitation". On the other hand, it was acknowledged that the concept of multi-functionality held potential for improving the forestry sector's position in the political arena. The concept was not only helpful in drawing attention to the importance of forests, it also provided the basis of an influential storyline for forest expansion and, later on, its "urban" variant.

Creating an Urban Forest Storyline

The integration of different land uses in the planning of open space in Flanders was the theme of a Green Space Strategy conference (1988). The timing of this platform for discussion was crucial, as it coincided with the preparation of significant changes in the planning system. The forestry sector advocated that the forest should be considered as a distinctive planning entity. As it was at the time, forests were divided into various planning zones, such as nature protection, recreation, and more. It was argued that planning was completely ignoring the intrinsic multi-functionality of the forest. In the foresters' view, the forest should be perceived as a physical entity with spatial impact, out of which a range of potential functions automatically follow (Dua, 1988). Thus, they were primarily concerned with valorizing the intrinsic multi-functionality of the forest itself, rather than encouraging the multifunctional use of open space in general.

The forest-based concept of multi-functionality also formed the cornerstone of the forestry sector's demand for forest expansion. A "forest expansion" group was formed with active FFA members. They assembled an appealing story line, beginning with the "fact" of the extremely low forest cover in Flanders, compared with several other European countries. This fact was put in contradiction to the exceptional importance of forests as they fulfil a multiplicity of functions, all of which are beneficial to society for several reasons. This led to the evident observation that as a result of the low forest cover these beneficial functions can only be fulfilled to a limited extent. The story then made the normative leap from facts to recommendations, indicating that in order to fulfil all these functions the forest cover should be expanded by 50 per cent by 2050 (1,000 hectares every year) (Muys et al., 1988). The strength of the storyline was not so much in the plausibility of facts and assumptions. It was the particular framing of multi-functionality in relation to space that created a representation of the forest as a norm on its own terms and, subsequently, forest expansion as the most logical decision.

Throughout the 1990s this storyline was used continuously to advance the cause of forest expansion, a prime example being the Long-term Forestry Plan (launched by the forest administration in 1993). On this occasion, the storyline was placed within the emerging sustainability discourse following the UN Earth Summit, 1992. Specifically, the Flemish forest situation was linked to industrial countries' responsibility for the decline of the world's forest cover. However, what is most critical here is that a forest expansion programme was outlined in maps of Flanders showing the priority areas for afforestation from a variety of functions. From this first "spatial translation" of discourse it became evident that afforestation near urban areas would sustain many forest functions and thus produce the most societal gain. Subsequently, the term urban or city forest was introduced and fitted perfectly within the existing forest-expansion storyline. However, apart from the precondition of a vast surface area—a minimum of 100 and preferably 500 hectares—the concept was not defined specifically. Despite its legal base (the 1990 Forest Decree), the Long-term Forestry Plan was never formally approved. Nevertheless, the urban forest idea became increasingly used. Later on, the sector's spatial vision of forest expansion was elaborated further in the 1996 "Desired Forest Structure for Flanders".

Institutionalization of Forest Expansion Discourse

The demand for forest expansion was actively pursued in the run-up to the Spatial Structure Plan for Flanders (SSP), whose aim was to provide a clear strategy and a frame of reference for future spatial development. Thanks to its prior visioning work, the forestry sector positioned itself as a well-prepared and convincing partner in the negotiation process. As a result, a forest expansion target of 10,000 hectares was inscribed in the formally approved SSP (1997). It stipulated that forest expansion should be principally linked up with existing forests in function of nature development or in the vicinity of urban areas (for instance, peri-urban forests in sparsely forested areas). The Flemish government was given the task of designating the areas for forest expansion in the regional land use plans (RUPs).

Whereas the claim for forest expansion was now formalized and translated into one of the SSP's binding provisions, this had more to do with the quantification of hectares than with the discourse in its entirety. The SSP focused on creating one coherent, structuring framework consisting of river valleys and well-connected areas of open space. The structure-based discourse implied that coherence and connection were the norms for open spaces rather than particular land uses. Furthermore, a largely nature sector-led view was followed which placed forests in different spatial categories such as "large units of nature" and "nature connection areas". This still left the forestry sector with great uncertainty. It was felt that the principle of multi-functionality was once again denied. It was feared, moreover, that both spatial and nature policies were going to deprive them of their competences (Vanhaeren in FFAs newsletter, January 1997). However, apart from that, the SSP was acknowledged as the starting point of serious attention being given to address the forest expansion issue. As a regulatory document, it also helped spread the urban forest concept throughout Flanders and even take root in local politics.

The subsequent establishment of a Forest Expansion Unit was another important step in the institutionalization of the forest expansion discourse. From the beginning it focused on the urban environment, "where the forest combines most of its functions" (Vitse in "Eigentijds", March 2001). It was its ambition to provide each city and town in Flanders with a forest. A three-step strategy was developed: (1) Location: where should the urban forest be located? (2) Concept: what should the urban forest look like? (3) Implementation: how should the urban forest be realized? It was expected that local politicians would more likely accept the urban forest idea if it was presented gradually. It was decided therefore to keep the first two steps of decision-making out of wider social debate, so as to enable "common interests" (i.e. urban forest creation) to prevail over the "personal interests" of local residents (Nachtergaele et al., 2002). Thus, the attribution of interests was another exclusionary strategy to anticipate and negate public criticism.

Reversing the Storyline

Meanwhile, the FFA and the spatial planning department at Ghent University were commissioned to undertake a pilot location study for creating a 300-hectare city forest near Ghent. This yielded an interesting combination of forest and planning discourse, also reflected in the study's report (Van Elegem et al., 1997). The text starts telling a structure-based "open space storyline", pointing to the loss of coherence and connection of open spaces under pressure of urbanization. It then continues with the urban forest storyline, framing the problem of forest shortage as the incapacity of the existing forests to provide the desired multi-functionality. Urban forest creation was legitimized further as it followed from both "scientific inquiry" (the Long-term Forestry Plan) and "societal consensus" (the SSP). Remarkably, the line of the story changed direction completely when it came to action. The city's, and not the forest's, multi-functionality was now the starting point (including the problems related to it, e.g. urban liveability). Urban greenspace—such as the city forest—was presented as elements that make up the urban image and structure. It was recommended that when creating new greenspace one should search for existing urban structures and "create proximity and connection" to them. In this way, a more pleasant living environment for urban dwellers could be created. So, the city forest was presented more as a possible means or solution than an end in itself. And, the urban-centred view urged to look from the city to the forest rather than the other way around (Van Herzele and van Woerkum, 2011). Reversing the storyline also implied that the forest was no longer acknowledged as a distinct entity, let alone a privileged one, and that other kinds of greenspaces might be part of the "solution". The study's engagement with searching for a forest location, however, did not allow such a conclusion.

In this case, the planners' concern with creating interconnected open-space structures was nicely reconciled with the foresters' desired image of a large entity, which would ensure the multi-functionality of the forest and its professional management. Hence, the 300-hectare "budget" was to be spent in order to achieve one unbroken forest unit. Furthermore, the study fostered a "city forest profile" that was a close copy of the close-to-nature forest image. For instance, rapid development of a "real forest climate" was to be enabled through planting fast-growing tree species. The forest-centred view also had great implications for the way in which acceptable usages were defined: "The target group mainly consists of urban dwellers searching for quietness and enjoying natural beauty. Active recreation should be avoided, yet, the creation of a limited space for play forest within the urban forest concept is within the bounds of possibility" (Van Elegem et al., 1997).

Multi-criteria analysis was used to test possible locations for their structure-strengthening, recreational, and ecological potential. Finally, a feasibility test was performed to deal with the acceptability of the suitable locations for other land-use sectors (agriculture, nature conservation, industry, etc.), which all were considered "competing categories". However, apart from proposing two best locations, no definitive choice was made and the results were still open for discussion.

Selling the Forest to Local Politicians

It became rapidly clear that the city forest was not an image that could easily be "sold" locally. The municipal government of Lovendegem was quick to oppose turning agricultural land into forest. Hence, further efforts in lobbying concentrated on the other location. This area—called Kastelensite—obtained a high ranking mainly because of its potential to keep separate the residential development of the city of Ghent and the municipality of De Pinte. However, politicians in De Pinte expressed fear that the forest would attract swarms of urban people. The city forest also evoked a strong connotation with "the city", of which local residents did not want to become a part. But that concern was counteracted in part by the argument that the forest could put up a barrier to the city (FFA personal communication, 2004). In a formal statement (March 1999), the city of Ghent declared its readiness for collaboration, but it also criticized the location study because it lacked any consideration of "how the forest project would be integrated into the present cultural landscape with its valuable landscape elements". Also in the newspapers there was much talk of "destroying the beautiful meadows" and "chasing away the farmers".

Nevertheless, the concept of a "massive forest core" (being a "true city forest") continued to be used in promoting the project (FFA press release, July 2000). In 1999, an EU-funded Life Environment project was started with the prime objective to create a firm societal support base for the "City Forest Ghent" (De Vreese et al., 2004). The project initiators (Flemish forest administration, province of East-Flanders, FFA) formed the "Bossanova" alliance to actively promote the city forest to the public and the local politicians. But despite intensive lobbying, the politicians remained reluctant. For example, during an election debate (September 2000) the Ghent deputy mayor of environment stated: "Just to be perfectly clear: city forests are not necessarily forests!" He opposed the idea of "building up" the area with trees, which would destroy its distinct character. In the same period, the Spatial Structure Plan for Ghent was in the making. The planners took the need for forest creation into account, but they started from a structure-based vision emphasizing the coherence and connection of open spaces (i.e. the open space storyline). The enhancement of the urban quality of life was a major concern. In this context, the concept of the four "groenpolen" (large multifunctional greenspaces in the urban periphery) was formed, as a main part of the city's green structure. The Kastelensite was included as one of these areas.

After the elections (October 2000), the new political coalition of Ghent declared its commitment to the realization of the four "groenpolen" in its governmental agreement (2001–2006). Although they aimed to create new forests in these locations, this was not the case in the Kastelensite, for which the "preservation of the present landscape values" was among the main action points. The local nature movement, the Minister of Environment, and the FFA reacted with disappointment, arguing that the location had been selected through scientific investigation and departmental budgets were already in place.

Changing the Forest's Image

Increasingly aware of the importance attached to the present landscape, Bossanova decided to change the name of the project from "City Forest Ghent" to "Park Forest Ghent" (December 2000). It was also realized that a more "consumer"-oriented view (focusing on scenic and recreational values) was necessary for promoting the forest more widely. This also implied a shift from avoiding active

Figure 4.2 The forest image was changed from one massive entity to an open concept of interacting land uses.

Source: Preparatory study for the RUP in 2001.

forms of recreation to actually promoting them. For instance, an article promoting the project made a new point when speaking of horse and mountain bike trails along and through the forest (Embo in the province's environmental magazine, November 1999).

In January 2001 the project was integrated in the planning process for the RUP. In this context, the extended area of Kastelensite-Scheldevelde was to become a "city landscape park": a multifunctional area of 1,200 hectares with a dominantly open-space character and including about 300 hectares of new afforestation. The process was coordinated and led by the regional Spatial Planning Division and was followed by a steering group consisting of representatives of various regional administrations, as well as the three municipalities involved.

The joint discussions led to a thorough revision of the initial plan: splitting up the forest into several units, ranging from three "core forests" to numerous small forest patches spread over the area. Thus, the urban forest concept was fundamentally changed, that is, from a massive single entity to a more open concept of interacting land uses. In the preparatory study for the RUP in September 2001, the new urban forest concept was presented in a structural sketch—showing a mixture of areas, including different types of forest, sustainable agriculture, habitat creation areas, and so on, but with smooth and fluent transitions between them. Special attention was given to scenic qualities, such as borders and gradients of transparency. The new image also infiltrated the project's campaigning. In Bossanova's Park Forest magazine, photographs with dense forest stands populated with squirrels and woodpeckers were replaced with pictures of meadows with trees and cows.

Growing Concern among Local People

While the RUP study brought the public administrations and politicians to agreement, concern among local people was growing. Farmers and their organizations continued to complain about the legal uncertainties caused by the project, which they thought would worsen by the fragmentation of the forest into multiple entities. Local residents welcomed recreational equipment, but the present landscape was a sensitive issue to them. Personal letters to Bossanova and interviews in the field revealed that they wanted to keep the landscape like it was. Creating a forest not only would change the landscape but also people's relationships with those who made and still maintain it, and create a dependency on those institutions being given the management task.

In October–November 2002, Bossanova organized a series of information meetings. The presentation of the structural sketch began with a shortened, more institutionalized version of the urban forest storyline, including the legitimization of both forest expansion by the SSF and site selection by the location study. Despite the detailed information given during these sessions, the farmers and local residents continued to question the very idea: Why is the forest being planned here? Is there a need for forest at all? Clearly, the logic of reshaping the open landscape into a forest was not fully understood. The residents were also concerned about the

Figure 4.3 Farmers' protest against the Park Forest project: "Where may I graze now?"

Source: photo by Griet Buyse.

practical implications: safety, property rights, privacy, tidiness, etc. However, their worries were often denied as too personal or less relevant or as something to be handled by the RUP's formal public consultation procedure.

From Structural Sketch to Land Use Plan

Early in 2002, the Flemish Land Agency became involved in the project. Consultations with individual landowners/farmers were held, in particular, for translating the structural sketch of the RUP into a detailed, parcel-wise land use plan. As a result, the agreed-on Park Forest concept was transformed into a segregated landscape with demarcated strips and parcels of land for singular land uses. For reasons of legal certainty, farmers wanted agricultural land use to be interpreted in its strict sense, implying that elements like "field forests" and "edge forests" were rejected. Thus, due to legally established procedures and farmers' expectations about them, the new urban forest concept was reduced to a juxtaposition of strictly delineated land uses. Remarkably, however, the initial afforestation target in terms of hectares of land remained largely untouched. The so-called "balance" was restored to a great extent by enlarging two of the core forests, resulting in a total of 285.5 hectares of afforestation. Thus, the spaces to be afforested were moved and rearranged or adjusted so that the desired forest expansion remained intact.

In December 2005, the Flemish government approved the RUP. The plan also included a regulation for the compulsory purchase of properties. In practice, however, land acquisition was provisionally settled out of court. In July 2007, a cooperation agreement was signed between the Flemish government and the Province of East Flanders. The former will acquire the land (between 2011 and 2014) and finance the project management, while the latter will coordinate the practical aspects of implementing the project on the ground (2012–2018).

Recent Developments

The Park Forest Ghent became a flagship for urban forest creation in Flanders. The project also received quite a bit of scholarly attention (e.g. Allaert and Leinfelder, 2005). Implementation practice, however, has had its share of ups and downs, and this was also the case in the rest of Flanders. Although nearly twenty urban forest projects were initiated between 2000 and 2003, political attention (at the regional level) slackened soon thereafter (Ledene et al., 2011). An unfavourable event was the abolition of the Forest Expansion Unit in 2005. Also, many urban forest initiatives remained on paper, partly because the projects were hampered by local politics and RUPs were still lacking. But there were also signs of progress. In the "Flanders in Action Pact 2020" it was claimed that at least half of the cities and towns should have an urban forest (or started a project) by 2020. The Flemish government agreement 2009–2014 was built upon this pact, and also the policy notes of several ministers referred to the need of urban forests. Meanwhile, municipalities increasingly started urban forest projects by themselves, eventually labelling these "play forests", "birth forests", "peace forests", etc. The government's mobilization of the "forest compensation fund" (revenues from licenced deforestations) was an important driver for such initiatives. Furthermore, there appeared several new supporters of the urban forest idea, such as private firms, public service agencies, and non-governmental organizations. It is notable, moreover, that the nature conservation movement is increasingly advocating the case of forest expansion, and urban forest creation in particular.

Concluding Observations on Discourse and Strategy

Throughout the years, the urban forest storyline has been told and retold, however, with some updates and adjustments. The most marked change is in the forest functions cited to promote urban forests. The new emphasis is on climate regulation and sustaining human health. Forest functions can indeed be flexibly adapted to the dominant discourses of the time. However, it is not just a tactic to reaffirm the importance of urban forests. The new functions appear to reflect a change in urban forest discourse towards a more people-centred concept. For example, a recent visioning report by the FFA (Ledene et al., 2011) defines (peri-)urban forests as being close to people, meeting the needs of urban dwellers, and forming part of their local community. (Peri-)urban forests are described as green areas with a balanced mix of closed and open landscape, including forests, nature areas, and/or "park structures". Thus, urban forest discourse is made accessible to a wide range of groups, is no longer disciplinary-controlled by forestry alone, and is more open for place-based interpretation. However, when it comes to achievements, the talk remains rooted in the old forest-centred image, for instance, evaluations are made in terms of "effectively afforested" land.

Likewise, in public campaigns the link with people is sometimes made explicit—e.g. "A Forest for Everyone"—but the numbers of planted trees are placed at the centre of the policy narration. At the ceremony of planting the millionth tree in

Flanders (October 2011), the Minister of Environment presented a new tool (the "Boswijzer") designed to "objectively measure" the evolution of forests in Flanders against the 2010 baseline. It is relevant to note here that numerical measurement also serves as a tool for political mobilization. Measuring a problem creates pressure to do something about it (Stone, 1988: 131). The selection of the measures can be seen as a strategic component of problem definition, especially when the outcomes are dressed up in a policy narrative (Fischer, 2003: 172).

In this regard, the old urban forest storyline is increasingly replaced by a new storyline, which has as its core the failure to realize ambitions (e.g. in Natuurpunt, 2013). Numerical measurements (hectares afforested/deforested, numbers of projects, euros spent) are confronted with the goals set out in the government's plans and agreements. The next move is to portray the gap between intentions and practice as a consequence of the failure of the government to provide the highly necessary forests. The story concludes with stressing the need to increase efforts and proposing regulatory ways to do so.

The shift in strategy is not surprising. The need for urban forests was already agreed, the discourse reached hegemony and became embedded in the institutions and practices of government. So, today's efforts for urban forest creation need to concentrate on implementation. However, discursive strategies are also suggestive of the dominant discourses they seek to appeal to. In this case, the institutionalization of urban forest discourse—and its subsequent translation into numerical targets, maps, budgets, regulations, etc.—has led to a formalization and standardization of discourse. The discourse has gradually become enclosed within the formal structure of institutions, including its sets of rules, competences, procedures, techniques, vocabularies, etc. These ultimately limit or condition the possible ways of looking at a problem or situation (Van Herzele and Aarts, 2013).

To conclude, the latest developments do raise some important questions about the effects of discourse institutionalization, in particular at the local level where urban forests are to be implemented. The Park Forest Ghent case revealed that the project's legitimation in terms of government plans and scientific study could not convince the local people. Likewise, figures of regional (or even local) afforestation targets and achievements would make no impression. Place-based approaches are required instead to make sense of urban forest projects, to mobilize locals and enable them to actively participate in the development of a discourse. In this regard, a further genealogy could focus on the urban forest discourses that develop and are distributed locally. But following Foucault (1981), it is very likely impossible to account for the positive and multiplication effects, if we also do not take into consideration the exclusionary and constraining function of the strategies in use. The interplay of mobilization and exclusion is a challenge for future research!

References

Allaert G. and Leinfelder H. (eds) (2005). *Parkbos Gent: over visievorming en beleidsnetwerking.* ·Academia Press, Ghent.

Buizer, M. and Van Herzele, A. (2012). 'Combining deliberative theory and discourse analysis to assess the deliberative incompleteness of centrally formulated plans'. *Forest Policy and Economics*, 16: 93–101.

De Vreese, R., Van Herzele, A., and Konijnendijk, C.C. (2004). Case study Ghent report. NeighbourWoods: EU Fifth Framework Programme QLK5-2001-00165.

Dua V. (1988). 'Meer ruimte voor bos?' In: *Ruimte voor Groen, deel I.* Vijfde Vlaams wetenschappelijk congres over groenvoorziening.Vereniging voor Groenvoorziening, Gent.

Fischer, F. (2003). *Reframing Public Policy: Discursive Politics and Deliberative Practices.* Oxford University Press, Oxford.

Foucault, M. (1972). *The Archaeology of Knowledge.* Routledge, Abingdon, UK.

Foucault, M. (1981). 'The order of discourse'. In: R. Young (ed.) *Untying the Text: A Post-structuralist Reader.* Routledge, London.

Graham, L.J. (2011). 'The product of text and "other" statements: discourse analysis and the critical use of Foucault'. *Educational Philosophy & Theory*, 43(6): 663–674.

Konijnendijk, C. (2003). 'A decade of urban forestry in Europe'. *Forest Policy and Economics,* 5: 173–186.

Ledene, L., De Pril, S., Couckuyt, L., and De Somviele, B. (2011). *Stads(rand)bosen: een randgeval?!* Vereniging voor Bos in Vlaanderen. www.bosplus.be/nl/kenniscentrum/publicaties/beleidsdossiers (accessed 23 March 2014).

Lust, N. (1980). 'Bosbouw op bio-ecologische basis'. In: *Natuurbehoud en landschapszorg in Vlaanderen* . Derde Vlaams wetenschappelijk congres over groenvoorziening. Vereniging voor Groenvoorziening, Antwerpen.

Muys, B. (spokesman) et al. (1988). 'Bosuitbreiding'. In: *Ruimte voor Groen, deel II.* Vijfde Vlaams wetenschappelijk congres over groenvoorziening. Vereniging voorGroenvoorziening, Gent.

Nachtergaele, J., De Vreese, R., Vanhaeren, R., and Van Slycken, J. (2002). 'Realizing urban forests in Flanders'. In: COST Action E12 Urban Forests and Trees, Proceedings No 2. Office for Official Publications of the European Commission, Luxembourg.

Natuurpunt (2013). Bosarm Vlaanderen. FACTsheet, April 2013.

Stone, D.A. (1988). *Policy Paradox and Political Reason.* Scotts Foresman & Co, Glenview, IL.

Van Elegem B., Embo T., and Allaert G. (1997). Studie van de bebossingsmogelijkheden en de afbakening van een regionaal bos en een stadsbos in de regio Gent, deel 1: het stadsbos. Vereniging voor Bos in Vlaanderen en Universiteit Gent, Gent.

Van Herzele, A. (2006). 'A forest for each city and town: Story lines in the policy debate for urban forests in Flanders'. *Urban Studies,* 43(3): 673–696.

Van Herzele, A. and van Woerkum C. (2011). 'On the argumentative work of map-based visualisation'. *Landscape and Urban Planning,* 100: 396–399.

Van Herzele, A. and Aarts N. (2013). '"My forest, my kingdom"—Self-referentiality as a strategy in the case of small forest owners coping with government regulations'. *Policy Sciences,* 46: 63–81.

Van Miegroet, M. (1971). 'De positive van het bos in Vlaanderen'. *Groene Band,* 1: 1–43.

Winkel, G. (2012). 'Foucault in the forests—A review of the use of "Foulcauldian" concepts in forest policy analysis'. *Forest Policy and Economics,* 16: 81–92.

5 Institutions, Law, and the Political Ecology of Urban Forests

A Comparative Approach

Blake Hudson

Introduction

The governance of urban forests involves a number of cultural, sociological, economic, political, and ecological influences. An additional influence is the institutional and legal context within which urban forestry is situated—a context that provides the overarching framework within which urban forests are created, protected, and managed, on the one hand, or mismanaged or entirely eliminated, on the other hand.

While the urban forest can take many shapes or forms, and no one group will agree on what an urban forest will or should be, one thing is clear—institutional and legal foundations influence the management and protection of the urban forest. At the same time, there is clearly a political element intertwined with institutions and law—an element that plays out significantly at the intersection of forests and urban development. There are important synergies between the institutions within which urban forest policy-makers are embedded and the political choices that drive the utilization and perpetuation of those institutions.

The legal institutions which political actors access both shape and are shaped by the management of the urban forest. Consider the example upon which this chapter primarily focuses, namely, how the constitutional structure of certain federal systems of government can complicate forest management on subnational scales, and in particular urban forest governance. Federal systems of government like Canada and the US divide regulatory authority over certain subject matter between the national government and the numerous subnational governments within their borders (states, provinces, and their constituent local governments). Written constitutions establish the legal framework within which these federal systems operate. This is, of course, an institutional component that can provide governance and resource-management strength or weakness depending on how that authority is allocated. A political component is how that legal framework is, or is not, utilized or adapted, or how laws arising out of the framework are, or are not, enforced and implemented to achieve policy and management goals. So consider the scenario presented by some federal systems whereby subnational governments maintain virtually exclusive control over forest resource management. While some subnational governments may undertake effective management of forests—in the urban context by providing greenbelts, open space, development "limit lines," treed parking lots and roadways, wildlife corridors and habitat, among a variety of other values and services—others may maintain little or no forest protection and encourage policies that replace forests with sprawling development. If there are gaps in subnational government policies related to forest protection and management, but no legal mechanism available to a higher-level governmental authority to more effectively coordinate or institute those policies, then forests in certain urban areas may succumb to increased development activities to the detriment of numerous valuable services they provide.

This institutional state of affairs can provide a useful lens through which to explore the influence of law in political ecology. Political ecology, of course, provides a broad perspective through which to study the management of urban forests. To be clear, this chapter does not seek to fully engage that broader perspective (see Perkins, 2011), but rather to take the approach of a legal scholar focusing on institutional influences in urban forest political ecology. The institutional perspective demonstrates how, once society makes a political calculation that a federal system of government, with a certain allocation of governance authority among levels of government, is the preferred institution, institutional inertia creates a feedback loop whereby the institution influences future cultural attitudes and related political action. In other words, law and policy related to urban forests may tend toward "over-decentralized"[1] institutions within federal systems without major shifts in cultural attitudes and political objectives that call for more coordinated action at higher levels of government. This "over-decentralization" may lead to the vast majority of urban forest policy being formulated—or more importantly not formulated—by state/provincial or local governments. These state/provincial governments, in turn, may leave urban forest management decisions solely within the hands of numerous local governments, who then may leave such decisions solely within the hands of varied private-property owners who may have interests at odds with maintaining and managing an urban forest. In this way, an institutional perspective also lays a foundation for assessing why in certain parts of the US and Canada private-property owners maintain the authority to appropriate urban forest resources, whereas in other parts of the country state or local governments take a more direct role in order to preserve the broader public's interest in the urban forest.

This chapter seeks not to answer the question of the "optimal" allocation of governance authority across levels of government in federal institutions, nor to prescribe normative solutions for the potential problems caused thereby. Rather, this chapter seeks to highlight the importance of institutions and law as one influence in the political ecology of urban forests. It thus grapples with central questions in the political ecology literature (McCarthy, 2006; Perkins, 2011), that is, who gets to make decisions regarding the management of the environment, why they make the decisions that they do, and in turn, who wins and who loses in each decision-making scenario that arises. In this way, hopefully society can consider the political *and* institutional changes that may be needed to preserve the varied urban forest landscapes in federal (and other) governmental systems.

Before highlighting some salient institutional dynamics related to urban forests in federal systems, a couple of points of clarification should be made. First, this chapter addresses a particular type of "urban forest"—a type that diverges somewhat from the usage of the term in other parts of the book. The chapter uses the term "urban forest" to signify the intersection between current urban areas and undeveloped rural lands. In other words, it is the urban forest of tomorrow upon which this chapter focuses (Figure 5.1). Under this conception, the term "urban forest" includes a key temporal component, because as populations increase and development expands society has opportunities *today* to establish laws and policies related to the shape and form of development that will impact the quantity and quality of forests within urban areas of the *future*.

Within the context of this chapter's case study, another way to think about this particular definition of urban forest is to ask: when does a lack of federal authority to coordinate subnational forest policy actually present a problem for urban forest management? Though debatable, one might argue that it is not such an important problem within the jurisdictional limits of current, high-density urban areas, as these are already in large part defined by development that replaced natural systems long ago. It might be difficult to forge a coherent policy at a national level to craft site-specific reforestation of areas already developed—compulsory ripping up of

Figure 5.1 Forests currently on the outskirts of urban areas will be the urban forests of tomorrow, and should be managed with that potentiality in mind.

Source: Blake Hudson.

sidewalks, buildings, roads, and other urban infrastructure, or coerced planting of abandoned city fields on an acre-by-acre basis by a national government in a federal system seems like an approach that could be neither particularly effective nor politically viable. This realm of urban forestry might very well be better left to subnational government policy without prescriptive federal inputs (though federal incentives may play an important role in spurring subnational action). Importantly, however, this institutional state of affairs has crucial implications for how the "frontiers" of the urban–rural forest intersect are managed. Though these frontiers are often ignored when discussing urban forestry, we must begin to look beyond the structure, makeup, and management needs of today's urban forest—i.e. trees in currently established urban areas—and envision the needs of tomorrow's urban forest. As municipalities continue to grow and expand, most often replacing natural capital with human-built capital, how will we integrate already standing, rural forestlands into new developments? And how do we prevent large swaths of standing forests in the path of rapid urbanization from being replaced with development in the first instance? These are perhaps the most important questions, especially since our failure to anticipate the need for forest integration into urban areas—and protection of some forests from urbanization—during the original development of many (if not most) lands is one reason we struggle to reintegrate forests into urban areas today.

Unfortunately, as evidenced by rapid development sprawl in urban areas in both Canada and the US, subnational governing bodies are failing to plan for future urban forests. Rather, closely tracking the decentralized allocation of regulatory authority present within these federal systems, urban forest policies are developing in a piecemeal manner, with some urban areas managing forests more responsibly than others. In a large part, society is ignoring the new frontiers of development and how progress might be made in a way that better incorporates existing forests into new urban land uses. The federal governments in Canada and the US, if given the proper institutional tools (i.e. legal authority), might very well be critical to establishing limit lines, as just one policy example, around major metropolitan areas so that sprawl does not proceed apace.[2] The establishment of controlled and planned development density requirements outside of those lines may more effectively maintain forest resources and their services. Of course, even when maintaining legal authority, national governments may politically refuse to act. Nonetheless, a prerequisite to the exercise of political will is maintaining the legal authority to take political action in the first instance, if and when the federal governments in these nations choose to act to protect forestlands from sprawling development. To be clear, a massive re-centralization of urban forest policy is not what this chapter advocates. Obviously, there is a wide variety of values provided by local governance, the involvement of private actors and non-governmental entities, the establishment of market mechanisms, and other non-federal, non-prescriptive methods of sustainably managing subnational forests, urban or otherwise. What this chapter is concerned with, however, are scenarios where subnational entities are not doing enough to protect important forest resources as we continue to develop lands on the frontiers of current urban areas.

A second point of clarification is an acknowledgement of the sheer complexity of forest policy in these nations. This chapter fully recognizes that there are many permutations of forest policies within scales of governance and across scales. For example, some forest policies seen at the state and provincial level are very basic, such as streamside buffer zones or clear-cutting requirements. These policies may tend to address (and capture) the extractive value of forest resources more than tending toward forest preservation policies that fully account for biodiversity, aesthetics, recreation, and other values. Other forest policies might take place at the municipal level, and might actually be more in the form of forest preservation, taking into account the many non-extractive values and services provided by forest resources. It may be that we need these latter policies to be more aggressively implemented at the state or provincial level, as are some urban growth boundary programs designed at the state level but implemented at the local level in the US. Or we may need local governments and municipalities to implement such policies more aggressively. We may even need state/provincial and local governments to implement the most fundamental, extractive forest policies if they are lacking. But regardless of whether state/provincial or local government forest policies take the form of basic extractive standards or forest preservation through land use planning, federal-level institutions may need to play a more robust role in setting at least a floor for those standards as each category of standards has an impact on the frontier forests that will constitute the urban forest of the future. Yet this federal role may be complicated by legal institutions and the role these institutions play in urban areas. So though this chapter focuses later primarily on examples of basic, extractive subnational forest policies, those policies are relevant to the more stringent forest preservation-type policies that may be necessary to forestall the threats that encroaching urban development poses to current rural forests and that states/provinces and their subunits may be unwilling to implement.

This chapter lays out briefly the legal landscape relating to urban forest policy in the US and Canada, before discussing the importance of urban forest resources and the threats to Canadian and US forests over the coming decades. Next, the chapter will discuss the politics at the subnational level that lead to a potential need for greater regulatory inputs at the federal level in order to avert these threats. The chapter concludes by highlighting how institutional components can inform a political ecology perspective and perhaps better equip society to improve upon the urban forest condition, especially at the frontiers of urban forests.

The Legal Landscape

Over some subject matters, federal system constitutions may allow concurrent regulatory jurisdiction whereby the national and subnational governments share legal regulatory authority, while on other subject matters constitutions may grant exclusive powers to the federal government and exclusive powers to subnational governments. In circumstances where subnational governments maintain exclusive constitutional authority to regulate resources, governance problems may arise. A "race to the bottom" may occur, whereby subnational governments compete over economic resources to the detriment of environmental values. An inability of the federal government to coordinate disparate, varied, or deficient subnational resource governance policies, or to address the lack of such policies where needed, can then lead to suboptimal protection or management of those resources. This constitutional state of affairs can ultimately preclude a national resource policy that takes into account a broader suite of forest values than subnational policies might alone, and can even complicate national government involvement in international resource governance, since the national government cannot bind subnational governments to international agreements (Hudson, 2011).

These legal constraints dividing governance between federal and subnational governance levels can have effects on local scales, as forests around urban areas may remain mismanaged or unregulated by states or provinces and the federal government, and in the absence of higher-level state/provincial or federal inputs, local governments may choose to increasingly facilitate the appropriation of forest resources by increased development and urban sprawl. These constraints can also have effects at the global level—at least to the extent that countries wish to harness national forest policies to combat climate change. Nearly 20 percent of yearly global carbon emissions over the last couple of decades have resulted from forest loss and degradation. This is an amount greater than that emitted by the global transportation sector each year (Myers Madeira, 2008). Not only is forest destruction a substantial source of atmospheric carbon, about one-third of global carbon emissions are absorbed by forests each year, making forest destruction a loss of the most significant terrestrial carbon sink (USDA, 2011) (Figure 5.2).

Canada and the US alone account for over 13 percent of the world's land base (List, 2013), over 15 percent of the world's forests (Siry et al., 2009), and maintain two of the top ten (carbon-driven) world economies as measured by annual gross domestic product (International, 2011). Even so, the involvement of these two federal systems in cultivating national forest policies to fill gaps where subnational, urban governments are failing to protect forest resources, or in participating in global climate governance related to forests, is potentially hindered by legal constraints that subnational governments may place on national government involvement in forest policy formation (Hudson, 2012).

So where exactly do these legal constraints originate? In the US, the Tenth Amendment of the US Constitution reserves powers not delegated to the federal government for the states. Though the US federal government has gained a foothold over many types of resource management under its Commerce Clause power, it has never attempted to craft direct forest management policy for subnational forests—whether extractive standards or forest preservation standards—because forest policy has been considered a land use regulatory role that is the responsibility of state and local governments (Rose et al., 2005). As a result, state and local governments are responsible for regulating the nearly 65 percent of US forests owned by subnational governments (5 percent) or private parties (60 percent) (UN, 2002). Consequently, the federal government is arguably limited in the types of forest management directives it might provide at the national level, only currently maintaining recognized ability to do so for the 35 percent of nationally owned

Figure 5.2 "Local" forest in south Alabama, USA. This forest is increasingly of global importance as the crucial role of forests in regulating climate change is becoming clearer.

Source: Blake Hudson.

forests over which it maintains constitutional control. Federal legislation aimed at certain types of management[3] would invariably be challenged by state governments and private-property owners as beyond the scope of federal power.[4]

The Canadian Constitution goes even further in demarcating the exclusive spheres of forest policy regulatory authority, explicitly granting the provinces exclusive control over non-federally owned forests (Constitution, 1985). Thus, the provinces maintain control over nearly all of the nation's forests since 84 percent of Canadian forests are owned by the provinces or private parties (Canadian, 2013). The provinces place even more constraints on the Canadian national government in the forest context than do the states in the US because while US state hegemony over subnational forest management has been understood as a "reserved power" for the states in the absence of a recognized enumerated federal power, Canadian provincial hegemony over forests is made explicit in the text of the Canadian Constitution. So while the US may have some flexibility to exert federal authority over subnational forests under the Commerce Clause or some other federal power, the Canadian federal government's options are expressly limited. In addition, and also unlike the US, Canadian courts have definitively found that the Canadian treaty power is constrained by reserved provincial powers.

The Importance of Urban Forests and the Threats to Them

With the above-described legal and institutional landscape in mind, consider the urban forest landscape embedded within that governance structure in Canada and the US, and the services provided by forests in and around urban areas (Figure 5.3):

- clean air services by filtering and trapping air pollutants;
- clean water services that prevent nutrient, chemical, and other non-point run-off from entering waterways;
- protecting fisheries by mitigating run-off eutrophication that leads to "dead zones" where fish cannot survive;
- flood control services;
- endangered and other animal species habitat;
- regulation of local ambient air temperatures in urban and rural areas during the summer;
- energy cost savings for households and businesses;
- a global climate regulator and major carbon sink/source of carbon sequestration;
- aesthetic values;

Figure 5.3 Forested watershed in the southern part of the US state of Alabama. This watershed cleans water, controls flooding, protects fisheries, and provides species habitat, among a variety of other services.

Source: Blake Hudson.

- cultural values;
- recreational values.

Yet the total values of these forest "ecosystem services" (Louman et al., 2009; Rasband et al., 2009) have only recently been seriously studied. Such services are under great strain not only in the developing world, as we so often hear about, but also in developed countries. For example, a recent US Forest Service Report, the "Forest Futures Project," details that by 2060 states in the southeastern US are estimated to lose up to 43 million acres of land to urban development, population growth, climate change, and invasive species, with total forest losses projected to be as high as 23 million acres, or approximately 13 percent of all forestland in the south (Wear and Greis, 2011). This is an amount equal to nearly all the forest acreage in the state of Alabama.

Similarly, though Canadian boreal forests store a great deal of carbon—an estimated 67 billion tons—deforestation of nearly 230,000 acres of forest a year due to cropland conversion and urbanization is a significant source of emissions (World, 2013). At this rate, over 11 million acres of Canadian forest may be lost over the next 50 years. One 2003 report detailed that:

> Winnipeg's urban footprint almost quadrupled between 1971 and 1991, yet its population only doubled. Calgary already takes up as much land as New York City with only a tenth of its population. In the thirty years before the last census, the Vancouver urban area population grew almost 70 per cent, with four-fifths of that growth occurring outside the core cities of Vancouver, Burnaby, and New Westminster. From 1979 to 1996, over 7,000 hectares of land were converted to urban uses in the region, an increase of over 21 per cent in urban land in only 17 years. In recent years 25,000 to 30,000 new housing units have been built annually in the Greater Toronto Area (GTA), most of them in the once rural fringes . . . About two million people, twice as many as 1961, live in the regions outside the city itself. This expansion converted farms and forests into low-density subdivisions. Another million are expected by 2021, which will mean further urbanization of the land.
>
> (Gurin, 2003)

Of course, forest gains may continue to be made elsewhere in both the US and Canada, such as the conversion of former agricultural lands to forests or in areas promoting policies that seek to capture non-market forest values. The key question yet to be answered, however, is whether the new pressures on forests in certain regions will cause deforestation to proceed at a rate that outstrips reforestation in other regions. If so, over the coming decades deforestation will no longer be only a developing-world problem. Without greater coordination of forest policy, either among the states and provinces or provided by some higher US or Canadian governmental authority, destruction of forests in the name of economic development may proceed apace as states and provinces jockey for economic growth and development.

Institutions, Law, and Political Ecology

To this point we have reviewed the legal landscape that complicates Canadian and US federal government ability to fill in gaps in subnational urban forest policies, especially at the frontiers of the presently sprawling urban–rural development intersect. We see that very similar legal restrictions on federal authority over subnational forest management currently exist in the two countries. Next, we highlighted some of the threats that may exploit this institutional state of affairs and which may cause great damage to urban forest resources over the coming decades. To gain a complete understanding of the implications of this institutional state of affairs, and the associated threats, we must understand how politics synergizes with law and institutions and how political choices drive the perpetuation of these institutions either toward or away from sustainable urban forest management policies. In other words, we must understand how institutions and law are a component part of the political ecology of urban forests in these systems.

The first way politics synergizes with the institution that is constitutional federalism in Canada and the US is that the political choices of the respective federal governments are limited to different degrees. In Canada, the express constitutional limitations on federal authority over forests make any political action at the federal level a virtual impossibility. Even if the Canadian federal government wanted politically to gain some foothold over provincial forest management there are no institutional (constitutional) provisions that even come close to allowing it to do so (Hudson, 2012: 947–48). Indeed, the Canadian federal government has to date been unable to even entice changes in provincial forest policy through incentives or subsidies. This leaves constitutional amendment as virtually the only way for the federal government to gain a foothold over provincial forest policy.

In the US, conversely, the current jurisprudential understanding of subnational forest policy as a matter of state and local regulatory authority is not made explicit in the US Constitution, but rather arises merely out of a Tenth Amendment reservation to the states of all powers not granted to the federal government. The US federal government has exercised the Commerce Clause power of its Constitution to gain regulatory footholds over virtually all other resources, from water, to air, to fisheries, to endangered species, to a variety of other resources. Yet, the federal government has never politically tested the extent of its regulatory authority under the Commerce Clause in the subnational forest context. Subnational forest management, more than any other resource category, has been jurisprudentially lumped in with general land use planning for so long—presumably because of the non-transient nature of forest resources—that the federal government seems to view it to be as far removed from federal authority as is a decision to zone part of an urban area commercial and another part residential (a typical example of state and local exercise of land use regulatory authority). In this way it is unclear whether the US federal government is as severely limited as Canada's federal government by constitutional law, or rather, whether it simply has not politically chosen to test the waters of its legal authority. In this way, legal perception is often a political reality in the US, as the federal government politically fails to act on crafting a subnational forest policy based upon perceived constitutional constraints (whether this perception is genuine or rather is a political choice is unclear). Ultimately, however, the US federal government may have more options available to it politically to fill in gaps in subnational urban forest policy and to reign in urban forest-threatening sprawl than does the Canadian federal government.

This distinction between Canada and the US, whereby the US federal government may be more free politically to act on subnational forest policies, could in fact be a benefit to urban forest management in the respective countries because of the second way in which politics synergizes with constitutional federalism in Canada and the US. The distribution of forest ownership between public and private landowners in the two countries may cause the US to be in greater need of federal inputs into subnational forest management than is Canada. Politically, subnational governments in the respective countries operate quite differently with regards to forest policy—at least with regard to the crafting of the basic, extraction-focused forest management standards highlighted earlier. Cashore and McDermott undertook a comparative study of global forest governance standards worldwide, and their results shed much light on how regional factors affect subnational forest policy (or the lack thereof). Cashore and McDermott (McDermott et al., 2010: 7–19) measured the stringency of forest regulations on a scale of 1–10 for countries and subnational governments all over the world, based upon management standards like riparian buffer zones, stand density requirements, clear-cutting restrictions, road clearing prescriptions, and afforestation and reforestation requirements. While the Canadian provinces, the US federal government, and states in the Pacific Northwest maintain fairly stringent basic forest management requirements for forests under their control—nearly an "8" on the Cashore–McDermott scale—the southeastern US states averaged a "1.2." This is a level far below that of even the developing world, where (at least on paper) forest regulations were far more stringent (McDermott et al., 2010: 327).[5]

So while Canadian provinces maintain relatively stringent management standards that are fairly consistent across the nation, US state forest policy stringency varies greatly by region. States in the Pacific Northwest, where 67 percent of forests are publicly owned, track the forest policy stringency of the Canadian provinces to a large degree. Southeastern US states, where 86 percent of forests are privately owned, maintain only weak, voluntary "guidelines." Governance culture in regions with low subnational forest policy standards is therefore greatly impacted by the distribution of forest ownership between private and public entities. Cashore and McDermott attribute this in part to a public forest management "spillover effect," whereby state governments in the US increase the forest policy stringency placed on private forests due to high percentages of federal forests within the state and the heightened management standards on those lands. Since 67 percent of forests in the US Pacific Northwest are publicly owned, the interests of a majority of civil society are aligned with the benefits that public forests provide the entire citizenry, as is also the case with Canadian citizens since 77 percent of forests are publicly owned across the nation. In the southeastern US, however, civil society consists mostly of private-forest owners resistant to government interference.

The fact that a vast majority of Canadian provincial forests are publicly owned—in stark contrast to the 60 percent of US forests that are privately owned—plays a large role politically in how their respective subnational governments manage forests, in both rural and urban areas. The provinces and northwestern US states seem to respond more readily to external demands for heightened forest management standards, such as those raised by environmental, scientific, or other members of civil society, than do US state governments in the southeast. This is likely due to the fact that a government will more readily respond to the demands of civil society regarding the management of its own lands than it will to the management of private lands.

In addition, most of the provincial and northwestern US states' electorate have no legally vested private property rights in the bulk of each region's forestlands. Rather, their interests are likely to be focused on the benefits that public forests can provide the subnational government's citizenry as a whole. In the US southeast, conversely, a higher proportion of civil society is made up of the very private-forest owners who would be regulated, and who are, therefore, more resistant to government interference and who elect politicians who feel the same. Canadian provincial and northwestern US state governments are simply more free to adjust the management of their own policies in response to the electorate's demands than are state governments in the southeastern US. It is true that the provinces may face countervailing pressure from the forest industry, as may the US federal government leasing timber rights on federal forests. Yet, Canadian provincial or US federal government interaction with a relatively unified, handful of industrial players leasing timber rights on public lands is a different matter altogether from southeastern US state governments, who interact with innumerable private-property owners, any one of whom may constitutionally challenge regulatory action as infringing private property rights.

What is going on at the state and local scale in regions like the southeastern US, with high private ownership of forests, may be the entrenchment of a neoliberal political philosophy. The region has never been subject to direct prescriptive regulatory efforts by federal or state governments in the forest sector, except for some requirements tangential to direct forest management under US federal statutes like the Clean Water Act or Endangered Species Act. Therefore, the region has always relied on the self-regulating market approach foundational to neoliberalism. In other words, because the forests of the southeastern US never went through the command and control era that subsequently spurred a shift toward neoliberal approaches to environmental management, these forests have long been de facto "neoliberalized"—though not intentionally as that term has come to be understood. Policy-makers in these areas have long relied strongly on market forces and metrics in the formation and evaluation of forest policy, prioritizing the benefits provided by private property rights, and avoiding command and control regulatory approaches (McCarthy, 2006: 99).

Specifically in the urban forest context, there is evidence that while a number of other regions of the nation tend to support direct government efforts to cultivate urban forests, the South prefers that such efforts be undertaken through voluntary means only (Perkins, 2011). Some inroads into the better management of urban forests by private-property owners might be made through tactics grounded in ecological modernism—through the promotion of ecosystem services and other market values provided by urban forestry (Perkins, 2011). Yet it seems clear that if non-governmental actors are not successful in such efforts, and state and local regulatory bodies continue to look with disdain on regulatory approaches, then the institutional ability of the federal government to overcome regional neoliberal strongholds and to provide a foundation of prescriptive protection becomes very important.

The low forest management standards found in southeastern states arise directly out of the current lack of national constitutional authority (or political will for that matter) to coordinate subnational forest policy combined with the southeastern US state governance philosophy and political culture regarding forests and land use. The US state of Alabama's perspective on voluntary "best management practices" is emblematic of this neoliberal governance philosophy. The Alabama Forestry Commission declares that it is the "lead agency for forestry in Alabama" but that it is "not an environmental regulatory or enforcement agency" and that it "[avoids] environmental problems through voluntary application of preventative techniques." (McDermott et al., 2010: 82–90). Yet when given the choice between preserving a forest (or managing it for the full range of ecological values) and cutting it down in the name of economic development and urbanization, voluntary choices do not always lead to "preventative techniques" that benefit forests, as evidenced by rapid urban sprawl in the southeastern US and the Forest Service's projected loss of up to 13 percent of the region's forests over the next 50 years.

To be certain, the forest management standards studied by Cashore and McDermott—riparian buffer zones, stand density requirements, clear-cutting restrictions, road clearing prescriptions, and afforestation and reforestation requirements—are aimed at the extractive use of forest resources, not the preservation of forests at the urban–rural forest frontiers. So it may be that though Canadian subnational governments seem to do a better job regulating extractive timber industries than US subnational governments, they may be in just as great a need of federal input into limiting development sprawl and integrating forests into new urban landscapes as US subnational governments are. It seems, however, that the private/public distinction highlighted by Cashore and McDermott may also play a role in the context of how willing subnational governments are to provide greenbelts, open space, development "limit lines," treed parking lots and roadways, wildlife corridors and habitat, among a variety of other values and services provided by urban forests. Indeed, in the US Pacific Northwest, where forest policies are more stringent, states like Washington and Oregon have been on the forefront of developing urban growth boundaries and limit lines to address sprawl, whereas states in the southeastern US are among the most rapidly sprawling regions of the country, with little or no use of limit lines and growth boundaries. And as highlighted by the Futures Report, perhaps not surprisingly, southeastern US states also contain forests facing the most severe threats in the US over the next 50 years.

Conclusion

The lens of political ecology typically fixes its gaze upon politics, culture, governance, and the environment. In this chapter, I argue that a deeper understanding of law and institutions can contribute to the political ecology perspective, as is illustrated by a comparison of the laws guiding urban forest management in Canada and the US. The national governments in these federal nations are limited, to varying degrees, by constitutional principles, and therefore may be unable to adequately fill in gaps in urban forest policy at the frontiers of new urban development. It seems that the Canadian federal government is more limited than the US federal government, as more political options are available in the US at the federal level to test the extent of their legal authority over subnational forests. This is fortunate, because the greater private ownership of forests in the US creates a unique situation that causes subnational forest policies to be extremely disparate, varied, and sometimes deficient when compared to Canadian provincial policies.

When thinking of the urban forest it is important to remember that urban forests are not only the forests that we see in cityscapes today, but also include future forests that will, or should, be integrated into the expanded cityscape of tomorrow. Places like the southeastern US are rapidly developing and urbanizing, and currently are doing so with little regard for protecting the forest resource. The same may be said of Canadian provinces and municipalities in some locations. We need to understand better both the political and institutional drivers at the intersection of forests and urban development so that we can plan better to harness the many values provided by forests in the urban areas of the future. Society must certainly continue to focus on the values that decentralization provides—harnessing the power of local governments to have access to the best information, to reduce central government meddling and bureaucracy, to maintain vested interests in the management of their resources because they exercise a high degree of control over them—while also preventing the many problems caused by the "over-decentralization" of forest policy. This is what national governments in federal systems were designed to do—to take action when subnational policies are so uncoordinated or so disparate as to cause harm to national interests. Urban forests, especially those on the frontiers of new development, are of both national and global interest. Though we have long understood forests to be a local government responsibility because of their non-transient nature, we now understand that the carbon they sequester and the climate they regulate are as transient and interconnected with global interests as a resource can be.

To the extent that developing urban areas are failing to protect these resources, the federal governments in these nations should be able to fill in the gaps. A political ecology perspective on urban forests, that takes into account the synergy between politics, culture, and the institutions that facilitate governance decisions, would seem to demand it.

Notes

1 *See* Hudson (2014) *Constitutions and the Commons.*

2 To be clear, some areas have had better experiences with government implemented growth boundaries than other areas. Certain municipalities in Oregon, like Eugene, for example, have long been considered a prime example of more environmentally friendly land use planning. Though even in Oregon, other municipalities, like Portland, are facing tough challenges to their growth boundary regulations. Similarly, in some urban areas of Canada "leapfrog development" has reduced some of the gains made by growth boundary regulation. For discussions of the gains made by, and the challenges to, growth boundary regulations in these divergent jurisdictions *see* Walker and Hurley (2011) *Planning Paradise: Politics and Visioning of Land Use in Oregon (Society, Environment, and Place)* and Sandberg et al. (2013) *The Oak Ridges Moraine Battles.*

3 For example, prescriptive prohibitions or limitations on clear-cutting and requirements related to stand density, forest road-building, riparian buffer zones, and afforestation or reforestation.

4 Such a challenge was recently lodged in the US Supreme Court case *Decker* v. *Northwest Environmental Defense Center* (133 S.Ct. 1326 (2013)) where the National Governors Association, National Association of Counties, National Conference of State Legislatures, International City/County Management Association, and Council of State Governments argued in an amicus brief that a federal requirement to obtain a permit under the Clean Water Act for nonpoint runoff from logging activities was unlawful because, among other things, the forest management activities in question were "traditionally regulated by state and local governments under their own laws." *See* Brief for the National Governor's Association et al. as Amicus Curiae, *Decker* v. *Northwest Environmental Defense Center*, 133 S.Ct. 1326 (2013), available at www.americanbar.org/content/dam/aba/publications/supreme_court_preview/briefs/11-338_petitioneramcungaetal.authcheckdam.pdf.

5 Regulations being more stringent on paper, of course, does not mean those resources are managed better, as enforcement and implementation of environmental laws may be weak.

References

Canadian Council of Forest Ministers. (2013) 'Sustainable Forest Management Policies in Canada', p. 1, www.sfmcanada.org/images/Publications/EN/Sustainable_Management_Policies_EN.pdf (accessed 6 May 2013).

Constitution Act, 1867, 30 & 31 Vict., c. 3 (UK), §§ 92A, reprinted in R.S.C. 1985, app. II, no. 5 (Can.)

Gurin, D. (2003) 'Understanding Sprawl: A Citizen's Guide', www.davidsuzuki.org/publications/resources/2003/understanding-sprawl-a-citizens-guide/ (accessed 15 July 2013).

Hudson, B. (2011) 'Climate Change, Forests, and Federalism: Seeing the Treaty for the Trees', *University of Colorado Law Review*, 82: 363.

Hudson, B. (2012) 'Fail-safe Federalism and Climate Change: The Case of US and Canadian Forest Policy', *Connecticut Law Review*, 44: 925.

Hudson, B. (2014) *Constitutions and the Commons: The Impact of Federal Governance on Local, National, and Global Resource Management.* RFF Press/Routledge, London and New York City.

International Monetary Fund. (2011) 'World Economic Outlook Database', www.imf.org/external/pubs/ft/weo/2011/01/weodata/index.aspx (accessed 6 May 2013).

Laitos, J.G., Zellmer, S.B., Wood, M.C., and Cole, D.H. (2006) *Natural Resources Law*, Thompson West, Eagan, MN, p. 849.

List of the Countries of the World. 'Countries of the World Ordered by Land Area', www.listofcountriesoftheworld.com/area-land.html (accessed 5 May 2013).

Louman, B., Fischlin, A., Gluck, P., Innes, J., Lucier, A., Parrotta, J., Santoso, H., Thompson, I., and Wreford, A. (2009) 'Forest Ecosystem Services: A Cornerstone for Human Well-Being', *International Union of Forest Resource Organizations World Series, Adaptation of Forests and People to Climate Change—A Global Assessment Report*, pp. 16–20, www.iufro.org/science/gfep/embargoed-release/download-by-chapter/ (follow "Download chapter 1" hyperlink), accessed 6 May 2013.

McCarthy, J. (2006) 'Neoliberalism and the Politics of Alternatives: Community Forestry in British Columbia and the United States', *Annals of the Association of American Geographers*, 96: 84–104.

McDermott, C.L., Cashore, B., and Kanowski, P.R. (2010) *Global Environmental Forest Policies: An International Comparison*. Earthscan, London.

Myers Madeira, E.C. (2008) 'Policies to Reduce Emissions from Deforestation and Degradation (REDD) in Developing Countries: An Examination of the Issues Facing the Incorporation of REDD into Market-Based Climate Policies', *Resources for the Future*, www.rff.org/rff/documents/rff-rpt-redd_final.2.20.09.pdf (accessed 5 May 2013).

Perkins, H.A. (2011) 'Gramsci in Green: Neoliberal Hegemony through Urban Forestry and the Potential for a Political Ecology of Praxis', *Geoforum*, 42: 558–566.

Rasband, J., Salzman, J., and Squillace, M. (2009) *Natural Resources Law and Policy*. Foundation Press, New York City, pp. 1206–1209.

Rose, G.A., MacCleery, D.W., Lorensen, T.L., Lettman, G., Zumeta, D.C., Carroll, M., Boyce, T.C., and Springer, B. (2005) 'Forest Resources Decision-Making in the US', In: C.J. Pierce Colfer and D. Capistrano (eds) *The Politics of Decentralization: Forests, People, and Power*, Earthscan, London, p. 239.

Sandberg, L.A., Wekerle, G., and Gilbert, L. (2013) *The Oak Ridges Moraine Battles: Development, Sprawl and Nature Conservation in the Toronto Region*. University of Toronto Press, Toronto.

Siry, J.P., Frederick, W., and Newman, D.H. (2009) 'Global Forest Ownership: Implications for Forest Production, Management, and Protection', XIII World Forestry Congress 3 tbl.1, www.pefc.org/images/stories/documents/external/global_forest_ownership_FD.pdf (accessed 6 May 2013).

UN Environment Programme. (2002) 'Global Environment Outlook 3: Past, Present and Future Perspectives, p. 110, www.unep.org/geo/GEO3/english/pdf.htm (follow "Forests" hyperlink), accessed 6 May 2013.

USDA Forest Service. (2011) 'US Forest Service Finds Global Forests Absorb One-Third of Carbon Emissions Annually', www.fs.fed.us/news/2011/releases/07/carbon.shtml (accessed 5 May 2013).

Walker, P.A. and Hurley, P.T. (2011) *Planning Paradise: Politics and Visioning of Land Use in Oregon (Society, Environment, and Place)*. University of Arizona Press, Tucson.

Wear, D.N. and Greis, J.G. (2011) 'U.S. Forest Service, the Southern Forests Futures Project: Summary Report', www.srs.fs.usda.gov/futures/reports/draft/summary_report.pdf (accessed 6 May 2013).

World Wildlife Federation. Canada's Boreal Forests, assets.panda.org/downloads/canada_forest_cc_final_13nov07_lr_1.pdf (accessed 6 May 2013).

6 Manufacturing Green Consensus

Urban Greenspace Governance in Singapore

Natalie Marie Gulsrud and Can-Seng Ooi

Introduction

Located in Southeast Asia, the densely populated island city state of Singapore is widely known as Asia's Garden City. Since its bid for a post-colonial identity in the 1960s, Singapore has actively envisioned itself as a clean and green garden city with the dual intent of "attracting foreign investment while also raising the morale of its citizens" (Lee, 2000). The city's green identity has served as a guiding vision of the city's development plan over the past five decades. Extensive urban biophysical greening measures have ensured the beautification of the rapidly urbanizing and industrializing infrastructure of the island while impressing and attracting "first-world" investors with an orderly and resort-like atmosphere (Ooi, 1992; Lee, 1998; Tan et al., 2013). The recognition of Singapore's green city vision is reflected in citizens' and visitors' perceptions of Singapore as a Garden City, with both groups ranking parks and greenery as one of the most important elements to Singapore's quality of life and as the number one thing that makes Singapore special (Tan et al., 2013; Hui and Wan, 2003). The global business community has eagerly praised Singapore's green reputation, naming Singapore as Asia's greenest city in the Economist Intelligence Unit Asian Green City 2011 Index (EIU, 2011). Singapore's green growth track record has caused academics and proponents to claim Singapore as a best practice case study in terms of green urbanism in Asia and abroad (Newman, 2010; EIU, 2011; Tan et al., 2013). What started as a green city vision in the 1960s is now a strongly established green city brand (Koh, 2011).

The success of the Garden City brand has not come without detractors. In 2012, when Singapore opened its showcase Gardens by the Bay complex complete with a vertical garden of "sustainable," concrete, photovoltaic-lit, musical "supertrees" covered in a colorful array of exotic plants and flowers (Figure 6.1), the Nature Society of Singapore (the country's leading nature conservancy group and champion of local biodiversity initiatives) challenged the National Parks Board to make the park more natural (Koh, 2011).

And despite the city's expansive vegetative cover encompassing over 56 per cent of Singapore's landmass in biophysical green, the actual space allocated to residents for recreational purposes in the form of parks and open spaces is lagging behind other global cities in Asia, Australia, Europe, and North America

Figure 6.1 Supertree Grove by day. Gardens by the Bay, Singapore.
Source: Natalie Marie Gulsrud, 2012.

(Tan et al., 2013). The Nature Society's objection to Singapore's latest version of urban greening—a keystone project as the city's newest vision to become a "city within a garden"—as well as the hard facts underlying the difficulty of providing adequate greenspaces for residents in such a densely populated city, highlights the contested claims of green city visions and place brands in general. Powerful terms such as "natural," "sustainable," and "green" often represent diverging meanings in a diverse urban context such as Singapore and underline the tension involved with place identity-making.

Singapore's long-standing green visioning efforts provide a fascinating longitudinal study in the processes and outcomes of green place identity-making. An analysis of the city's green brand since the 1960s shows how key terms such as "green" and "sustainable" have been applied in diverse and contradictory ways, shifting and developing in line with political leadership and pressure from civil society (Lee, 2000). The dueling visions of economic prosperity and social reform encased in the city's Garden City identity have often clashed, producing unexpected results. This tension, however, should not be surprising. Urban political ecologists have challenged the notion of the sustainable city, arguing that cities are defined by urban and environmental processes that benefit some groups while negatively impacting others. Seen from an urban political ecology perspective, Singapore's green city identity-making is a historical reflection of intricately interwoven socio-environmental urban processes with serious political ramifications. And while

Singapore has gained international renown for its green image, there is limited critical analysis of the political and social construction of the city-state's green identity over time (see Neo, 2007).

This analysis of Singapore's green city visioning project from the 1960s to the present aims to uncover the historical trajectory and impacts of the city's green identity-making process as seen through an urban political ecology lens. The study is directed by questions of which socio-environmental narratives are revealed in the green visioning processes and who has benefited and lost in the act of green identity-making.

Singapore Through an Urban Political Ecology Lens

As cities such as Singapore compete to differentiate themselves, many of them are turning to marketing themselves and crafting green or environmentally sustainable profiles (Jonas and While, 2007). Green place branding is a form of image-making, a process steeped in historical, political, and cultural discourses (Morgan and Pritchard, 1998; Pritchard and Morgan, 2001). The sustainability narratives presented in a green brand are a by-product of the same historical, political, and cultural discourses found in a community. By framing and presenting selective images of local character, green place brands affirm and reproduce an understanding of local identity both for insiders and outsiders (Moilanen and Rainisto, 2009). Green city brands have become a global tool for municipal leaders to promise a better quality of life, promote sustainable development, and increase their competitive advantage. Green campaigns can pursue goals of what Harvey (1996) calls ecological modernization by prioritizing the maintenance of existing values, patterns, and social relations based in capital accumulation. One consequence of green place branding is that some ideas of local authenticity are left out of the image-making process (Govers and Go, 2003). Sustainability campaigns have also been cited with the power to provide new opportunities for disempowered groups by reshaping urban environments and thereby making them more equitable places (Krueger and Gibbs, 2007: 5). The exclusion or inclusion of certain aspects of community identity in green place brands can influence decision-making power and thereby resource allocation (Govers and Go, 2003). In this sense, green city visioning is a community agenda-setting tool establishing political and cultural norms.

In the case of Singapore, green city visioning has served not only as a tool for nation and community building but also as a tool for development and land use allocation (Ooi, 1992). The People's Action Party (PAP), the ruling political party in Singapore since its independence in 1959, has deftly programmed the city for economic development in part through top-down urban planning policies and social engineering processes (Grice and Drakakis-Smith, 1985). The PAP's leadership philosophy, which has centred on "strong, wise and far-sighted government," has viewed public participation as "a process of mass education," and has therefore implemented a top-down process of establishing and implementing policy goals in a large part through strategies and slogans such as the Garden City vision

(Grice and Drakakis-Smith, 1985: 348; Ooi, 1992). Singapore's Garden City campaign has been an "intrinsic feature in its land use planning and development policies" over the past decades, informing citizens of their duties to cultivate and civilize the islands post-colonial jungle environs (Ooi, 1992: 65). Additionally the Garden City campaign has boosted Singapore's international reputation as a clean and green oasis in Southeast Asia, and therefore a reliable place for foreign investors and companies. In this sense, Singapore's green city vision can be understood not only as a marketing tool but also as a strategy for socio-economic development.

Singapore's claim to be Asia's greenest city makes for excellent political rhetoric but calls for more rigorous attention to the concepts used in the construction of such green city brands as well as their engagement with potential social and political changes resulting from the city's green reputation (Krueger and Gibbs, 2007). Urban political ecology provides a helpful lens to uncover and analyze these changes. Heynen et al. assert that urbanization occurs within the context of complex and intrinsically interwoven social, political, and ecological processes (2006; Harvey 1996). Throughout urbanization these processes play out in the reproduction of urban environments that "embody and reflect positions of social power" (Heynen et al., 2006). In this sense urban political ecology takes to task the nature/culture logic suggesting that there is nothing unnatural about human-produced environments because cities are specific historical results of socio-environmental processes (Davis, 1996; Harvey, 1996; Heynen et al., 2006; Wachsmuth, 2012). Cities are also by-products of capitalism and in this sense urban nature is as much a commodity as steel, glass, and concrete are because urban nature is produced under "capitalist and market-driven social relations" (Heynen et al., 2006: 5). In a capitalist city, such as Singapore, urban political ecology argues that the urban environments of the city are "controlled, manipulated and serve the interests of the elite at the expense of marginalized populations" (Swyngedouw, 2004). This theoretical frame encourages questions regarding how urban environments are shaped and reshaped: "Who produces the political, economic, and social configurations of a city and who benefits from these configurations? Who produces what kind of socio-ecological configuration for whom?" (Heynen et al., 2006: 2). Or in other words who has produced Singapore's green city vision, how has the green vision discourse impacted the political and physical landscape of the city and whom has benefited from this green identity making?

Liveable Singapore, a Legacy of Clean and Green

Singapore's garden city legacy presents a unique opportunity to explore the political and social consequences of the city-state's green identity-making process over time. The case study is temporal in nature tracking three notable shifts in Singapore's green city identity from the inception of Singapore as an independent city-state in 1965 to the present time.

Singapore's progression from post-colonial slum to Asia's greenest and most liveable city has been carefully documented in newspaper articles, political speeches, and government planning documents. Since 1959 Singapore has been governed by a single ruling party, the PAP, and the newspaper articles, political speeches and planning documents analyzed for this article have been read and analyzed as a reflection of PAP policies and political discourses. The PAP perspective is triangulated with peer-reviewed articles written specifically about Singapore during this time period in the fields of urban planning, urban greenspace governance, biodiversity and nature conservation, and neoliberal governance. In addition, this study draws on field interviews conducted in the winter of 2011 with officials from Singapore National Parks Board (NParks), Singaporean academics involved in urban greenspace governance, as well as citizens engaged in community gardening. The second author of this paper is a citizen of Singapore and was personally involved in challenging the government's greening initiatives in the early 1990s. All data collected have been analyzed qualitatively to examine which socio-environmental narratives are revealed in the green visioning processes and to note who has benefited, or not benefited, from the act of green identity making.

Singapore as a "Clean and Green" Garden City: 1959–1987

> A well kept garden is a daily effort and would demonstrate to outsiders the people's ability to organise and to be systematic . . . The grass has to be mown every other day, the trees have to be tended, the flowers in the gardens have to be looked after so they know this place gives attention to detail.
>
> Lee Kwan Yew, 1965 (Kwang et al., 1998: 12)

In 1963, when Prime Minister Lee Kwan Yew began a tree-planting campaign to green and beautify the island, he set in motion not only an effort to biophysically alter the landscape of an island where more than 95 per cent of the original temperate jungle vegetation had been eliminated through colonial agricultural pursuits but also to morally reform and discipline the citizens in a campaign of economic development (Grice and Drakakis-Smith, 1985; Ng and Corlett, 2011). When Singapore gained self-governance from the British in 1959, after 140 years of colonial rule, the landscape was ravaged by drought and the city's inhabitants were suffering the effects of rapid urbanization and industrialization with high rates of unemployment and deplorable slum-like living conditions in the city center (Hassan, 1969; Lee 2000). The Garden City campaign, officially announced in 1968, was a government-driven, top-down, economic development plan to "achieve First World standards in a Third World region" with the ultimate aim of attracting direct foreign investment to boost Singapore's shift from an agricultural to an industrialized economy (Lee, 2000: 174). At the heart of the campaign was the concept of "cleaning and greening," not only the landscape but also the "rough and ready ways of the people," to encourage "considerate and courteous behaviour" (Lee, 2000: 173). In this sense the act of greening manifested not only biophysical results but economic reforms as well.

Not all participants in the Garden City campaign were equally pleased with the greening results. Beginning in the early 1960s, citizens complained about the lack of native species chosen to green the island. One opposition member in parliament stated, "The Chief Parks Officer had probably learnt a lot from the West, and was planting trees and flowers of Western colours and not those reflecting the East . . . Tourists to Singapore find that they have no need to come to Singapore to enjoy our scenery because these are very westernized" (The Strait Times, 1967). Another opposition member attacked the Parks Department for beautifying outlying rural villages, in no need of tree planting, with native jungle trees. When impoverished citizens in these villages asked for water they received trees instead (The Strait Times, 1967). Even the Singapore Horticultural Society was displeased with the initial results of the greening campaign, stating "this official planting program does not represent nature dominant in the extravagance of the tropics, but nature controlled and cringing, told not to interfere with street lighting and the vision of motorists" (The Strait Times, 1967). What was natural and what was green was called into question by citizens, gardeners, and politicians alike. By 1980, over one million flowering shrubs and instant trees were planted in the city transforming Singapore from what Lee once called "a mudflat swamp" into a modern green city (Kwang et al., 1998: 12).

To encourage the citizens of Singapore to green and clean their environs into a Garden City became an act of equal parts discipline, education, and behaviour modification. Speeches held by government officials on tree-planting days were laced with cautionary tales about the dire consequences of citizens losing confidence in Singapore's ability to urbanize and industrialize as well as an emphasis on the critical importance of well-behaved citizens in this process. One such speech in the name of greening and economic development given by Lee, as reported by *The Strait Times* in 1968, spelled out Lee's so-called "formula for success":

> When morale is down people become apologetic and the place is in shambles. Singapore will not be allowed to go thus. We will keep it trim, clean and green. Flowers bloom and ferns will grow where there was dirt and tarmac. Other governments can give you fountains or stadiums or monuments. But they can't give you the capacity to organize and discipline yourselves. No donor country can give . . . what you must have in yourselves: the self-discipline to keep in good condition what you own. Now workers and the unions must enter into the spirit of it. When word gets around that there are keen and striving workers in Singapore, then we shall blossom as the workshop, the dynamo, of South East Asia.
>
> (The Strait Times, 1968)

According to Lee, greening demonstrated to the international finance world Singapore's desire to be taken seriously in the labour market and to partake in capitalist modes of accumulation. Citizens no longer were allowed to live "lazy" rural lifestyles. As members of a developing nation, citizens were reminded to "enter the spirit" of capitalism not only by disciplining and cultivating the biophysical landscape of Singapore but also by applying similar acts to their bodies. The act of planting and cultivating trees and shrubs instilled discipline and pride in citizens and therefore a willingness to resign to the transformation of their selves as instruments of the economy. The radical alteration of Singapore's landscape under the clean and green Garden City campaign created a favorable environment for foreign-financed industrialization. Under this campaign city slums were removed, the river dredged and purified, farm animals removed from the city centre and housing developments decentralized to create a green suburban layout and more efficient use of land and space for factories and businesses (Lee, 2000). Ultimately, Lee constructed a "stable and docile population" to support and partake in his new industrialized economy (Grice and Drakakis-Smith, 1985: 348).

"The Next Lap: Tropical City of Excellence": 1988–2004

The second phase of Singapore's green city vision, Tropical City of Excellence, can be seen within the context of the government's shift from a focus on strictly functional greening to recreational and lifestyle greening. By the early 1990s, standards of living in Singapore had improved markedly with the country posting an increasing per capita gross national product (GNP). As a result, the green politics of Singapore emphasized a greater quality of life (Savage and Kong, 1993). The 1991 Concept Plan published by the Urban Redevelopment Authority (URA) highlights Singapore's need for more open space and recreational green areas, the conservation of old city areas, as well as the conservation of natural areas (URA, 1991). URA's emphasis on Singapore's tropical identity reflected the fact that the city's altered landscape was a top source of concern for citizens and government officials alike in the 1980s and 1990s. Citizens continued to voice displeasure at the lack of local identity represented in the built and biophysical environment of the city while government officials were alarmed by the increasing numbers of educated citizens emigrating from Singapore, thus threatening the country's economic competitiveness and productivity (Geh and Sharp, 2008). Greening became an act of local identity-making in the name of economic competitiveness and environmental responsibility.

Environmental conservation played a key role in domestic and international politics during this period. The 1992 Earth Summit in Rio, for example, ushered in a string of environmental conservation initiatives outlined in Singapore's visionary 1992 Green Plan, "The Singapore Green Plan—Towards a Model Green City" and subsequently the 1993 "Singapore Green Plan—Action Programmes" (Neo, 2007). Nature conservation figured in the plan with a commitment that "5 per cent of the land after reclamation to the full will be set aside for nature conservation to promote the appreciation of nature and interest in the country's natural resources" (SGP-1992: 29). Environmentalists had recently convinced the government to designate Sungei Buloh wetlands as a bird sanctuary and—in combination with the election of Goh Chok Tong in 1991, who promised to lead Singapore in a kinder and more consultative manner—there was great potential for non-governmental groups to pursue their own brand of greening. National Tree Planting Day in 1991, for the first time, focused on a public reforestation effort at a nature reserve where 300 saplings of native Singaporean species were planted as an act of greening instead of cleaning (The Strait Times, 1991).

Despite this apparent broadening of green governance, the green politics in Singapore of the 1990s illustrated ever-contested concepts of greening, nature, and conservation. The greater role of citizen consultation and civic engagement in the management of Singapore's biophysical resources was tempered by the government's commodification of sensitive ecosystems and wildlife habitats. The case of the Marina South Duck Pond illustrates this tension between engaged citizen conservation efforts and the government's entrepreneurial approach to greening. Situated along the southern shoreline of Singapore, Marina South was an "idle" 11-hectare piece of land-turned-roosting-ground for rare migratory ducks, of which some species had not been spotted previously in Singapore, and thus acclaimed by birdwatchers and the Nature Society of Singapore (The Strait Times, 1992a). When plans were announced by the Ministry of the Environment to reclaim and pave over the ponds, the Nature Society of Singapore launched a lobbying effort to mediate the government's actions, which was to no avail (The Strait Times, 1992b). Government officials claimed the site was too commercially valuable to be set aside for a bird sanctuary and that the pond presented a public health problem in terms of the potential for mosquito breeding (The Strait Times, 1992b). Additionally, officials maintained that there was nothing natural about the site and therefore the site did not qualify for conservation status (Neo, 2007). While environmentalists savoured the Marina South ponds for their rare and vulnerable biodiversity, as well as the added beauty in the city, the government measured the worth of the site not only in terms of its commercial value but also in terms of what it deemed to be natural. By deeming the ponds—originally dug out by humans—unnatural, officials limited the definition of natural and, thereby, of nature to habitat not touched by humans. In this sense the vast majority of Singapore's habitat was deemed unnatural and therefore unworthy of conservation according to the Ministry of the Environment's number one criterion for selecting conservation sites: "sites must be natural and ecologically stable" (Neo, 2007: 191). Citizens' interpretations of greening the city fell short of the government's definition of natural.

Following the government's reclamation of the land in 1992, the site lay undeveloped until 2005 when Prime Minister Lee Hsien Loong announced the creation of Singapore's premier urban outdoor recreation space, Gardens by the Bay, on the site of the former Marina South ponds (Interview, 2011). Greening in the "Tropical City of Excellence" was not about nature conservation or promoting biodiversity but the commodification of biophysical landscapes for recreational opportunity and, ultimately, garden-theme-park-based tourism (Figure 6.2).

City in a Garden: 2004 to Present

If Singapore positioned itself to increase quality of life and authenticity of local identity through greening as a "Tropical City of Excellence," Singapore as a City in a Garden now envisions the city as an innovative, environmentally sustainable

Figure 6.2 Tourists in front of green wall and waterfall. The commodification of biophysical green resources is here illustrated by tourists paying to pose in front of a gigantic green wall and waterfall in a re-enacted jungle at the Gardens by the Bay, Singapore.

Source: Can-Seng Ooi, 2013.

playground enveloped in a garden (Ministry of the Environment, 2009). Citizens and the private sector have actively been involved in this latest green city vision. Since 2005, over 600 community gardens have been established throughout the city giving the public—predominantly retirees and school children—a direct outlet to greening and tending their own gardens (Interview, 2011) (Figure 6.3). In addition, since 2011 NParks actively has solicited feedback from community members on shaping the city in a garden vision through road shows, focus groups, interviews, and surveys (Interview, 2011). The private sector has contributed to NParks' vision of the city in a garden through donations and financial partnerships to support the city's green heritage (Interview, 2011). NParks' officials are quick to point out how more community members than ever before are engaged in realizing

Figure 6.3 Citizens engaged in community gardening, Singapore.

Source: Natalie Marie Gulsrud, 2011.

the city in a garden vision. This is both a concession to increasing concern from community members regarding the sustainability of Singapore's green heritage as well as a response to current public administration theory that suggests good governance is based on public participation (Interview, 2011).

The experience of the garden—living in it, working in it, playing in it, and creating it—is the current vision of the city with the end goal of incorporating even more biophysical infrastructure into the densely populated city. Ministry of the Environment officials are proud to report, "between 1986 and 2007, the green cover in Singapore grew from 36% to 47% despite a 68% growth in population . . . Ten per cent of Singapore's land is dedicated as greenspace, 5% of which is protected as nature reserves . . . The city also is home to 2,900 species of plants, 360 species of birds and 250 species of hard coral" (Ministry of the Environment, 2009: 29). What the ministry neglects to tell in this story is that "more than 95% of Singapore's original forest cover has been cleared and less than 10% of the remaining 24 km^2 of forest is primary . . . The remaining forest reserves occupy only 2.5% of Singapore's land area and contain over 50% of the remaining native forest biodiversity" (Ng and Corlett, 2011: 20). Marine biodiversity is also in decline. Therefore, the concept of Singapore as a city in a garden calls directly into question the city's notion of environmental sustainability as specific indigenous landscapes are silently eliminated to actively promote "supertrees" and recreational greening.

Nowhere in Singapore is this narrative more typified than in the city's iconic Gardens by the Bay theme park located in the Marina Bay area of the island, once home to extensive marsh and wetland habitats supporting rare dragonfly species as well as a large swath of secondary growth forest and critical habitat for a host of bird and butterfly species (NSS, 2010). Gardens by the Bay is a park spanning 250 acres of reclaimed land in central Singapore established by NParks in 2011 as the iconic project of the city in a garden vision. The park is characterized by a variety of themed vertical gardens and conservatories as well as a collection of 100-foot "supertrees" that, as *The New York Times* recalls, "resemble oversize stone palms, each dripping with ferns, orchids and bromeliads" serving as the backdrop of a nightly laser show (Graham, 2013). Nestled in between the Mediterranean conservatory, the cloud forest, and the grove of supertrees is a quaint reminder of Singapore's past, a reconstructed Kampong (village in Malaysia) representing local vegetation grown in Singapore's rural villages, all but one of which suffered the fate of the bulldozer during the original clean and green Garden City campaign of the 1960s. While the remaining settlement—Kampong Buangkok—is also a popular tourist destination and icon of local food culture in Singapore, it is sadly under constant threat of demolition as the commercial value of the estate is extremely valuable (Graham, 2013). In their article on changing landscapes in Singapore, Chang and Huang remind us that "as with other landscapes of redevelopment, the built landscape at the Singapore River has been reconstructed as a (com)modified version of local history" (2005: 270). Gardens by the Bay, located not along the river but by the marina, is the city's latest project to reinvent and revitalize the urban landscape, "to erase its darker histories, utilitarian geographies and uneconomical spaces" (Huang and Chang, 2002). In erasing the past the city is able to selectively present narratives of greening and recreation—ultimately the sustainable Singapore experience keeping in line with the city's vision to transform Singapore into a garden (Chang and Huang, 2005).

Manufacturing Green Consensus

Contemporary neoliberal urban governance has embraced a somewhat paradoxical politics of economic growth tempered by a reduction in negative environmental externalities, such as carbon outputs, and an increase in environmental benefits, such as the growth in urban greening campaigns (Krueger and Gibbs, 2007; Jonas and White, 2007). Green city brands, in this context, suggest that a city can have it all: economic growth and environmental sustainability, entertainment-oriented urban greenscapes, and increased urban biodiversity. But Singapore's green imaging campaign offers a different interpretation of the neoliberal promise, revealing contested socio-environmental narratives of nature, greening, and sustainability. These conflicts provide poignant examples of how an urban greening campaign can directly impact the biophysical, social, and political fabric of a city in an uneven manner and point to broader lessons that can be learned from green city imaging and branding campaigns.

Singapore's green image was founded in the production, distribution, and exchange of biophysical green resources (Smith, 1984) in order to boost economic prosperity and modify a Third World landscape into a First World one. In this sense, the biophysical landscape of Singapore became a means of profit accumulation whereby landscapes with low economic value were replaced with landscapes that, literally, could pay the rent (Smith, 2007). The city's biophysical landscapes continue to be determined by their economic value and revenue-generating potential as illustrated by the city's most recent greenspace investment, Gardens by the Bay. One obvious consequence of this pattern is decreased local biodiversity as cement supertrees are prioritized over wetlands. As experts look to Singapore as a model green city, more communities might choose green city images with high-earning, tourism-based, biophysical landscapes over lower-earning, less-entertaining landscapes.

Singapore's clean and green campaign produced a greener landscape and a more stable and docile population. Under the guise of the clean and green campaign in the 1960s, the PAP government linked an ideology of survivalism and economic nationalism with an attack on "radical trade unionism, community subversion, and racial strife," usually associated with Malaysian settlements and cultural circles (Tan, 2012). Razing "slum housing" or village settlements often inhabited by one dominant ethnic or cultural group was another tactic used by the government to reduce racial tension in the 1960s while also cleansing the landscape of unsanitary, cramped living conditions. The physical act of planting and tending to trees and shrubs was a way for these newly displaced citizens to embrace the disciplined spirit of capitalism and demonstrate their allegiance to the entrepreneurial state. It was also a measure by which the government demarcated productive and unproductive citizens, or, in other words, good and bad citizens.

Cleaning and greening in Singapore has come at a cost, not only of biodiversity but also the public celebration of ethnic, cultural, and political diversity. City imaging campaigns function not only to unite communities in a shared vision but also serve to alienate and eliminate identities of local authenticity that do not align with that vision. Green city visions such as Singapore's are a promise of a better quality of life, sustainable development, and an increased competitive advantage, but these promises might not be available equally to all citizens of the community.

Perhaps the most poignant lesson we can learn from Singapore's green city experience is the consequence of prioritizing effectiveness and efficiency above all other goals. Singapore's current green vision is a brand, City in a Garden, which aims to enhance urban quality of life by increasing urban greenery and flora in the city. In as much as a place brand, such as City in a Garden, could inspire a sense of collective belonging focused on biophysical identity, only certain versions of biophysical identity are allowed in the brand. Gardens by the Bay is indicative of the effectiveness and efficiency with which a place brand can dictate the economic and social values of a community. What is natural, what is green, what is sustainable is actively and selectively defined through the commodification and marketization of Singapore's green resources. Aronczyk and Powers remind us that to fully understand the role of brands and branding in public life we must look beyond the "marketization of institutional practices" to the "commodification of public discourse itself" (2010). While the brand's reputation might be important to its users, it is the owners of the brand that accrue the economic value of the product and therefore actively control and limit the social and political potential for participation in the creation of the brand (Aronczyk and Powers, 2010). Singapore's City in a Garden brand manufactures and maintains one variety of green governance.

Singapore's green imaging experience is not unique. Global cities are actively competing for resources through green city profiles, marketing themselves as environmentally sustainable places. While Singapore is an extreme case in the longevity and authoritarian governance of its campaign, it provides nonetheless a fascinating study in the potential consequences of green city imaging and branding. Other developing cities in Asia and Africa are currently looking to Singapore for clean and green solutions to urban development and expansion. The Singapore Cooperation Enterprise was established in 2006 to market and export the country's public sector expertise in land use and transportation planning, amongst other things. Singapore's green city model might yet be replicated in other contexts. Ultimately the case of Singapore raises questions regarding the meaning of urban sustainability and what being a green city actually entails. Globally, green city branding campaigns are driven by entrepreneurial governance initiatives based on efficiency and effectiveness. Certain definitions of green are privileged over others. Green city brands promise something unique but as more and more cities profile themselves as green in an entrepreneurial fashion those unique characteristics of biodiversity and local identity, foundations of urban sustainability, might be lost.

Interviews

Anonymous, local community in bloom manager of Singapore NParks, interview February 22, 2011

Mr Kong Yit San, Assistant Chief Executive Officer of Singapore NParks, interview February 23, 2011.

Ng Cheow Kheng, Deputy Director of Horticulture and Community Gardening of Singapore NParks, interview February 22, 2011.

Philip Li, local community gardner at Hort Park, Singapore, interview February 22, 2011.

Dr Tan Puay Yok, Deputy Director of the Center for Urban Greenery and Ecology (CUGE) Research, interview February 24, 2011.

Dr Wong Nyuk Hien, Professor in the Department of Building/School of Design and Environment at the National University of Singapore (NUS), interview February 23, 2011.

References

Aronczyk, M. and Powers, D. (2010) 'Introduction: Blowing up the Brand', in Aronczyk, M. and Powers, D. (eds) *Blowing up the Brand: Critical Perspectives on Promotional Culture*, Peter Lang, New York City.

Chang, T.C. and Huang, S. (2005) 'Recreating Place, Replacing Memory: Creative Destruction at the Singapore River', *Asia Pacific Viewpoint*, 46(3): 267–280.

Davis, M. (1996) *Ecology of Fear: Los Angeles and the Imagination of Disaster*. Metropolitan Books, New York City.

EIU (2011) Asian Green City Index: Assessing the Environmental Impact of Asia's Major Cities. Prepared by the Economist Intelligence Unit, 2011. www.siemens.com/press/pool/de/events/2011/corporate/2011-02-asia/asian-gci-report-e.pdf (accessed 31 March, 2014).

Geh, M. and Sharp, I. (2008) 'Singapore's Natural Environment, Past, Present and Future: A Construct of National Identity and Land Use Imperatives', in Wong, T.C., Yuen, B. and Goldblum, C. (eds) *Spatial Planning for a Sustainable Singapore*, Springer-Science + Business Media B.V., Netherlands.

Govers, R. and Go, F.M. (2003) 'Deconstructing Destination Image in the Information Age', *Information Technology and Tourism*, 6(1): 13–29.

Graham, A. (2013) 'Green Acres by Singapore's Skyscrapers', *The New York Times*, 14 July, pp.TR4. www.nytimes.com/2013/07/14/travel/green-acres-by-singapores-skyscrapers.html (accessed 16 June, 2014)

Grice, K. and Drakakis-Smith, D. (1985) 'The Role of the State in Shaping Development: Two Decades of Growth in Singapore', *Transactions of the Institute of British Geographers*, 10(3): 347–359.

Harvey, D. (1996) *Justice, Nature and the Geography of Difference*, Blackwell Publishers, Oxford.

Hassan, R. (1969) 'Population Change and Urbanization in Singapore', *Civilizations*, 19(2): 169–188.

Heynen, N., Kaika, M., and Swyngedouw, E. (2006) 'Urban Political Ecology: Politicizing the Production of Urban Natures', in Heynen, N., Kaika, M., and Swyngedouw, E. (eds) *In the Nature of Cities*, Routledge, New York City.

Huang, S. and Chang, T.C. (2002) 'Selective Disclosure. Romancing the Singapore River', in Goh, R.B.H. and Yeoh, B.S.A. (eds) *Theorizing the Southeast Asian City as Text: Urban Landscapes, Cultural Documents and Interpretive Experiences*, World Scientific, Singapore.

Hui, T.K. and Wan, T.W.D. (2003) 'Singapore's Image as a Tourist Destination' *International Journal of Tourism Research*, 5(4): 305–313.

Jonas, A.E.G. and While, A. (2007) 'Greening the Entrepreneurial City?: Looking for Spaces of Sustainability Politics in the Competitve City', in Krueger, R. and Gibbs, D (eds) *The Sustainable Development Paradox: Urban Political Economy in the United States and Europe*, The Guilford Press, New York City.

Koh, B.S. (2011) *Brand Singapore: How Nation Branding Built Asia's Leading Global City*, Marshall Cavendish Business, Singapore.

Krueger, R. and Gibbs, D. (2007) 'Problematizing the Politics of Sustainability' in Krueger, R. and Gibbs, D. (eds) *The Sustainable Development Paradox: Urban Political Economy in the United States and Europe*, The Guilford Press, New York City.

Kwang, H.F., Fernandez, W., and Tan, S. (eds) (1998) *Lee Kuan Yew: The Man and His Ideas*, Singapore Press Holdings, Singapore.

Lee, K.Y. (1998) *The Singapore Story: Memoirs of Lee Kuan Yew*, Prentice Hall College Div, New York City.

Lee, K.Y. (2000) *From Third World to First: Singapore and the Asian Economic Boom*, Harper Collins, New York City.

Ministry of the Environment (2009) *A Lively and Liveable Singapore*, Ministry of Environment, Singapore.

Moilanen, T. and Rainisto, S. (2009) *How to Brand Nations, Cities, and Destinations: A Planning Book for Place Branding*, Palgrave Macmillan, New York City.

Morgan, N. and Pritchard, A. (1998) *Tourism Promotion and Power: Creating Images, Creating Identities*, John Wiley and Sons, New York City.

Nature Society Singapore (NSS) (2010) 'Comments on Garden by the Bay', Nature Society Singapore, Singapore.

Newman, P. (2010) 'Green Urbanism and its Application to Singapore', *Environment and Urbanization Asia*, 1(2): 149–170.

Neo, H. (2007) 'Challenging the Development State: Nature Conservation in Singapore', *Asia Pacific Viewpoint*, 48(2): 186–199.

Ng, P.K.L., Corlett, R.T., and Tan, H.T.W. (eds) (2011) *Singapore Biodiversity: An Encyclopedia of the Natural Environment and Sustainable Development*, Editions Didier Millet, Singapore.

Ooi, C.S. (2011) 'Paradoxes of City Branding and Societal Changes', in Dinnie, K. (ed) *City Branding: Theory and Cases*, Palgrave Macmillan, New York City.

Ooi, G.L. (1992) 'Public Policy and Park Development in Singapore', *Land Use and Policy*, 9(1): 64–75.

Pritchard, A. and Morgan, N. (2001) 'Culture, Identity, and Tourism Representation: Marketing Cymru or Wales?' *Tourism Management*, 22(2): 167–79.

Savage, V.R. and Kong, L. (1993) 'Urban Constraints, Political Imperatives: Environmental design in Singapore', *Landscape and Urban Planning*, 25(4): 37–52.

SGP-1992 (1992) 'Singapore Green Plan—Towards a Model City', Ministry of Environment, Singapore.

Smith, N. (1984) *Uneven Development: Nature, Capital, and the Production of Space*, The University of Georgia Press, Athens.

Smith, N. (2007) 'Nature as Accumulation Strategy', *Socialist Register*, 43: 19–41.

Statistics Singapore (2013) 'Annual GDP at 2005 Market Prices and Real Economic Growth' Prepared by Statistics Singapore. www.singstat.gov.sg/statistics/browse_by_theme/economy/.../gdp1.xls (accessed June 13, 2013).

Swyngedouw, E. (2004) *Social Power and the Urbanization of Water. Flows of Power*, Oxford University Press, Oxford.

Swyngedouw, E. and Kaika, M. (2000) 'The Environment of the City or . . . the Urbanization of Nature', in Bridge, G. and Watson, S. (eds) *Reader in Urban Studies*, Blackwell Publishers, Oxford.

Tan, K.P. (2012) 'The Ideology of Pragmatism: Neoliberal Globalisation and Political Authoritarianism in Singapore' *Journal of Contemporary Asia*, 42(1): 67–92.

Tan, Y.P., Wang, J., and Sia, A. (2013) 'Perspectives on Five Decades of Urban Greening in Singapore', *Cities*, 32: 24–32.

The Strait Times (1967) 'People Ask for Water, They Get Trees', *The Strait Times,* 20 December, p.4.

The Strait Times (1968) 'Lee's Success Formula in a Word—Confidence', *The Strait Times,* 2 September, p.13.

The Strait Times (1991) 'Just Keeping Singapore Clean to Making it Green', *The Strait Times,* 2 October.

The Strait Times (1992a) 'Rare Ducks' Pond Being Filled at Marina South', *The Strait Times,* 20 April.

The Strait Times (1992b) 'ENV Fills in Ducks' Pond Despite Appeals', *The Strait Times,* 12 May.

Urban Redevelopment Authority (1991) '1991 Concept Plan', Urban Redevelopment Authority, Singapore.

Wachsmuth, D. (2012) 'Three Ecologies: Urban Metabolism and the Society–Nature Opposition', *The Sociological Quarterly,* 53: 506–523.

7 The Places of Trees in Honduras

Contributions of Public Spaces and Smallholders

J. O. Joby Bass

Introduction

Revelations of increases in tree cover in places throughout the developing world in recent years have led many scholars to reconsider commonly held assumptions about widespread deforestation (Zimmerer, 2004). During previous research, I have found increases in vegetation across much of Honduras between 1957 and 2010 (Bass, 2004; 2006a; 2006b; 2012). Using repeat photography, I found that these increases were remarkably complex in site, context, and formation—ranging from settlement forests, public plantings, spontaneous dense growth, random dispersed growth, and forest patches (Bass, 2004). The factors that explain these increases are likely as numerous and complex as the landscapes they have helped shape. Practically all of these trees are associated with people, if they are not explicitly culturally derived. Tree increases appear particularly prevalent in explicitly cultural settings, such as in towns and around houses, both rural and urban.

This chapter considers two of the types of settings in which increases in trees have been commonly observed: public spaces, specifically urban plazas, and the immediate surrounding of houses. As two types of landscapes that are explicitly associated with the human-settled landscape, public open space and the built environment are also two of the foundations of urban landscapes. A political ecology framework points toward a central state policy of promoting forest conservation and a more diffused perception of small holders to mitigate risk as a couple of ways to conceptualize these changes.

Trees, Conservation, and Risk

Political ecology offers useful and productive perspectives to the study of trees in urbanized places. As the domain of most of the world's people, these landscapes are inherently political, with multiple interests and forces operating upon them and shaping and reshaping them. These forces exist on many different scales. Particularly in urbanized areas, biophysical conditions and landscapes are all situated within the many different interests—political, economic, social, and ecological—at work (Pelling, 2003). This chapter looks at two types of urbanization landscapes and addresses how the observed changes may be related to state forest policy and the work of western conservation non-governmental organizations (NGOs) and how both may be tied to shaping perceptions and, ultimately, to changes in landscapes. This points to potential links between policy and the state and efforts to mitigate risk among smallholders and the landscapes they make and live in (see Blaikie et al., 1994; Sandoval Correa, 2000; Bass 2006b).

In recent years, the fields of cultural and political ecology have broadened, increasingly embracing urban areas as fertile ground of study (Pelling, 2003). They have also broadened conceptually, for example, by successfully embracing the ontological nature of landscape (Zimmerer and Bassett, 2003: 3). Landscapes teach and the shapes they take communicate values, messages, and agency of people. Like monuments in a museum, landscapes not only reflect but they reinforce normative values (Robbins, 2003). The presence of trees in a park, near a house, or on the landscape, may not only reflect what people do but also how they interpret and interact with the environments around them (Bass, 2006b). The trees may also indicate, perhaps, the presence of a political discourse that sees trees as valuable tools for averting risk in the landscape.

Forest transition theory suggests that, as regions become developed, rural populations are drawn to urban centers for economic opportunity in various economic sectors, allowing agriculture to be concentrated and intensified on the most arable lands and leaving marginal rural areas to secondary forest succession (Mather, 1992; Rudel, 1998; Rudel et al., 2002; Rudel, 2006; Aide and Grau, 2004; Farley, 2010).

Political ecologists understand risk as vulnerability that is both biophysical and socio-political (Davis, 1998). Urbanized areas in Latin America are generally highly vulnerable (Pelling, 2003). Steep slopes with inadequate infrastructure, a large, poor population, and very little planning make many places in the region risky. The landscapes and forms of vegetation featured here can be seen as related to risk-mitigation strategies, and the increased population and built environment are associated with urbanization. The trees observed here provide for people. Food, shade, firewood, construction materials, habitat, and carbon sequestration associated with climate change are all products these trees provide. In this way, planting or encouraging tree growth is a way of mitigating future risk. As this chapter shows, multiple adaptive strategies associated with multiple perceived risks contribute to a cumulative landscape of increases in trees associated with settled areas. Further, the discourse of conservation expressed by people across the study area indicates that, over time, state forest conservation policy may have had a role in shaping both perception and landscapes in the area.

Repeat photography is the process of replicating a photograph or set of photographs for purposes of assessing change. First used over a century ago to monitor glaciers, its use and popularity have expanded in recent years (Bass, 2004; see Webb et al., 2010). By using multiple photographs of a geographical locale separated by time intervals (usually of years or decades), the method allows the researcher to observe local patterns of change in oblique landscape photographs that might not be, for example, observable through aerial photographs or satellite imagery. Further, because it requires extensive fieldwork, it allows the researcher to gain better understandings of conditions, relationships, and changes that are engendered through them.

This chapter looks more closely at two of the many contributions to this increase. The re-landscaping of town plazas and other public spaces into small symbolic forests and the increases in trees associated with individual households both contribute significantly, I argue, to the apparent increases in overall tree cover observed to have taken place over the past half century. Seen as examples, these landscapes make up parts of the complex mosaic that shapes cultural landscapes and urban environments.

Public spaces and domestic surrounds are both exemplary forms of landscape associated with urbanization. Both are directly associated with the population density and built environments of settled, or urban, areas. Public spaces are inherently tied to multiple politically charged processes of state policy, urban living, change, and growth. Like urban spaces, even rural agricultural households can be seen through a lens of the forces and processes at work shaping urban landscapes.

Especially since the 1970s in Honduras—with road construction and the subsequent diffusion of schools, electricity, internet, television, and telephones—life in small rural areas has become more similar to life in urban areas. The traditional rural-to-urban migration associated with urbanization has been accompanied by the movement of urban amenities to rural areas, or "the capture of the countryside by the city" (Cosgrove, 2006: 64). Recent years have shown greater interconnectedness between long-isolated people and distant places (MacDonald and WinklerPrins, 2012). Much of this is characterized by changing information flows and generally associated with changes in consumption and production patterns that seem to affect many things, from social institutions (Tucker, 2008) to environmental conditions (Aide and Grau, 2004), from foodways to the production of waste (Moore, 2009) and levels of economic productivity (Houghton, 2009). In this way, both landscape types here can be associated with the processes tied to urbanization.

Public Spaces

As perhaps the quintessential public space or place in Latin American towns, the plaza is both a social and civic area (Herzog, 1999; Low, 2000). A sort of central design element of spatial organization, it is a place of gathering. In some cases, the plaza remains an important site for political engagement (Masuero, 2003; Crossa, 2004; Smilde, 2004). These public spaces are important. As a consistent part of the landscape, the plaza, according to Miles Richardson (1974), helps direct behavior and activity. Additionally, like monuments—perhaps even as monuments—they teach.

Plazas in Latin America appeared early in the colonial era. The 'Law of the Indies,' actually a series of edicts spanning over 200 years, mandated particularities in the founding of settlements in the Spanish New World (Stanislawski, 1947; Crouch et al., 1982; Veselka, 2000). Many of the design aspects came from the Spanish landscape (Stanislawski, 1947; Smith, 1955), though some argue that the plaza design is an articulation of the tensions between indigenous American ideas and European ones (Low, 2000). The grid-pattern town with a central main plaza and adjacent church was the basic form. This settlement form likely came to Spain with the Moors. From there, it was carried to the Americas where it formed the basis for new Spanish settlements. From Texas to South America, most towns in Latin America still maintain this design.

The plaza in Latin America is a public space (Richardson 1974; 1982; Veselka, 2000). It forms a public social nucleus and helps order space and life (Richardson 1974; Low, 2000). The plaza town may be seen as a representative form of the Latin American settlement landscape. The plaza as a public space is also a repository of messages. Some of these messages are in the form of monuments; the mother breast-feeding her child, the important historic leader, such as Central America's Francisco Morazan or South America's Simon Bolivar. In Gracias, Lempira, Honduras' legendary Lenca leader of native resistance to the Spanish stands prominently in the plaza. Nearby signs announce Lempira's virtue, "*Nuestro Gran Cacique*" ("Our Great Leader"). They teach. Bolivar, Morazan, and Lempira all, then, through their specific representations become objectified, important, and virtuous; perhaps they also even become simplified or reduced.

Through repeat photography, I discovered that plazas and public spaces throughout the country have been changed in recent years. What once were open areas of public space have become spaces of trees. Plazas have been reordered. One of the most common aspects of this is the presence of trees. Plazas have become tiny representative public forests. They have been consistently planted with trees, filling the open, sunny gathering space with shade and greenery (Figure 7.1). This presence may itself also be a message.

When West visited and photographed Honduras in 1957, plazas there, as seen in his photographs, were generally still the open areas they had long been. John B.

Figure 7.1 Left (a) (1957) and right (b) (2010) Santa Lucia, Francisco Morazan, former mining center north-east of Tegucigalpa, Honduras. The town has become a weekend getaway and bedroom community for more affluent residents of the capital city. The increases in vegetation are clearly evident but difficult to classify and explain. Domestic gardens and orchards, public plantings, and random dispersed tree growth all contribute to this cultural forest.

Source: 7.1(a) Robert C. West Latin American Photograph Collection, Louisiana State University; 7.1(b) J. O. Joby Bass.

Stephens (1969a and b [1841]) traveled through Central America in the early 1800s.The many plazas he encountered and described were typically open places, often used for both political and religious gathering. Residents would gather in plazas for religious events (Stephens 1969a: 254) or for markets (Stephens,1969a: 192). Others would gather as participants in various military alliances and activities (Stephens, 1969a: 74, 230; Stephens,1969b: 76, 85). The plaza landscapes he described were open, treeless places (Stephens, 1969a: 192). When West visited Honduras a century later, he photographed places and landscapes throughout the country, including many of the country's plazas. His photographs show essentially the same plaza landscape that Stephens described over a century before.

It was through the process of repeat photography that I became aware of, among other things, a change in plaza landscapes. Most plazas that I encountered in 2001 in Honduras were *not* the open, treeless spaces that West and Stephens had seen. Most were, rather, places filled with trees. They had become essentially small civic forests. This is true for many municipal parks and open public spaces too. In short, a very different kind of place was created by official action.

The 'landscaping' of the plazas apparently occurred primarily in the 1980s and 1990s, much of it during the 1990–1994 presidency of Rafael Callejas. These changes also took place on the heels of an increase in attention to forests and deforestation issues in the country (Sandoval, 2000). This is reaffirmed by the presence of signs in the plazas. The signs are usually on the trees and typically advertise the virtues of trees, forests, and environment, and the need to protect them. This message is seemingly ubiquitous. It is throughout the country; painted on walls as murals, on the walls of schools, on T-shirts, on NGO-sponsored posters, on national automobile license plates. And, again, hanging on the trees in the plazas.

The trees in the plazas and the signs on them can perhaps be tied to a larger environmental discourse that has taken place in Honduras; and perhaps throughout much of the hemisphere or even the world (see Bass, 2006b). The discourse is one of environmental concern with a focus on forests and deforestation and the pertinence of protecting the environment (Downing, 1992; Utting, 1993; Painter and Durham, 1995). It also takes place within a complex set of interests, endeavors, and outcomes (Brosius et al., 1998; Escobar, 1999). The concurrence of the development of specific government conservation policies, the appearance of the trees, and the signs hint that they are related.

Households

Chance encounters with a Lenca man (see West 1998 for an overview of the Lenca culture region) and others while I was en route to a camera station (Figure 7.2) illustrate another cultural aspect of repeat photography. I heard his metal hoe chinking against rocks before I saw him. I topped a rise in the giant, dibble-planted cornfield, and saw the Lenca man hoeing up feral mustard greens. He stood up and leaned on his hoe, smiling, as I approached. I stopped near him and we exchanged greetings, briefly touching our palms together in a gesture of shaking

Figure 7.2 Left (a) (1957) and right (b) (2010) Llano Yarula, La Paz, Bolivia. Trees on distant slope are primarily agroforestry patches where residents are growing construction materials, domestic kitchen orchards around houses, and fencerows. The junipers in the foreground are growing along a barbed wire fence that did not exist in 1957. Though very rural, evident increases in trees are associated with increases in houses ("capture of the countryside by the city"). New tree patches in the distance have houses in them—they are essentially householder versions of what appears to have become an urban forest in many towns.

Source: 7.2(a) Robert C. West Latin American Photograph Collection, Louisiana State University; 7.2(b) J. O. Joby Bass.

hands. We briefly discussed his corn and how lucky the people in the area were to have been spared most of a recent drought. Then I asked him about the small oak forest patch behind him. He nodded toward it and confirmed for me that it was indeed bigger than it used to be. "It's a small forest. The trees, the forest, are good. Good for the air, for the animals, for the rain, the temperature." He told me that, owned by a local man, the forest was used partially to supply firewood and construction materials for locals. I asked him if he thought it would be all right for me to go and check it out. He nodded assuredly, "Of course." Then, looking back toward it, he continued, "It's beautiful, the forest. We need forest here. It's good."

I continued my walk to the tree patch. It was a small, fenced oak–laurel forest patch with only minimal underbrush and a few birds. I stood in the shade at its edge and looked over at another hill. There at the top was the new patch of trees that appears in the top center of the new photograph (Figure 7.2b). I headed for it.

At a barbed wire fence, I looked up to see a big dog hurrying across a packed earth yard, hopping a handmade palmetto broom as he came after me. From a hole in an adobe house, a woman's voice yelled at him to stop and then at me to tell her what I wanted. I crawled through the barbed wire, watching the dog, and crossed over to her window. I told her that I had old photographs of the area and tried to show one to her. She glanced at it, but quickly. I told her how the older photograph appeared to indicate that this area had more trees than in the 1950s and I asked her about the two new patches near her house.

She nodded across the barren earth of her yard toward the oak–laurel stand and said that the man who owned the land there had let the trees grow up. He had fenced it off to keep animals out and had essentially raised a forest. She said that he sells the wood for both firewood and construction materials "because we are poor here—*pobrecitos*. We are poor so we still have to cook with firewood." In 1995, 95 percent of rural Hondurans still used firewood as their primary source of energy, making up 60 percent of the country's total energy consumption (UNPD, 2000). The Lenca highlands are among the country's poorest areas and remain very rural in character. Many people in the area still depend on firewood for cooking energy, though the number is rapidly decreasing.

I asked about another, smaller tree patch that I had passed on top of the hill. Her face lit up. "That's mine," she smiled, "I planted it." She told me that she had planted the patch of *Perymenium strigillosum* (tatascan) for fence posts. Standing in even rows, the trees were young and not very big. She said that they need lots of fence posts because they have so much barbed wire fencing. Though fences often follow old "living fences," barbed wire is the rule now, and this requires fence posts, again, "because we are poor—*pobrecitos*. We have to make everything. This house, our meals, firewood, everything. We are poor here." Though self-deprecating, she smiled a lot. She left the dusty windowsill and came outside to hold the dog so I could leave.

In most cases, I found that areas of settlement have seen an increase in trees—settlement forest—over the past few decades. This was consistent across much of the country and was true for both urban and rural areas (Figure 7.3). Much of this is made up of trees associated with individual houses, some providing shade, others providing food (Figures 7.1, 7.2, and 7.3). Walking through the town of Oropoli,

Figure 7.3 Left (a) (1957) and right (b) (2010) El Chichicaste, El Paraiso, Honduras. A small town near the border with Nicaragua, El Chichicaste has grown over recent decades. As this growth has taken place, an urban forest has also grown. Domestic gardens and orchards, public plantings, and random dispersed tree growth all contribute to this cultural forest.

Source: 7.3(a) Robert C. West Latin American Photograph Collection, Louisiana State University; 7.3(b) J. O. Joby Bass.

El Paraiso, having just re-photographed a scene overlooking the town that showed an increase in both settled area and trees, I saw a man standing against the front wall of his cement-block house. He greeted me as he repaired a mahogany slingshot that he uses to kill rabbits and raccoons out at his *milpa* (swidden field). He grinningly said that, of course, as God's creatures, they have to eat too but that he just can't let them have it all. I showed him the early photograph and asked him about the town and its trees. His face lit up and he pointed to the side of his house to his fenced yard. "Want to see my trees?" he asked.

We stepped through the house, my appearance surprising his wife, and into the backyard, where I got a tour of the yard. The dark, fenced area was filled with trees, an arborescent kitchen garden. As he had only moved into town from his rural land 12 years before, many of the trees were still small. We waded through the brushy shade as he grabbed a limb on each tree and identified it. All were fruit trees. The soil beneath us was dark and damp, an anomaly in the area. This small kitchen forest was made up of *Citrus* spp. (orange and lime), two *Mangifera* spp. (mango), *Carica papaya* (papaya), *Musa* sp. (banana), *Cocos nucifera* (coconut), *Tamarindus* sp. (tamarind), and *Crescentia cujete* (calabash), in addition to several ornamental bushes and plants. Many other houses had similar plantings.

The role of smallholders in contributing to the overall increase in trees across Honduras is difficult to quantify but clearly significant. Across the full photograph sets, smallholder intensification is the most common transition pathway clearly related to visible increases in vegetation. It is also related to a lot of vegetation that exists in a wide variety of contexts and formations. Unlike the public forests of the plazas, these trees may or may not have been planted intentionally. Like the public forests of the plazas, though, these trees represent a change in human–environment relationships tied to many complex processes.

Further, as previously discussed, even rural smallholders in places like Honduras live within the context of urbanization. If they don't have daily or regular access to an urban agglomeration, they certainly are influenced by the goods, services, and ideals coming from there, an urban hegemony along the lines of dependency theory (Frank, 1967). In this way, smallholder, or individual household, landscapes can and should be seen as situated within many of the same processes shaping the settlement forests that are so common and widespread in urban areas, including central policy and civic discourse.

Discussion

Increases in trees, vegetation along fencerows, in towns, around houses, along roadsides, as well as in forest patches have characterized landscape changes across the past half-century in Honduras. The variety of patterns of increases observed likely mirrors the variety of factors affecting them. Plaza and settlement forests have increased as villages, towns, and cities have grown. Understanding these changes within a context of the 'pathways' that may be linked to them may help overcome some of the difficulties that have been met trying to explain their patterns (see Fairhead and Leach, 1995).

Conceptual pathways to forest transition seek to elucidate primary affective processes (see Farley, 2010). Of five potential pathways of economic development, forest scarcity, state forest policy, globalization, and smallholder tree-based intensification, the last appears to explain more than the others the increases in tree cover I have observed. However, I will argue that these changes ultimately can be linked to state forest policy and the discourse of conservation. Smallholder intensification observes that increases in tree cover can be related to many other factors and result in a variety of vegetation increases such as some forms of agroforestry, windbreaks, woodlots, etc. (Farley, 2010). This includes a variety of processes and vegetation types that occur in a variety of contexts, forms, and places. One reason for this—and problem with the conceptualization—may be the breadth of the category. Again, increases in tree cover can be related to many factors. Improvements in prevention of diseases, such as malaria, may be related to this change, as the thick vegetation providing habitat and refuge for mosquitoes may have been kept down in areas near houses and settlements to avoid disease in the past. Rural-to-urban migration, by agriculturalists, likely contributes. The public discourse on forest and conservation issues appears to also contribute (Peet and Watts, 1996; Bass, 2006b). This is almost a catch-all category for processes that cannot be identified clearly within other conceptual pathways. In many cases, the processes in this pathway appear to be most significant in affecting change at the local level. However, its potential importance lies in the fact that the myriad processes collected in this category may be among the more important in helping understand how people shape and perceive landscapes.

But the issue of perception should be seen within a context of public discourse and state forest policy. State forest policy is an important factor in understanding rural Honduran landscapes, at least where there are trees. The formation of modern national forest policies and agencies took place basically across the same time period as this study. In the 1950s, the first concepts of so-called "professional forest management" arrived in Honduras, essentially directly from the US under the auspices of the Food and Agriculture Organization (FAO) and United States Agency for International Development (USAID) (AFE-COHDEFOR, 1996: 88–89). Professional Honduran foresters came to universities in the US to be educated in the field. The forestry school, *La Escuela Nacional de Ciencias Forestales* (ESNACIFOR) was founded in 1969 to promote and expand a professionalized forestry sub-sector (AFE-COHDEFOR, 1996: 127). In 1971, the *Ley Forestal* introduced ideas of multiple use, sustainability, and conservation (Sandoval, 2000: 277). In 1974, Honduras formed the now-defunct *Corporacion Hondurena de Desarollo Forestal* (COHDEFOR) (Sandoval, 2000).

COHDEFOR was formed with the economic and professional assistance of Germany, the US, and the United Nations Program for Development (AFE-COHDEFOR, 1996: 39). At first, the organization focused on organizing and managing timber resources. Reforestation and protection were not really part of the operation.

Eventually, though, toward the late 1980s, a larger ecological movement began to take hold. Interest in conservation and protection of resources, in biodiversity and ecosystem management began to gain strength. Reforestation began to receive attention, as did the formation of protected areas, causing approximately 24 percent of the country to be eventually put under a variety of forms of protected status, ranging from national parks to forest reserves to inhabited forest areas with regulations on their use (Sandoval, 2000: 280–284; AFE-COHDEFOR 1996: 150–153).

With the neoliberal economic restructuring of the 1990s, COHDEFOR received different degrees of influence over a significant portion of Honduras' land, public and private. In 1993, the Congress passed laws that, officially at least, assured the protection of the environment and guaranteed that forests would be protected, managed, and, if need be, replanted, formally 'concretizing' the growing global preoccupation with the environment (Sandoval, 2000: 280–284). Progressively, conservation concerns increased, coincidentally with increases in involvement by foreign agencies, mostly NGOs (see Sandoval, 2000: 287, 331–315, 409–414; AFE-COHDEFOR, 1996: 39, 155, 185, 199). Social and educational programs were put in place throughout the country, largely with the help of, if not directed by, foreign interests from the US or Europe (Sandoval, 2000: 298–302; AFE-COHDEFOR, 1996: 185). As this was happening during the same period that the government and NGOs alike had mounted environmental education campaigns throughout the country (Inter-Hemispheric Education Resource Center, 1988), the impact on perception has been potentially tremendous.

The official view of Honduran forests in recent years could perhaps be best summed up in this statement by AFE-COHDEFOR:

> The forests of Honduras constitute an important source for the generation of economic resources to finance government programs. The same is true for many property owners of private forest areas. However, for other sectors of the population, the forest is seen as an obstacle to the expansion of their agricultural and ranching activities, thus ignoring the high value and significance they have for society through the protection of soils, fauna, the regulation of water sources, the moderation of climate, the purification of the air and recreation.
>
> (1996: 141) (author's translation)

Combined with the work of NGOs (mainly western), Honduras' state forest and conservation policies and ideals appear to have affected environmental conditions, and the presence of trees. Indeed the conservation message is ubiquitous. In addition to the signs in plazas, conservation messages appear on billboards, public murals, and even the national license plate. Cumulatively, perhaps, then the observed increases in trees combined with the consistent message I have heard from residents regarding the virtues of trees indicate that state policy has had an impact, both on landscapes and on people's perceptions. Following the notion that the control of discourse shapes actions (Peet and Watts, 1996) and landscapes (Velasquez Runk et al., 2010), perhaps we see here a direct link between policy and public discourse that ultimately has reshaped both cultural perceptions and cultural practices, evidenced through the landscapes they make.

Conclusion

Analyses that have historically utilized national scale data have probably obscured variations in vegetation change that have been taking place at more local scales. In some places, forest increases have actually been observed concurrent with population increases, challenging some traditional assumptions (Tucker, 2008). In many countries, both deforestation and afforestation can be observed simultaneously in different regions (Batterbury and Bebbington, 1999; Perz, 2007; Redo et al., 2009; Lambin and Meyfroidt, 2010). Increases in trees in some places actually seem to be associated with the presence of people. As urban areas grow in many places, so do their urban forests.

As populations have grown, both in the countryside and towns, so have the trees around homes. In rural areas, this appears in photograph pairs as small clusters of trees—practically all fruit bearing—in formerly open areas. Occasionally, the houses within the trees are visible. In towns, this appears as the growth of an urban or settlement forest. It can be associated with different conceptual pathways, as it probably has taken place due to a variety of factors. Immigration by rural people, changes in conservation ethics and changes in the geography of disease are all likely tied to this. In another way, this can be seen as the other end of the traditional set of forest transition processes, whereby rural folk move to town and change the way they use land, not only in the rural areas but in the urbanized spaces to which they move.

The kinds of vegetation that are significant here have elsewhere been described as *tree resources outside forests* (Herrera-Fernández, 2003). The vegetation appears to be potentially significant, as it is particularly widespread, is thought to have high conservation value, and often involves native species (Herrera-Fernández, 2003; Farley, 2010). These trees are also often found in explicitly cultural landscapes, often in towns and/or around houses. A more nuanced perspective on what is meant by smallholder intensification would likely shed more light on the ways that individuals collectively shape landscapes regarding trees in our settled worlds.

Both settlement forests and symbolic forests are increasingly a part of urban life across the world. Documented increases of trees in settled landscapes are now common. The public use of trees in landscaping is also a common part of urban life. Together, these trees represent something about the value of trees in the minds of people. The shade, food, carbon sequestration, soil retention, and wood resources they provide are represented by their increasing presence, both practically and symbolically. That the practical and symbolic both appear together is likely not a coincidence, as perception shapes agency. State forest policy and centralized conservation discourse appear to be effective in shaping perception and, perhaps, landscape conditions, even at the household scale.

Further research should inquire more deeply into the relationships between political discourse and landscape (Velasquez Runk et al., 2010). As ethnographic work indicates, a consistent message regarding the value of trees seems to exist among residents. This message is similar to that found in Honduras' state forest policy as developed over the later part of the twentieth century (Sandoval Correa, 2000). What, if anything, does this perception have to do with the recent increases in trees? Landscape analysis hints that control over the political discourse, important for shaping what takes place (Peet and Watts, 1996), favors the "tree-centric" currently. Trees are seen and talked about as virtuous, as ways to mitigate climate change, as sources of food and fuel and construction materials. Their benefit is touted in public spaces. They are never associated, for example, with mosquitoes that cause disease. Though this arboreal bias has been addressed in rural landscapes of conservation (Velasquez Runk et al., 2010), more work should be done.

As populations grow and gain increasing access to more and more information, communication and transportation opportunities, economic options and activities, conservation ideals, land use, and settlement patterns are all changing, influencing each other as well as landscape conditions across the world. Honduran landscapes contain material manifestations of a remarkably high number of different processes seen to shape environmental conditions. Increases in vegetation that are occurring in such varied arrangements and places are reflections of the many reasons that are behind them. Seeing this at the scale of oblique photography allows a more intimate perspective on an issue typically informed by more distant imagery or maps that use coarse data relative to the types of changes seen in many places. This more humanized scale also perhaps offers more insight into the true complexity of understanding what we are up to and how it shapes our landscapes.

Acknowledgment

This chapter is derived in part from Bass (2006b; 2010; 2012). Many thanks to the respective publishers for permission to reprint portions of that body of work.

References

AFE-COHDEFOR. (1996) *Analisis del Sub-Sector Forestal de Honduras*. Tegucigalpa: Cooperacion Hondurena-Alemana, Programa Social Forestal.

Aide, T. M. and Grau, H. R. (2004) Globalization, Migration, and Latin American Ecosystems. *Science*, 305(24): 1915–1916.

Arreola, D. D. (1992) Plaza Towns of South Texas. *Geographical Review*, 82(1): 56–73.

Bass, J. O. J. (2004) More Trees in the Tropics. *Area*, 36(1): 19–32.

Bass, J. O. J. (2006a) Forty Years and More Trees: Land Cover Change and Coffee Production in Honduras. *Southeastern Geographer*, 46(1): 51–65.

Bass, J. O. J. (2006b) Message in the Plaza: Landscape, Landscaping, and a Forest Discourse. *Geographical Review*, 95(4): 556–577.

Bass, J. O. J. (2010) Learning Landscape Change in Honduras: Repeat Photography as Discovery. In: R.H. Webb, D.E. Boyer, and R. M. Turner (eds). *Repeat Photography: Methods and Applications in the Natural Sciences*. Washington DC: Island Press.

Bass, J.O.J. (2012) Trees, Repeat Photography and Pathways to Landscape Transition in Honduras. *RevistaGeografica*, 150: 55–74.

Batterbury, S. P. J. and Bebbington, A. J. (1999) Environmental Histories, Access to Resources and Landscape Change: An Introduction. *Land Degradation & Development*, 10(4): 279-289.

Blaikie, P., Cannon, T., Davis, B. I., and Wisner, B. (1994) *At Risk: Natural Hazards, Peoples, Vulnerability and Disasters*. London: Routledge.

Brosius, J., Tsing, A. L., and Zerner, C. (1998) Representing Communities: Histories and Politics of Community-based Natural Resource Management. *Society and Natural Resources*, 1(2): 157–168.

Cosgrove, D. (2006) Modernity, Community, and the Landscape Idea. *Journal of Material Culture*, 11: 49–66.

Crossa, V. (2004) Entrepreneurial Urban Governance and Practices of Power: Renegotiating the Plaza in Mexico City. *Antipode*, 36(1):127–129.

Crouch, D. P., Garr, D. J., and Mundigo, A. I. (1982) *Spanish City Planning in North America.* Cambridge, MA: MIT Press.

Davis, M. (1998) *Ecology of Fear.* New York City: Metropolitan Books.

Downing, T. E. (ed) (1992) *Development or Destruction: The Conversion of Tropical Forest to Pasture in Latin America.* Boulder, CO: Westview Press.

Escobar, A. (1999) After Nature: Steps to an Antiessentialist Political Ecology. *Current Anthropology*, 40(1): 1–30.

Fairhead, J. and Leach, M. (1995) False Forest History, Complicit Social Analysis: Rethinking some West African Environmental Narratives. *World Development*, 23(6): 1023–1035.

Farley, K. A. (2010) Pathways to Forest Transition: Local Case Studies from the Ecuadorian Andes. *Journal of Latin American Geography*, 9(2): 7–26.

Frank, A.G. (1967) *Capitalism and Underdevelopment in Latin America: Historical Studies of Chile and Brazil.* Berkeley, CA: University of California Press.

Grau, R. H. and Aide, M. (2008). Globalization and Land-use Transitions in Latin America. *Ecology and Society*, 13(2): 16.

Herrera-Fernández, B. (2003) *Classification and Modeling of Trees Outside Forest in Central American Landscapes by Combining Remotely Sensed Data and GIS.* Doctoral dissertation. University of Freiburg, Germany.

Herzog, L.A. (1999) *From Aztec to High Tech: Architecture and Landscape Across the Mexico–United States Border.* Baltimore, MD: The Johns Hopkins University Press.

Houghton, D. (2009) *Cell Phones and Cattle: The Impact of Mobile Telephony on Agricultural Productivity in Developing Nations.* Unpublished honors thesis, Durham, NC: Trinity College, Duke University.

Inter-Hemispheric Education Resource Center. (1988) *Private Organizations with US Connections, Honduras.* Albuquerque, NM: Inter-Hemispheric Education Resource Center.

Lambin, E. F. and Meyfroidt, P. (2010) Land Use Transitions: Socio-ecological Feedback versus Socio-economic Change. *Land Use Policy*, 27: 108–118.

Low, S. (2000) *On the Plaza: The Politics of Public Space and Culture.* Austin, TX: University of Texas Press.

MacDonald, T. and Winklerprins, A. (2012) The Best of Both Worlds? The Case of Peri-Urban Migration in Western Pará State, Brazil. Presented at Conference of Latin Americanist Geographers, January 13, 2012, Merida, Mexico.

Masuero, A. (2003) Santiago's Civic Square, Plaza de la Constitucion. *Arq*, 53: 58–61.

Mather, A. S. (1992) The Forest Transition. *Area*, 24: 367–379.

Moore, S. A. (2009) The Excess of Modernity: Garbage Politics in Oaxaca, Mexico. *The Professional Geographer*, 61(4): 426–437.

Painter, M. and Durham, W.H. (1995) *The Social Causes of Environmental Destruction in Latin America.* Ann Arbor, MI: University of Michigan Press.

Peet, R. and Watts, M. (1996) *Liberation Ecologies: Environment, Development, Social Movements.* London: Routledge.

Pelling, M. (2003) *The Vulnerability of Cities: Social Adaptation and Natural Disaster*. London: Earthscan.

Perz, S. G. (2007) Grand Theory and Context-specificity in the Study of Forest Dynamics: Forest Transition Theory and other Directions. *Professional Geographer,* 59(1): 105–114.

Redo, D., Bass, J. and Millington, A. (2009) Forest Dynamics and the Importance of Place in Western Honduras. *Journal of Applied Geography,* 29(1): 92–111.

Richardson, M. (1974) Spanish-American Settlement Pattern as Societal Expression and Behavioral Cause. *Geoscience and Man,* 5: 35–51.

Robbins, P. (2003) Fixed Categories in a Portable Landscape: The Causes and Consequences of Land Cover Categorization. In: K. Zimmerer and T. Bassett (eds) *Political Ecology: An Integrative Approach to Geography and Environment-development Studies*. New York City: Guilford Press.

Rudel, T. K. (1998) Is There a Forest Transition? Deforestation, Reforestation, and Development. *Rural Sociology,* 63(4): 533–552.

Rudel, T. K. (2006) Shrinking Tropical Forests, Human Agents of Change, and Conservation Policy. *Conservation Biology,* 20(6): 1604–1609.

Rudel, T., Bates, D., and Machinguiashi, R. (2002) A Tropical Forest Transition? Agricultural Change, Out-migration, and Secondary Forests in the Ecuadorian Amazon. *Annals of the Association of American Geographers,* 92(1): 87–102.

Sandoval Corea, R. (2000) *Honduras: su gente, su tierra y su bosque*. Tegucigalpa: Graficentro Editores.

Smilde, D. (2004) Popular Publics: Street Protest and Plaza Preachers in Caracas. *International Review of Social History,* 49: 179–195.

Smith, R. C. (1955) Colonial Towns of Spanish and Portuguese America. *The Journal of the Society of Architectural Historians,* 14(4) Town Planning Issue: 3–12.

Stanislawski, D. (1947) Early Spanish Town Planning in the New World. *Geographical Review,* 37(1): 94–105.

Stephens, J. L. (1969a [1841]). *Incidents of Travel in Central America, Chiapas, and Yucatan, Volume 1*. New York City: Dover Publications.

Stephens, J. L. (1969b [1841]). *Incidents of Travel in Central America, Chiapas, and Yucatan, Volume 2*. New York City: Dover Publications.

Tucker, C. (2008). *Changing Forests: Collective Action, Common Property and Coffee in Honduras*. New York City: Springer Academic Press.

Utting, P. (1993). *Trees, People and Power: Social Dimensions of Deforestation and Forest Protection in Central America*. London: Earthscan.

United Nations Program for Development (UNPD). (2000) Informe sobre desarollo humano, Honduras: 2000. Programa de las Naciones Unidas para elDesarollo, Tegucigalpa, Honduras.

Velasquez Runk, J., Ortiz Negria, G., Pena Conquista, L., Mejia Pena, G., Pena Cheucarama, F. and Cheucarama Chiripua, Y. (2010) Landscapes, Legibility, and Conservation Planning: Multiple Representations of Forest Use in Panama. *Conservation Letters,* 3: 167–176.

Veselka, R. E. (2000) *The Courthouse Square in Texas*. Austin, TX: University of Texas Press.

Webb, R., Boyer, D., and Turner, R. (2010) *Repeat Photography; Methods and Applications in the Natural Sciences*. Washington DC: Island Press.

West, R. C. (1998) The Lenca Indians of Honduras: A Study in Ethnogeography. In: R.C. West *Latin American Geography: Historical-Geographical Essays, 1941–1998*. Geoscience and Man, Vol. 35: 73–86.

Zimmerer, K. S. (2004) Cultural Ecology: Placing Households in Human Environment Studies—The Cases of Tropical Forest Transitions and Agrobiodiversity Change. *Progress in Human Geography*, 28(6): 795–806.

Zimmerer, K. and Bassett, T. (eds) (2003) *Political Ecology: An Integrative Approach to Geography and Environment-development Studies.* New York City: Guilford Press.

Part 2

Arboreal and Greenspace Agency in the Urban Landscape

8 (Urban) Places of Trees

Affective Embodiment, Politics, Identity, and Materiality

Owain Jones

> [A] part of the world is what sets up, drives and energizes our emotional experience.
> (Slaby, 2014: 1)

Introduction

The novelist Richard Ford opens his 1995 novel *Independence Day* thus:

> In Haddam, summer floats over tree-softened streets like a sweet lotion balm from a careless, languorous god, and the world falls in tune with its own mysterious anthems. Shaded lawns lie still and damp in the early a.m.

'Tree-softened streets'—does that not beautifully capture the kind of impacts trees can sometimes have in a city? The 'softening' may come through multiple, interleaving registers, some more obvious than others; visual, auditory, temperature wise, aspiratory, and culturally—through memory—and even literally as tree roots loosen paving and street substrates.

This then is about the affects which stream between the arboreal and the social, through the senses and other 'skills of the body' in everyday practices—about exchanges between trees and people as they both go about their daily business in the city. That immediate becoming-in-place is a complex-enough process, but humans, being what they are, absorb these streams of 'data' through layers of memory, culture, and the like. Trees can set out urban space, its tone and feel; mark them as socio-economic–cultural–ecological space of some particular kind or other. Trees can be *lead players* in the city performed.[1]

Consider what could be said of a very ordinary city square with a population of mature and younger trees in it—Victoria Square, Bristol (UK).[2] The trees here make, quite literally, the space by enclosing it in vertical and horizontal dimensions. Other trees are busy in their own ways, for example, carrying plaques of remembrance for local citizens who had campaigned for the square's arboricultural heritage. And as they live their everyday lives, they generate all the sensory, material, and cultural richness that differing trees bring to a place. And people inevitably engage with richness as sensing moving beings as they use the square in many ways—to stroll through, to sit in at lunch break, to meet their lover (or estranged partner), to hang out after school, to walk the dog, to play football or Frisbee, to watch out for whatever wildlife is hanging on in the city—to remember, relax, forget, to steel oneself. The square has a rich (often surprising and contested) aboricultural history, which is always on the move. Since I studied it intensely, some 15 years ago, a few of the large trees have been reduced, or felled, and a new generation introduced, including five new species of oak tree donated by two local residents' groups (the Clifton and Hotwells Improvement Society and the Clifton Garden Society) to mark the Queen's Golden Jubilee in 2002. One large Tree of Heaven (*Ailanthus altissima*), which was reputed to be the 'finest in Bristol'[3] spectacularly fell in 2013 crushing cars and closing a road for 14 hours.[4] The giant beech tree (*Fagus sp.*) which dominates the centre of the square has lost some major limbs, making it look and feel like 'half a tree', the absent parts acting like an absent echo of the half that remains (Figure 8.1).

The basic aim of this chapter is to suggest that the turn to affect in various parts of the social sciences and arts and humanities, and the deepening understanding in the natural sciences of how organisms—be they human or trees—affectively exchange information and substance (e.g. chemical emanations) in deeply ecological retaliations, should be of central interest to those studying various aspects of tree society relations, be they urban trees or otherwise; or single trees, or populations of trees in place. Beyond that, I delve into more detail about trees and their affective capacities which in large part means paying attention to them as palpable living things, not just as 'trees' but also as types of trees and individual trees in particular places.

To expand this introduction a bit further; our previous work on trees and place[5] focused very much on the agency of trees, their lively and recalcitrant materiality, and how they operate in specific place and time—specific local circumstances of culture, politics, management, economy, soil type, and climatic conditions. I am pleased to say that others too are addressing the liveliness of trees and other 'plant actors'.[6] In our 'trees work', emotions and feeling were to the fore in so far as it addressed how people engaged with particular trees in particular places, and that, in effect, we put the specific embodied agencies of trees at the centre of place narratives, *and* how these entangled with the social. This is not to imply a natural determinism, or one-way flow of agency. There are always multidirectional flows of actions and meanings and *feelings*—as communities and agencies respond to trees and act with and upon them. What I now feel is needed, in order to deepen these *narratives of exchange*, are more fully developed notions of the *affective energies of collective life.*

New affect-based understandings of place and landscape (which will be set out below) can be of use in the understanding of how individuals and communities engage with treed spaces in imaginative and in embodied (practical) terms, and also in how the complex composition of such spaces (as cultural, ecological, political, economic, and living entities) can be appreciated. These ideas are not only connected to basic ontological understandings of how the world unfolds, but also to questions of well-being and social and ecological benefit through the ways people construct their individual and collective identities in terms of senses of self and senses of place.

Figure 8.1 The giant beech tree with missing half of canopy.

Source: Owain Jones, 2012.

There are a number of intersecting ecological, cultural, political, and economic dimensions to this. What kind of trees should we be planting in cities? What about the agency of established (often large and powerful) trees which can be unruly? What about the contested relationships between different people's varying needs and expectations of the city? See, for example, David Ley's great paper (1995), on the cutting of two 30-metre sequoias in a suburban garden in Vancouver, which opens up vistas of migration, colonial power, and cultural difference and conflict in the city. How should we imagine (and manage) the future of city trees in relation to climate change? Should iconic species be abandoned for new ones that might survive future conditions better?

What I am proposing here is sustained attention on the transmissions between trees and people. Such investigations of feelings and values are important, but need to sit alongside thinking about how they are performed, how such feelings come about, and where they are located. The chapter will briefly review innovative, theoretical and methodological approaches to place, landscape, and heritage which have been developed in recent years in geography, anthropology, sociology, and elsewhere, and then move on to sketch out various dimensions of people–tree transmissions in urban settings.

City–Trees

Trees and the city, what a combination! Possibly one of the ultimate interpenetrations of culture and nature. In his famous work Robert Pogue Harrison (1992) said that civilization (the settlement-the city) is, in a very real sense, a clearing in the primeval forest. The city—*supposedly* the apotheosis of culture—nature erased both spatially and temporally. The forest—the apotheosis of nature, a space beyond culture—offering possibilities of escape (Thoreau, 1864); refuge from power (Schama, 1995); or of temptation, otherness, and loss of self (see Jones, 2011 for summaries). So then, how do we read the many cities that are lucky enough to have large tree populations? I was always struck by the idea that parts of London are de facto forests because of the number of trees in place; street trees, garden trees, park trees, trees in office and mall landscaping, 'wild' trees on railway margins and brown field sites. Given the size (an important affective issue) that some trees can reach, they can come to dominate a landscape even if densely packed with buildings. They can override the complex symmetries of grand architectural spaces and facades as in Victoria Square.

Of course, the divide between nature and culture is a false one, as Ingold (2011) and many others have argued. (Human) bodies are ecologies of differing life forms, forces, circulations, and substances. So too are cities. They have never been 'without nature', they always teem with animals, microbes, lichens, and the like (Amin and Thrift, 2002). But when trees are in place, the ecological, non-human production of urban space is *writ large in the amazing material forms, and in the daily and seasonal doings of trees.* This is why trees are often one of the key mediums for green city planning. To separate out one set of actors (trees) for special attention is to risk losing sight of that ecological richness of all life. But I feel there are many compelling reasons for paying close attention to city trees. One is the fact that now, (in)famously, more than half of the world's populations live in cities and thus have particular and varying relationships with 'nature'. Another is that urban biodiversity can, in certain circumstances, flourish, while that in rural/wilderness areas is declining. A third is that trees clearly *can* contribute to the sustainability and resilience of cities (see below). The reverse may well also be true.

How are we to read, practice, and manage the city as a forest? What are the implications of urban trees for society? What are the implications of the city for trees? These questions are particularly pressing given the uncertainties and risk of coming decades, and demand concerted and shared work to answer them. In their comprehensive urban forestry reference book Konijnendijk et al. state that 'urban forestry is facing very significant challenges, and the availability of sound knowledge across cities, regions, and counties is crucial, as is the identification of good practice' (2005: 5).

To continue to generate 'sound knowledge' and to push deeply into the extraordinary richness of trees–cities–peoples' performativities (i.e. the lived ongoing moments of relational life, which spring from habit and novelty in fluxing relations), we need not only to consider the lively, embodied emplaced agencies of specific trees in action in place, but also how people (and other beings and processes) receive, exchange, and act with trees. This has of course been considered before but it now needs to heed the affective realm of becoming.

Political ecology has been prominent in addressing urban natures. As Smith (2005) summarizes:

> Political ecology provides a powerful means of cracking the abstractions of this discussion about the metabolism or production of nature. Rooted in social and political theory it is also grounded in ecology and has an international scope. When complemented by an environmental justice politics, which is less internationally focused and less theoretical but more politically activist in inspiration, political ecology becomes a potent weapon for comprehending produced natures.
>
> (2005: xiv)

Urban trees have been a focus within urban political ecology focusing, amongst other things, on important questions such as the uneven distribution of social well-being or the reverse, associated with urban trees in relation to other forces such as ethnicity and income (Heynen et al., 2006). Peckham et al. (2013) summarize that:

> To be successful, sustainable urban forest management must reflect urban society's values which essentially strike at the heart of what we want and what we believe is important . . . Focusing on a narrow range of physical and material benefits of the urban forest betrays a lack of appreciation of other values of urban trees that are independent of human material interests, such as love of nature, reverence of its beauty, or being inspired by its spiritual qualities. *Our research confirms that urban trees, as citizens understand the trees through their senses and knowledge-building processes, help improve personal outlooks, attitudes, and senses of well-being.* As revealed by citizen expression in our data, the social values of urban trees are foremost on people's minds.
>
> (2013: 8, emphasis added)

To dig deeper into those processes of sensing and valuing and into the transmissions which take place, we need to heed the affective realm. But also, importantly, affect-based approaches would question the clear splitting of the physical and material benefits that trees bring from those more associated with feeling, values, and emotions. Affect places emotion at the centre of the physical/ecological body.

Finally in this section comes quite a big question about affect, trees, and city life. Writers, such as Irus Braverman, have cast a much more quizzical and critical look at city trees (US) and the politics that they are entangled in. This analysis is claimed to be a counter to the generally 'romantic' values that underpin seeing trees as an inevitable good in the city. Rather, 'trees are made and used as symbols of class, race, and status' (Braverman, 2008a: 45), and can be actants deployed and/or enrolled into the governance and policing of US streets (Braverman, 2008b) to suit particular (capitalist, elitist, neo-liberal) ideologies.

Such critiques are clearly worth considering. No-body can be politically neutral—be it a person or a tree, and politics and economy always entwine together through such processes as property rights and real estate value. In terms of economy, Luttik (2000) shows that the presences of trees (and other agents of nature) can affect property prices. So the question is, might such political economy energies be affectively transmitted between trees and people (perhaps trees helping to gentrify a district which some then feel excluded from)? Or does affective transmission *short circuit* economy and politics and set up an (alternative) ethics/politics of bodies in relation (to each other)? My tentative answer is that it is not 'either or'—they both go on. But here I am focusing more on the latter.

The Affective Turn

The 'affective turn' (Dewsbury et al., 2002; Clough and Halley, 2007; Stewart, 2007; Thrift, 2008; Cadman, 2009; Pile, 2009; Anderson and Harrison, 2010; Woodward and Lea, 2010) has generated profound challenges to the social sciences and related disciplines in terms of *what the social is, how it operates, and how it can be engaged with through ontology, epistemology, and methodology* (see Law, 2004 on methods). It is the outplaying of an alternative philosophical/scientific pedigree (Spinoza, Darwin, Deleuze) (see Thrift, 2008 for details) to that of modernism/rationalism, which never bought into the whole Cartesian (enlightenment) derived notion of the self as pure, separate, 'rational' mind which the modern subject is more or less trapped in. Questions of the body are to the fore as various disciplines—cultural geography, social anthropology, archaeology, cultural studies, history, literature, visual culture, memory studies, and psychoanalysis—consider how lives in place and space are practised as cultural, political, psychological, and material configurations in which identities (individual and collective) are conjured, and power and politics are exercised.

Affects are inherently relational—about exchange between and through bodies (McCormack, 2007). They are about how bodies function moment-by-moment and interact with each other and with(in) the environment through panoplies of senses, movements, contacts. It is important to realize that these affective registers lie largely outside the realms of language, thought, rationality, and reflexive consciousness. The social sciences which operate in and through representations, constructed in and of language and reflexive thought, thus struggle to 'see' and work with affective life. This is, in part, Thrift's (2008) justification for non-representational theory which attempts to extend social science work into the affective spectrum. It is important to note that ecofeminist thinkers such as

Plumwood (2002) have long argued that we need to re-enter realms of feeling in order to build effective knowledge of human–nature/animal relations.

Affective registers are the bodily systems and processes which make life, and which run pre, within, and beyond reflexive consciousness, and through which we live the bulk of our lives as beings-in-environment on a moment-to-moment basis. Memory functions, emotions, motor movements (e.g. balance), and sense/response systems are key but not exclusive parts of affect. Key scientific work (e.g. Damasio, 2003, 1999, 1993) has shown that our conscious, reflexive, rational, language-using self (the focus of most social science) is underpinned, enframed, *and coloured* (in terms of mood) moment-to-moment by these processes. These insights into the biological (Thrift, 2008), bodily, basis of individual and collective social life challenge the Cartesian settlements of identity and agency, and the key dualisms of modernism: mind/body, nature/culture, subject/object.

However, past these basic points, affect is a complex and contested term. Thrift (2008) sets out three differing accounts and legacies of affect as a concept. Here I work with the idea that affect comprises: (a) the systems of the body which enable us to function as meaningful beings *and to engage with the world and each other*, most of which are pre/beyond consciousness and reflexive thought/language; (b) the 'channels' that these engagements occur through; and (c) the traffic on those channels. Emotions are vital elements of affective processes but *the two are not synonymous*. Affect leads us to see human (selves) as complex embodied animals not as virtual minds, and as 'dividuals' rather than 'individuals'. There are plenty of obvious and important reasons for thinking that this is a good way to take understandings of the social. These reasons are essentially ecological, and span from the current regimes of ecocide being wrought on planetary biodiversity to failure to heed the fact that humans themselves are a community of bacteria and other biotica, that share, and make possible, the body.

This quote from Anderson captures the essential relational, processual, and *emergent* flows of affect:

> The capacities to affect and to be affected that enact the life of everyday life do not, however, simply emerge from the properties of the humans or nonhumans that exist 'in' Euclidean space or 'in' linear time. That is, affect does not reside in a subject, body, or sign as if it were an object possessed by a subject . . . 'Being affected – affecting' emerge from a processual logic of *transitions* that take place during spatially and temporally distributed encounters in which 'each transition is accompanied by a variation in capacity: a change in which powers to affect and be affected are addressable by a next event and how readily addressable they are' (Masumi, 2002: 15) . . . What is at stake is the composition of harmonious or disharmonious relations amongst diverse collectivities of humans and nonhumans that produce 'rises and falls, continuous variations of power . . . signs of increase and decrease, signs that are vectorial (of the joy–sadness type) and no longer scalar like the affections, sensations or perceptions' (Deleuze, 1998: 139).
>
> (Anderson, 2006: 736, emphasis as original)

There have been extensive debates over the affective dimensions of politics (Amin and Thrift, 2013; Barnett, 2013) and the politics of the affective turn. As Barnett (critically) summarizes, for Amin and Thrift political judgements 'are not made in rational or deliberative ways; they follow key lines of emotion' (2013: 3). So it thus follows that (for Amin and Thrift) '[m]any political impulses are contagious and require only momentary thought' (Barnett, 2013: 14). This turn away from what could be termed deliberative politics aggravates those who still invest in the ideological mega structures of Marxism. But the affective turn does not deny or ignore the significance of economy, culture, rationality (management), language, and reflexive thought (which are reached by standard social science methods) but stresses that much of the richness and import (power) of human interaction within place, landscape, and heritage is articulated through a range of other embodied processes and practices. Here I summarize five things which I think give affect-based approaches particular purchase in relation to thinking about (urban) trees and forests.

1. Beyond social constructionism and towards a full(er) politics of material rationality. The social cultural turn told us, rightly, that we can never free 'objects' from cultural constructions. This is particularly so for trees for pretty obvious reasons. But equally 'objects' have their own energies—'Whilst material objects, including nature, are a screen for our projections, they are also much more than this. Objects affect us, announce their presence to us, insinuate themselves upon us, *call to us*' (Hoggett, 2009, emphasis added). So this is neither cultural nor natural determinism but about relational exchanges flowing into on-going co-becoming. Nature has agency much of which is transmitted (to us) and into complex webs of relational ecological becoming through affective registers.

Thinkers such as Latour (1993) argue that any politics that ignores this living fabric of the world is not fit for purpose and is, instead, masking our journey to destruction. Sadly most politics does exactly this at present. The affective politics that Amin and Thrift want to develop asks us to ponder the political implications of these extensions of the scope of the social: "what if we were to admit actors who traditionally have been regarded as objects of political attention but rarely as subjects?" (2013: 40).

2. The half-second delay (and the corporations). The half-second delay is a well-established (scientific) fact; that the sub-conscious mind works in front of the conscious mind. Our trajectories of thought and action are set in place before thought and consciousness kick in. This space of 'raw life' is where much affective business occurs. It is important to note that this is not just some esoteric, arcane, philosophical curiosity, or a puzzling challenge to notions of free will. This gap is absolutely vital in social, political, and economic terms. The corporations know this. Collectively they spend billions on branding, design, and marketing so that they 'get at you' through the half-second delay—so they get at you affectively. Your body, your emotions want it—once that is set up the mind (and wallet) follows along. Consider our city square with its framing community of trees. They provide a setting for 'raw city life' as it is practised in all the ways already alluded to. They are, to use the ideas of Slaby (which are developed below), actually in the (sub-conscious) mind and body as a matrix as thought and practice unfold.

3. 'The feeling of what happens'. Closely linked to the above is the fact that the *primary* forces/systems in our brains/minds are pre-subconscious as well as articulated in the electro-chemical processes of the emotions and related systems. These enable, enframe, and exceed rational thought/language. As Damasio famously has explained 'feelings of pain or pleasure or some quality in between are the bed-rock of our mind. We often fail to notice this simple reality . . . But there they are, feelings of myriad emotions and related states, the continuous musical line of our minds.' (Damasio, 2003: 4). This kind of work draws upon Charles Darwin's other great work, *The Expression of the Emotions in Man and Animals,* in which the embeddedness of the human as an organism in evolved nature is clearly expressed.

Now the trees in the square which are extending into people's moment-by-moment becoming will generate and/or colour conscious feelings, through the senses, memory, culture, and so on. This is where the boundary between the cultural and the biophysical blur. The giant beech tree which holds pride of place by the path in the centre of the square, might be generating cool shade on a hot day, and a sub-aural white noise of leaf rustle in a gentle breeze, and complex light patterns might be playing in the canopy and be projected as animated light and shade. These might well be barely perceived or felt, but they, and memories of trees as culture, will delicately trace out in the convection currents of a subconscious mind.

4. Art and music essentially operate within the affective realm. One way to appreciate the implications of affect—or aspects of it as a process—is to consider the artist realm. The impacts of music both on the individual, and collectivity, operate at a set of levels which span a whole range of registers. There is of course a huge range of art and music which 'speaks' of nature, place, and landscape and, within, those (urban) trees. Ottorino Respighi's symphonic poem *Pines of Rome* (1924) is one notable example of music that explores the multiple placings and materialities of trees in urban space and time. One implication of all this is that the affective turn in the social sciences has involved a growing interest in performance and arts as ways (methods) of thinking about, and researching in a non-representational mode, people's engagement with the world, place, and landscape.

5. Methodological implications. The affective turn does not simply require a casting out of established research methods (be they quantitative or qualitative); rather it requires an extended palette of methods, in range, how things are done, and the attitudes through which they are applied. For example, ethnography can be extended to consider visual (Pink, 2007) and sensual (Pink, 2009) registers. Participant observation adjusts to a more performative form (see below), ethology and performative/action research techniques (Law, 2004; Pearson, 2006) can all be developed in general for addressing affective becoming in place and for the particular business of researching relational life. There is an impulse for experimental approaches, particularly when seeking to 'research' with non-human others such as trees. See, for example, Roe and Greenhough's (2014) proposal for 'experimental partnering' which suggests that 'established research techniques can be developed in new directions by becoming attentive to the ways in which novel epistemological and ontological frameworks can shape the production of research knowledges with assemblages of non-humans' (p. 45).

Affecting Places

Firstly we need to consider notion of places—as they are not simply spaces, but rather the bedrock of everyday becoming (Casey, 1993). Places are (taken to be) complex outplays of intersecting flows of material and agency, where the human and non-human (materials, technologies, plants, and animals) combine and recombine in and through series of registers in cycles of comings and goings. Temporal dimensions are critical insofar as some comings and goings will be fleeting while others—such as that of tree growth (and economic cycles)—operate over decades and even centuries (as well as their seasonal travels). This complex temporal ecology of 'timescapes' brings multiple rhythms to places which are central to their affective life (Lefebvre, 2004). The Bristol square can be fruitfully envisaged in this way. It is spatially grounded but also an intersection of rhythmic and novel flows.

Urban places are particular forms of entanglements (always unique but with some shared characteristics). Materiality and non-humans (of all kinds, animals, plants, technologies) play their full part in the production of space which, in some instances, can become culturally, politically, ecologically, and economically critical and/or contested. Ingold's (2011) notion of meshwork captures this living complexity of becoming-landscape. If we take Chambers' assertion that 'tree-planting *makes landscapes happen*' (1997: 5, emphasis added), we get a glimpse of relational agency distributed between people, trees, soil, air, water, and so on, and the sense of landscapes (and places) as happenings.

So places/landscapes are complex temporal–spatial–material processes into which social, cultural, and symbolic meaning entwine. People's engagement with (differing) places is articulated through a range of *affective bodily practices* (commuting, working, resting, playing, eating, sitting); sensing (touch, sight, sound, smell); and a range of non-cognitive affective processes (feelings, emotions). Thus these approaches can be seen as post-phenomenological as set out in Ingold's (2011) dwelling approach and Thrift's (1999) 'ecologies of place', and in a whole range of other recent writing in anthropology and cultural geography (e.g. Ingold, 2000; Massey, 2005; Pearson, 2006; Wylie, 2010).

This very recent (at time of writing) conference-session abstract shows how affect, place, and materiality are being fused into new forms of conceptualization and analysis:

> Place can be understood as a type of situated affect or feeling, a mode of active, sensory engagement. By making sense of place we shape how we dwell in the

> built and natural environment, how we understand and appreciate its sounds, sights, textures, flavors, and scents. By making sense of place we orient ourselves to speeds and rhythms, as well as to movement, rest, and encounter. Thus place is the sum total of the sensations that it gives rise to, the cumulative incorporation of those feelings carved into soils, skies, and shores, and the embodiment of its affective intensities on its dwellers.
>
> (Vannini, 2013)

Affective life is very much about embodiment, relationality, and materiality as it seeks to deal with the self as a *performative entity always in specific space-time circumstances of the now which are melding with habits and remembrances of past experience*. In this there is great complexity, uncertainty, and even unknowability. This is how Stewart describes the affective/performative view of the social:

> To attend to ordinary affects is to trace how the potency of forces lies in their imminence to things that are both flighty and hardwired, shifty and unsteady but palpable too. At once abstract and concrete, ordinary affects are more directly compelling than ideologies, as well as more fractious, multiplicitous, and unpredictable than symbolic meanings. They are not the kind of analytic object that can be laid out on the single, static plane of analysis, and they don't lend themselves to a perfect, three-tiered parallelism between analytic subject, concept, and world. They are instead, a problem or question of emergence in disparate scenes and implement strict forms and registers; a tangle of potential connections.
>
> (2007: 4)

The specific and material forms places take (such as that in Victoria Square) have a 'liveliness' which pushes onto the social (as well being constructed by the social). This might well be particularly significant for thinking about people's engagements with trees and wooded environments both individually and collectively because it stresses the textures of space, and that is what trees provide. 'Advocates of affect offer it up as a way of deepening our vision of the terrain we are studying, of allowing for and prioritizing its "texture" . . . This texture refers to our qualitative experience of the social world, [and] to embodied experience that has the capacity to transform as well as exceed social subjection' (Hemmings 2005: 459).

To engage fully with these textures we need to *immerse ourselves in the site*:

> The idea is to get embroiled in the site and allow ourselves to be infected by the effort, investment, and craze of the particular practice or experience being investigated. Some might call this participation, but it is a mode of participation that is more artistic and, as with most artistic practices, it comes with the side-effect of making us more vulnerable and self-reflexive. It is not however an argument for losing ourselves in the activity and deterritorializing ourselves completely from our academic remit, but nor does it mean sitting on the sidelines and judging. Rather the move, in immersing ourselves in the space,

> is to gather a portfolio of ethnographic 'exposures' that can act as lightning rods for thought. [Such] a method produces its data: a series of testimonies to practice. This is of course the flipping over of 'participant observation' to 'observant participation' . . . to emphasise the serious empirical involvement involved in non-representational theory's engagement with practices, embodiment and materiality.
>
> (Dewsbury, 2009: 326–327)

Transmissions

> Mind is (a continuation of) life: a continuation of—and thus structured in some respects similar to—the metabolic process of a system's self-organization in exchange with appropriate environments. Mental processes are thus essentially active, performative sequences of organism/environment interaction.
>
> (Slaby, 2014: 37)

In the book *Tree Cultures* we talked about not only the agency of trees, but also the agencies of specific types of trees, and individual trees in particular settings. Forests and woodlands are very highly charged affective spaces because of their spatial, material, and cultural qualities. These qualities and their habits (seasons, ecologies) are engaged with/through all the affective channels (such as sensing, movement, memory, emotion). Slaby's (2014) phenomenologically based notion of 'extended emotion' seems useful in thinking this through. That is, the way that modes of becoming are enacted within and through the self and materiality and landscape. In this approach 'phenomenologically, it can seem as if an environmental structure creeps in upon us, fills our experiential horizon and affects our bodily poise, our posture, our readiness and potentialities to act, and even the execution and style of our agency' (2014: 36). This seems to offer a way of addressing how the materiality and agencies of nature become affective in everyday life and also notions of the 'ecological self'.

Affective processes are both enlaced with the legacies of memories (of varying kinds) yet also have a potential of creative departure (see Jones, 2011). Thus, they are of great political import, as non-representational theory (e.g. Amin and Thrift, 2013) argues. The affective transmissions trees are enmeshed in are complexly derived and articulated. Below I sketch out a few key aspects of this.

Materiality, Sensory Richness, Be(longing), and Identity

Senses and practices of place are key to the successful performance of identity and well-being at the individual and collective level. These are always implicated in the bricolage of particular places as made up by disparate and entangling materials in process. Trees can be very powerful in this respect. They 'locate us in time and place' (Sinden, 1989 cited by Rival, 1998: 19). Harrison (1991), talking about the huge old trees in his local landscape, says, 'To stand beneath one of these maimed colossi is to be overwhelmed by its powerful, resonant presence' (p. 135). These oak trees are 'the living tissue of time' (p. 135) and therefore (as do other trees, to differing extents) they meet a need, which, Harrison says, he 'believe[s] to be indispensable, for parochial monuments, landmarks, milestones and other points of reference by which each person can take his or her own bearings in time and place' (p.139). This applies to all places, be they urban, suburban, or rural. Trees can deliver the need, which is, as Harrison has it, 'a demand [for a] delicate and penetrating relationship with our physical world, our environment' (p. 139). They do so because of their rich material, spatial, temporal, sensory, and cultural registers. These are clearly 'visible' in the town square, with the complex depths of views and sound, offered to those using it in some way. Because of all these capacities, great strides have been made in planting mature trees in the city (Figure 8.2); a process of considerable expense and technical challenge.

Figure 8.2 Newly planted tree in Munich, Germany. Swathed in fabric and supported with multiple cables.

Source: Owain Jones, 2013.

Trees and the Body

A number of writers and artists have explored the relationships between the human body and tree bodies. Both are 'vertical beings' with arrangements of trunk and limbs. The sheer size of (some) trees is remarkable but often taken for granted. The verticality of the tree body (and control of it) is a critical factor in the deployment of street trees. The verticality and spreading of trees has long been associated with, and considered a precursor to, religious architectures (e.g. the great Christian gothic cathedrals of Europe). But it can also be deployed in the creation of grand urban order (Figure 8.3). Furthermore, the tree body is so expressive of life, in so far as, as Hughes-Gibb has it, '[a] tree is really a great community of life with hundreds of growing points all combined into one' (undated, p. 141).

Figure 8.3 The verticality of many tree species makes them an ideal spatial medium for city streets (Munich).

Source: Owain Jones, 2013.

Some suggest that modern, commoditized, consumption-based, industrial, mechanized life is impoverished in terms of affective engagement with landscape (nature). Psychogeography seeks to combat this through walking in the landscape (and other means) where the body, senses, mind, emotion can re-engage with the textures of places. Forests and woods might offer themselves as therapeutic affective landscapes.

'Other' Space

As I have discussed elsewhere (Jones, 2011), there is a long tradition of seeing treed spaces as potentially 'other' where identity can be unsettled, where there are portals to other realms, where dangers and temptations lurk (as in fairy stories). Forests have been constructed as other to civilization. Harrison writes that:

> Western civilization literally cleared its space in the midst of forests. A sylvan fringe of darkness defined the limits of its cultivation, the margins of its cities, the boundaries of its intuitional domain; but also the extravagance of its imagination . . . The governing institutions of the West—religion, law, family, city—originally established themselves in opposition to the forests.
>
> (1992: ix)

The question then arises, can wooded spaces become 'other space' *within* the city? Perhaps it is simply a question of scale and density of wooded space. If a wooded place can close around you, and you can feel alone, or lost—or actually be lost, if lines of sight are closed off—then perhaps yes. Having said that, I think that spaces that are enclosed by trees such as Victoria Square do offer a break in the psycho-geography of the city where filaments of otherness can find a space.

Passions and Practices

As suggested above, there are plenty of clues in art and literature and psychological/ (eco)feminist approaches to trees/forests about the passions people hold for—have with—trees. From their ethnographic research on trees and people in and around UK urban areas, Macnaghten and Urry suggest that 'the value people appear to attach to woods and forest arises from the specific 'affordances' that the latter could offer for particular bodily *desires*' (2000: 180, emphasis added). Furthermore 'the affordances of woods tend to be local, in that specific 'local' circumstances and experiences shape people's sense as to what is necessary or desirable for their bodily engagement with such spaces' (Macnaghten and Urry, 2000).

Recent examples of performance and installation art continue art's long interest in trees as makers of spaces and landscapes. For example in Canberra (Australia) the *City of Trees* is a multi-disciplinary project by UK artist Jyll Bradley 'reflecting upon Canberra's 100 years through an exploration of the city's remarkable tree-scapes'. Seven audio pieces, related to specific trees and mapped locations, are online and also listenable to in the landscape.[7] In Coventry (UK) creative instigator

Anne Forgan has conducted an art project on *Trees in the City*, commissioned by Artspace in Coventry, a part of their ARC project, as a way of thinking about creative engagement with communities. There is also a group dedicated to raising awareness of, and collecting stories and images of the city's trees. Both of these projects talk of a local famous black mulberry tree in a local park.[8] Thus certain city trees become public figures in the life and history of the city. And in London, to celebrate their 20th year of working, the organization *Trees for Cities* celebrated the city's 'unique tree heritage', by delivering a programme of tree-related, community-engagement activities as part of the City of London Festival. As part of these activities, they commissioned the international artist Konstantin Dimopoulos to create *The Blue Trees* environmental art installation in the heart of London.[9] All these projects are examples of passions and trees and the city coming together in ways which not only explore existing urban arboriculture but also extend it in creative ways in affective registers.

To return to Victoria Square, Bristol, one more time. Last year I got a rather surprising, but pleasant email, from a resident of the square who happened to be a retired drama lecturer from one of the city's universities. He told me that he had written a short drama based upon the chapter about the square in the 'trees book' and this was to be performed at the square's annual community day. Would I like to attend? Of course I took up the kind offer, and prior to the event it was arranged to have a display of the archive of material I and the organizer had about the square's arboricultural history. The short play involved a series of voices—the trees, residents past, resident present, the square's owners, the corporation—in conversation about their entanglements together as the square moves on as a living eco-social space. Many of the current residents were in attendance, watching and picnicking beneath the canopy of the square trees, including the lovely cedar of Lebanon (Figure 8.4). This small drama was an example of, *and* spoke of, the on-going affective life of this urban square with its changing communities of people and trees.

Conclusions: Materialized Hope for Shared City Futures

The central argument of this chapter is, in essence, pretty simple. It is that if we are to take trees in cities seriously we need first to see trees as agents—as specific active makers of place always in co-constructive relations with other agents, such as people and technology, and second to appreciate and investigate how trees and people (and others) are bound up in relationships of affective (emotional/feeling) exchange.

These needs are both general and specific in so far as—general—these are necessary ways of reading the becoming-world whatever particular issue or focus one is concerned with—specific—because, as discussed, trees are particularly powerful generators of effective traffic.

To put it another way, there is an extraordinary richness to people's engagement with places and landscapes, and trees compound such richness. And much of that

Figure 8.4 The Party in the Square within the spaces formed by the trees.
Source: Owain Jones, 2012.

richness is articulated in affective registers. Perlman (1994) talks very convincingly about the 'power of trees' and the 'negotiations' which take place between trees and people, with the trees taking part through 'their rich resonant particulars' and 'actual natural life' (p. 2). Coming from an environmental psychological background he was really, I think, talking about affect but only under differing methodological and disciplinary terms.

Trees work in their own ways and in their own time. We need to heed that when we come to consider them in urban settings. In terms of management, how people live with them, and how we study such 'life under the canopy'. Anderson (2006) discusses hope as a very particular and important dimension of affect. In a world beset by very horrible forms of ecocide (of cultural, physiological, and ecological diversity) planting trees in the city, or letting them grow wild, and caring for them, seems to me like some kind of materialized hope in action.

As stated earlier, trees in the city can be seen as one of the ultimate interpenetrations of nature–culture. But as Ingold says—the nature–culture divide is a falsity. So rather we need to see cities as assemblages, and we need to plan them so they are places of ecological flourishing. And flourishing as in Guattari's sense of ecology—individual psyche, community, cultural, and ecological in its more conventional sense of biodiversity. It is not hard to see how trees can contribute to all three in human terms. Konijnendijk et al. conclude their reference work on European urban forests with the assertion that opportunities abound for urban forestry to assume the role of a 'regenerative tool, a catalyst for change and improvement, quality of the landscape and investment decisions and most importantly, the prime driver behind pushing the new urban agenda of livability forward' (2005: 501). Having said that, this optimistic view of urban forests needs to be leavened with the more critical stance of scholars such as Braverman.

Finally this must also not be judged in terms of human well-being. To do so is to re-inscribe the oh-so-damaging nature–culture division and hierarchy. We need to keep imagining how the city is co-constructed and shared by a whole host of living things, trees and humans being just some of them. We need to think how humans can serve tree well-being in the city as well as vice versa. Maybe trees take priority in at least some decision-making. The United Nations has previously said that children are a good 'indicator species' for city health. I think the same can be said for trees—if a city is a good place for tree becoming it most likely is working well in other ways, for others, too.

Notes

1 Performed is used here and elsewhere instead of 'practiced' because it reflects how practice has elements of habit, rehearsal, and self-scribing in it.
2 This square and its arboriculture history is described in detail in Jones and Cloke (2002).
3 www.cliftonhotwells.org.uk/chis_tree.html.
4 www.bristolpost.co.uk/Clifton-residents-evacuated-tree-tumbles/story-19239539-detail/story.html.
5 With Paul Cloke (see Jones and Cloke 2002 and other output).
6 E.g. Sandilands (2014).
7 www.canberra100.com.au/programs/city-of-trees/.
8 coventrytrees.blogspot.co.uk/p/about-us.html.
9 www.treesforcities.org/about-us/20th-birthday/the-blue-trees-of-london/.

References

Amin, A. and Thrift, N. (2002) *Cities: Reimagining the Urban.* Cambridge: Polity.

Amin, A. and Thrift, N. (2013) *Arts of the Political: New Openings for the Left*, Durham, NC: Duke University Press.

Anderson, B. (2006) Becoming and Being Hopeful: Towards a Theory of Affect, *Environment and Planning D: Society and Space*, 24: 733–752.

Anderson, B. and Harrison, P. (2010) *Taking-place. Nonrepresentational Theories and Geography.* London: Ashgate.

Barnett, C. (2013) Book Review Essay: Theory as Political Technology, *Antipode*, http://radicalantipode.files.wordpress.com/2013/07/book-review_barnett-on-amin-and-thrift.pdf (accessed 3 December 2013).

Braverman, I. (2008a) Everybody Loves Trees: Policing American Cities through Street Trees. *Duke Environmental Law & Policy Forum*, 19: 81–118.

Braverman, I. (2008b) Governing Certain Things: The Regulation of Street Trees in Four North American Cities, *Tulane Environmental Law Journal*, 22(1): 35–60.

Cadman, L. (2009) Nonrepresentational Theory/Nonrepresentational Geographies. In: R. Kitchin and N. Thrift (eds) *International Encyclopaedia of Human Geography*, Amsterdam: Elsevier, pp. 456–463.
Casey, E. (1993) *Getting Back into Place: Towards a Renewed Understanding of the Place-world*, Bloomington, IN: Indiana University Press.
Chambers, D. (1997) Introduction. In: G. S. Thomas *Trees in the Landscapes*, London: John Murray, pp. 5–8.
Clough, P. T. and Halley, J. (2007) *The Affective Turn: Theorizing the Social*, Durham, NC: Duke University Press.
Damasio, A. (1993) *Descartes' Error: Emotion, Reason, and the Human Brain*, New York City: Penguin Putnam.
Damasio, A. (1999) *The Feeling of What Happens: Body and Emotion in the Making of Consciousness*, London: William Heinemann.
Damasio, A. (2003) Mind over Matter, *Guardian, Review*, 10: 4–6.
Deleuze, G. (1998) Spinoza and the Three Ethics, *Essays Critical and Clinical* translated by D. Smith and M. Greco, London: Athlone Press.
Dewsbury, J.-D. (2009) Performative, Non-representational, and Affect-Based Research: Seven Injunctions. In: D. DeLyser, S. Atkin, M. Crang, S. Herbert, and L. McDowell (eds) *Handbook of Qualitative Research in Human Geography*, London: Sage.
Dewsbury, J.-D., Harrison, P., Rose, M. and Wylie, J. (2002) Enacting Geographies, *Geoforum*, 33: 437–440.
Harrison, F. (1991) *The Living Landscape*, London: Mandarin Paperbacks.
Harrison, R.P. (1992) *Forests: The Shadow of Civilization*, Chicago: University of Chicago Press.
Hemmings, C. (2005) Invoking Affect: Cultural Theory and the Ontological Turn, *Cultural Studies*, 19(5): 548–567.
Heynen, N., Perkins, H.A., and Roy, P. (2006) The Political Ecology of Uneven Urban Green Space: The Impact of Political Economy on Race and Ethnicity in Producing Environmental Inequality in Milwaukee, *Urban Affairs Review*, 42(1): 3–25.
Hughes-Gibb, E. (Undated) *Trees and Man*, London: Alexander Moring Ltd.
Hoggett, P. (2009) *Psychic Transactions with the Material World*, paper presented at Exploring Subjectivity, Politics and the Natural Environment—A day of presentations and discussion, Cardiff University, www.cardiff.ac.uk/socsi/newsandevents/events/0605 2009subjectivity.html (accessed 3 December 2013).
Ingold, T. (2011) *Being Alive: Essays on Movement, Knowledge and Description*, London: Routledge.
Ingold, T. (2000) *The Perception of the Environment: Essays in Livelihood, Dwelling and Skill*, London: Routledge.
Jones O. (2011) Forest Landscapes: Identity and Materiality. In: E. Ritta and D. Dauksta (eds) *Society, Culture and Forests: Human–Landscape Relationships in a Changing World*, Guilford: Springer, pp. 159–178.
Jones, O. (2008) Stepping from the Wreckage: Geography, Pragmatism, and Anti-representational Theory, themed issue: Pragmatism and Geography, *Geoforum*, 39: 1600–1612.
Jones, O. and Cloke, P. (2002) *Tree Cultures: The Place of Trees, and Trees in their Place*, Oxford: Berg.
Konijnendijk, C., Nilsson, K., Randrup, T. B., and Schipperijin, J. (eds) (2005) *Urban Forests and Trees: A Reference Book*, Berlin: Springer.
Latour, B. (1993) *We Have Never Been Modern*, Hemel Hempstead: Harvester/Wheatsheaf.
Law, J. (2004) *After Method: Mess in Social Science Research*, London: Routledge.

Lefebvre, H. (2004) *Rhythmanalysis: Space, Time and Everyday Life*, London: Continuum Books.

Ley, D. (1995) Between Europe and Asia: The Case of the Missing Sequoias, *Cultural Geographies*, 2: 185–210.

Luttik, J. (2000) The Value of Trees, Water and Open Space as Reflected by House Prices in the Netherlands, *Landscape and Urban Planning*, 48: 161–167.

McCormack, D. P. (2007) Molecular Affects in Human Geographies, *Environment and Planning A*, 39: 359–377.

Macnaghten, P. and Urry, J. (2000) Bodies in the Woods, *Body & Society*, 6(3–4): 166–182.

Massey, D. (2005) *For Space*, London: Sage.

Massey, D. and Thrift, N. (2003) The Passion of Place. In: R. Johnston and M. Williams (eds) *A Century of British Geography*, Oxford: Oxford University Press.

Massumi, B. (2002) *Parables for the Virtual: Movement, Affect and Sensation*, Durham, NC: Duke University Press.

Pearson, M. (2006) *'In Comes I': Performance, Memory and Landscape*, Exeter, UK: University of Exeter Press.

Peckham, S. C., Duinker, P. N. and Ordónez, C. (2013) Urban Forest Values in Canada: Views of Citizens in Calgary and Halifax, *Urban Forestry & Urban Greening*, Articles in Press.

Perlman, M. (1994) *The Power of Trees: The Reforesting of the Soul*, Woodstock, CT: Spring Publications.

Pile, S. (2009) Emotions and Affect in Recent Human Geography, *Trans Inst Br Geogr*, NS 35: 5–20.

Pink, S. (2009) *Doing Sensory Ethnography*, London: Sage.

Pink, S. (2007) *Doing Visual Ethnography*, London: Sage.

Plumwood, V. (2002) *Environmental Culture, The Ecological Crisis of Reason*, London: Routledge.

Rival, L. (1998) 'Trees, from Symbols of Life and Regeneration to Political Artefacts'. In: L. Rival (ed) *The Social Life of Trees: Anthropological Perspectives on Tree Symbolism*, London: Berg.

Roe, E. and Greenhough, B. (2014) Experimental Partnering: Interpreting Improvisatory Habits in the Research Field, *International Journal of Social Research Methodology*, 7(14): 45–57.

Sandilands, C. (2013) Dog Stranglers in the National Park? Vegetal Politics in Ontario's Rouge Valley, *Journal of Canadian Studies*, 47(3): 93–122.

Schama, S. (1995) *Landscape and Memory*, London: Fontana Press.

Sinden, D. (1989) 'Orchards and Places', in Common Ground, *Orchards: A Guide to Local Conservation*, London: Common Ground.

Slaby, J. (2014) Emotions and the Extended Mind. In: M. Salmela and C. von Scheve (eds) *Collective Emotions*, Oxford: Oxford University Press, pp. 32–46.

Smith, N. (2005) Foreword. In: N. Heynen, M. Kaika and E. Swyngedouw (eds) *The Nature of Cities: Urban Political Ecology and the Politics of Urban Metabolism*, London: Routledge: xii–xv.

Stewart, K. (2007) *Ordinary Affects*, Durham, NC: Duke University Press.

Thoreau, H. D. (1864) *The Maine Woods*, New Haven, CT: Yale University Press.

Thrift, N. (2008) *Non-Representational Theory: Space/Politics/Affect*, London: Routledge.

Thrift, N. (1999) Steps to an Ecology of Place. In: D. Massey, P. Sarre and J. Allen (eds) *Human Geography Today*, Oxford: Polity: pp. 295–352.

Vannini, P. (2013) Making Sense of Place (Conference session abstract), ISA Conference, Yokohama, Japan, July 13–19, 2014. www.isa-sociology.org/congress2014/# (accessed 28 November 2013).

Woodward, K. and Lea, J. (2010) Geographies of Affect. In: S. Smith, R. Pain, S.A. Marston and J.P. Jones III (eds) *The Sage Handbook of Social Geographies*, London: Sage: pp. 154–175.

Wylie, J. (2010) Non-Representational Subjects? In: B. Anderson and P. Harrison (eds) *Taking-Place: Non-representational Theories and Geography*, Oxford: Ashgate.

9 Order and Disorder in the Urban Forest

A Foucauldian–Latourian Perspective

Irus Braverman

> [T]ell an expert how big the building is and he will tell you exactly where to place the exits: it's disciplined even at the very earliest part of the design. . . . But a sidewalk is managed chaos: there is nobody controlling this. Trees happen to be [on] the sidewalk. The course of accommodating the trees will bring some discipline and some rigor to how we manage the sidewalks.
>
> (Simon, interview)

Introduction

We pass by street trees every day. Their existence as well as their particular location in the city seem obvious, innocuous, natural. But, as is the case with most taken-for-granted "things" (Brown, 2001), some excavation is bound to reveal a more complicated and even ideological story. This study focuses on such a story: the story of the clandestine governance of nature and of humans by way of nature—all through the construction and regulation of city street trees. This story problematizes the mundane display of urban space in general, and of urban street trees in particular, as technical and apolitical, and instead promotes an understanding of nonhumans and humans as constantly negotiating spatial order and disorder through law.

Specifically, this chapter proposes that the "art of governance" is relevant not only to human populations, but also to nonhuman things and networks. It suggests that legal norms and practices must, and in fact do, take physical matters into account. The chapter is organized to correspond with the social stratification of streetscape into the bifurcated places of aboveground and underground, and the "in-between" place of ground level. While these strata, along with their binary juxtaposition, are socially constructed, they are also constrained by material and mental conditions, such as visibility and usability. Operating through regulations and guidelines, professional practices and everyday acts, a detailed bureaucratic apparatus attempts to know and govern these places by managing nonhumans into a certain order that both serves and controls humans. But such prefixed orderings seldom work. Instead, various dynamics flow among and between the street's strata, between humans and nonhumans, and between living and nonliving things.

The dynamics of tree governance are explored here from the perspective of three spatial technologies: the grid, the grate, and the Dig-Safe procedure. Whereas the grid demonstrates the governance of aboveground things and places, Dig-Safe is a story of underworld governance, and the grate exemplifies management on the interim level of the ground. Accordingly, the construction of these spatial technologies brings to the surface the potentially varied legal approaches toward matter. Relatively speaking, trees in the aboveground are susceptible to tight levels of management, while on the level of the concrete their materiality is negotiated more fluidly. Finally, in their underground manifestation as roots, the trees are mostly left unregulated, because the Dig-Safe procedure ignores their existence altogether. This stratification highlights what largely goes unnoticed, that law and matter, nomos and physis, are inseparable and intertwined, both physically and discursively.

The perspective offered here draws on two scholarly traditions: governmentality (Foucault, 1991) and Science and Technology Studies: especially Actor Network Theory (ANT) (see, e.g. Akrich, 1992: 205; Callon, 1986: 196; Callon and Law, 1995: 481; Johnson, 1998: 298; Latour, 2004; Latour, 1997: 63) and Thing theory (Appadurai, 1986; Brown, 2001: 1; Latour, 1986; Mitchell, 2001: 167–184; Pels, 1998; Pels et al., 2002). While studies of governmentality do not explicitly take up ANT's call to consider the actancy of things, there is an affinity between Bruno Latour's theory in *We Have Never Been Modern* (1993), which suggests that nonhumans exert inherent control over humans, and Michel Foucault's theory, which suggests that material structures have specific political effects, quite apart from the class or other interests of the people controlling them (Rose et al., 2006: 83).

This chapter is based on ethnographic research carried out between May and November 2005 in four North American cities: Toronto and Vancouver in Canada, and Brookline and Boston in the United States. It relies on twenty-four in-depth interviews with city officials, mostly urban planners, city engineers, and urban foresters that operate within local governments. The interviews are supplemented by direct observations of various tree sites and other practices (coalition meetings, for example), as well as secondary data such as state and federal statutes, municipal by-laws and policies, environmental reports, and newspaper articles.

Treescaping: From the Ground Up

Literally and figuratively, trees—especially in their presence from the ground up—stand at a major crossroad. On the one hand, trees are conspicuous signifiers of nature in the city. But while they are perceived as belonging to the realm of nature, they are also routinely categorized as nonhuman entities, as things, or, in the case of urban life, as street furniture. In *The Order of Things*, Foucault depicts the binary between living and nonliving things as central to natural history (1970: 68). Latour's work challenges another binary: the binary between humans and nonhumans (Latour, 2004: 62–82; Latour, 1991). The dialogue between trees as living organisms and trees as things, has exerted a myriad of tensions into the management of street trees, also enabling certain forms of governance to emerge. In this sense, the tree is situated at the nexus of Foucauldian and Latourian discourses.

From Regulation of Tree Distances to Regulation of Human Movement

Vancouver's city arborist Paul Montpellier emphasizes the tree's viability: "It's not like managing park benches. Trees are alive and they're growing, and they relate to the other trees and to birds and squirrels and insects and everything else—you're trying to manage a living system" (Montpellier, interview). According to Bill Brown, a scholar of Science and Technology Studies, a tree is not an object and cannot become one (Brown, 2001: 3). The tree's status as a "living image" (Mitchell, 2001: 177) distinguishes it from other street things, making it both more and less governable at the same time. Its thingness is an embodiment of the liminality of artifice and nature, a representation of the boundaries between the urban environment and wilderness. The street tree is a living testimony of the human's desired otherness, a desire both expressed and constrained by law, which pretends to extend itself beyond the domesticated order over a surface of chaos that needs to be disciplined (Pels, 1998: 113).

At the same time, the tree's viability is often neglected by city managers, who design treescapes to resemble other sidewalk amenities. Accordingly, the Boston guidelines list trees alongside mailboxes: "Sidewalk elements like trees, plants, light fixtures, benches, kiosks, mail boxes, and newsstands" (Boston Transportation Department: 19). Treescaping is described by the Boston streetscape guidelines as an inherent part of an urban order intended to "[d]evelop a pedestrian friendly environment which encourages sidewalk activity and is both pleasant and comfortable for users" (17).

The design and management of the public urban street is facilitated by the application of rigid distance calculations. In Vancouver, the thirty-foot-distance rule between individual trees is but a fraction of a much larger body of "distance rules" that pertain to trees. These guidelines also require a twelve-foot separation between the building line and the curb, with a minimum of six-foot width reserved for sidewalks. Curran explains that this distance allows "[two] wheelchairs to pass" so that "they don't have to be juggling and squishing, or . . . waiting to go around the tree" (interview). Similar considerations prevail in Boston. Boston's landscape architect refers to the "clinical requirements" defined by the Americans with Disabilities Act (ADA) such as a "four-foot clearance for a person with a wheelchair to navigate down a sidewalk" (Anonymous, interview), and a one-foot "shy distance" is designed on each side of the zone (Boston Transportation Department: 17).

Urban trees are excellent classification technologies. Through placing street trees between the sidewalk and the road, pedestrian traffic is funneled into the fixed corridor between buildings and curb lines. Curran explains that planting trees on grass boulevards "helps divide the vehicles from the pedestrians," creating "a bit of a safe haven and a corridor" (interview). The trees function, essentially, as a nonhuman policeman, physically restricting the movement from sidewalk to road and vice versa. In other words, the placement of trees in the streetscape restricts the mingling of humans and machines, pedestrians and cars. Although the direct objectives of these regulations are things (trees and curbs, building lines and wheelchairs), they mostly target human behavior and movement. The strict boundary established by the linear alignment of trees in relation to curbs and building lines not only produces a sense of order in public space, but also conceals the policing nature of this order behind the innocuous presentation of trees. Trees thus classify, police, and normalize.

"Lollypop" Trees and the Natural Grid

The modern grid is commonly attributed to Baron Haussmann's design of Paris in the 1860s, which was intended to prevent civic unrest in the city (Sennett, 1970: 87–91). In *The Conscience of the Eye* (1992), Richard Sennett regards New York City as an example of the construction of neutral spaces for the efficient advancement of capitalist interests "to be played upon as a chessboard" (55). While everything "is graded flat" in the "natureless" part of the city, Sennett claims, Central Park is configured conversely as a Nature without a City (Sennett, 1970: 61), exemplifying the human production of nature and city as binary figures of chessboard extremity.

Instead of juxtaposing the linearity of the grid with the fluidity of nature, as Sennett suggests, urban forestry provides a two-in-one solution: the natural grid. The transformation of Paris occurred not only by carving straight-lined streets, but also through the configuration of tree-lined boulevards (Miller, 1997: 48). The tree's alignment in relation to other trees, and to building lines and curbs, fills the width of a sidewalk, which can then be designated solely to humans. This structure reinforces the horizontal street grid. Ian Buchanan, York Region's Manager of Natural Heritage and Forestry Services, refers to this horizontal alignment as "lollypop trees" (interview). The trees contribute to the already-grid-shaped street by both intensifying and softening the mechanical features of the modern grid. If the forest was once the enemy of civilization (Schama, 1995: 83) and outside of law (Harrison, 1992: 62), it is now partitioned into highly regulated bodies of individual trees that are fixed in the concrete, watered through complex irrigation systems, and separated by thirty feet from other trees to prevent any sort of "natural" revolt. Urban trees are thus monuments of human dominance over nature. Simultaneously, a field of knowledge is produced to make urban forestry into a science that can manage trees en masse, rather than in their singular formulations.

Trees, like skyscrapers, also reinforce a vertical grid (Sennett, 1992: 57). Garry Onysko, one of Vancouver's tree inspectors, explains that every tree gets "pruned once every seven years. . . . [This way] they are [all] inspected and have a work history" (interview). Street trees are classified into trees higher than thirty feet, defined as "SYS large," and trees below thirty feet are defined as "SYS small." "The purpose of deciding if they're small or big," Onysko explains, "is to determine which truck to send out, either a boom-truck or a man-cab . . . [T]his division [of] trucks is standard in this profession in North America and I am sure throughout the world" (Onysko, interview).

Figure 9.1 The natural grid, Buffalo, NY, USA.
Source: Irus Braverman.

A complex network of things and humans is therefore engaged in the management of trees: inspection crews are organized according to truck type, which are in turn built to fit various tree heights. Pruning machines not only reflect but also affect tree height, which is manufactured to fit "system size." Vancouver's Street Tree Guidelines includes both a "Preferred Street Tree Species List," which states the "system size" of each species, as well as a parallel "Unsuitable Trees" list (Street Tree Guidelines: 12). Both lists offer the following general instruction: "Remember to always plant the right tree for the right place" (Street Tree Guidelines: 12).

The aboveground visibility of the street tree renders it a suitable object for the scientific, ordering gaze. The next section explores another street technology, this time one that is situated on the ground. Through on-the-ground technology, city officials negotiate humans and things in a much more fluid fashion than that performed aboveground.

The Grate: Governance on the Ground

Experts have designed various techniques to address the tree's special need for soil around its roots when surface paving city streets. One of the more widespread techniques utilized in North America is the grate. Typically, the grate comes in two pieces that form a collar around the trunk of the tree.[1] Its advantage is that it mitigates between the tree's need for soil, water, and air and the human need for a compact surface to advance predictable walking.[2] By providing a thing that is both solid and also melts into holes and openings, and that is transient enough to be placed and replaced to facilitate the (re)location of trees, the materiality of the grate solves a specific managerial problem in that it negotiates the materialities of humans and trees. Specifically, the grate balances the protection of trees from humans and the protection of humans from trees. But as is often the case, this technology is also not immune to complications: the holes pose an obstacle for humans who use canes to read their way through street space. "If the holes are

Figure 9.2 Trees, grates, and other street furniture, Cambridge, MA, USA.

Source: Irus Braverman, 2007.

governed they're not gonna get their canes stuck in the holes," explains Boston's landscape architect about the city management of holes (Anonymous, interview).

Alongside canes, holes, engineers, and sight-challenged people, trees also perform an active role in the making of the grate. Trees grow. By growing their trunks into the grates, trees continuously confound the plans of grate engineers and kill themselves in the process. Because it would be both time-consuming and economically inefficient to expand the diameter of the grate's central hole every time the tree grows into it, the only prefixed solution is to design a grate that perfectly balances the diameters of canes and tree trunks together with the required compactness of the soil. Because such perfect balance is rarely a thing of this world, the solution to this material problem must come through the techno-legal regulation of grate holes. Indeed, regulation typically kicks in when things start causing trouble, and "it is only once most of these anti-programs are countered that the path taken by the statement becomes predictable" (Latour, 1991: 105). Technical objects and people are brought into being through a process of reciprocal definitions in which objects are defined by subjects and subjects defined by objects (Akrich, 1992: 222).

The grate is but one example of an on-ground street thing that is designed and managed to negotiate the relationship between humans and nature, and between the particular mobilities of trees and humans. While the aboveground management of trees demonstrates a tight form of governance, mostly for the sake of managing humans, governing through grates presents a much softer and reciprocal negotiation between humans and things. On the ground level, the physical thingness of the tree is part of the program rather than a re-programming. But such human negotiations are much more difficult under the compressed concrete than either above or on the ground. Roots—the tree's presence underground—are not only invisible to the human eye but are also too messy and unpredictable to correspond with aboveground grid requirements. Under the ground, then, the order of certain things gets much murkier.

Underground Governance

Tight tree management aboveground stands in stark contrast with the strong disregard for tree management underground. Underground space is not only less visible to city bureaucrats and experts, but is also less visible to most other city dwellers. Such physical invisibility is the prime reason for the regulatory neglect of this space. The main technology for translating the underworld into a more legible on-the-ground map is the Dig-Safe procedure. Dig-Safe illuminates the ways in which human relationships, in this case the relations between city experts, not only manage but also create space. Moreover, it demonstrates that law's non-management of trees underground can be as consequential, if not more, than the most intense form of regulation aboveground.

Roots vs Pipes, Engineers vs Foresters

My interviewees describe underground city space as embodying a spatial battle between pipes and trees. "Utilities don't usually conflict with each other," Brookline's senior civil engineer Tom Condon tells me. "The biggest [problem] with utilities is their effect on . . . the roots of existing trees" (interview). However, my interviewees have different views on who should have the upper hand in this battle. While Condon's "engineering" perspective depicts trees as messing around with pipes, the urban planner laments that "[t]he tree is an orphan" (Simon, interview), explaining that "Toronto's community council is "generally not in favor of trees . . . plumbing or water [is] more fundamental for the city officials" (Condon, interview). "Not to be overly cynical, but trees don't vote, trees don't talk, right?" a Torontonian tree activist asks me (Weinbaum, interview).

Engineers and foresters have developed a distinct vocabulary to address what they see as the self-interest of their respective object, be it pipe or tree. Through the process of representation, foresters and engineers articulate what "their" things say or want, why they act the way they do, and how they associate with each other, posing themselves as "spokespersons" for the trees or the pipes. A binary divide is thereby constructed between those who speak for the trees and those who speak for the pipes. Bruno Latour uses the term "translation" to describe this process (Johnson, 1988: 306). Translation is the mechanism by which certain actors, in this case human experts, control others, in this case trees and pipes, through representing "the many silent actors of the social and natural worlds they have mobilized" (Callon, 1986: 224). Boston's landscape architect describes the work of translation in the city council: "we go into a meeting and there are eight people around the room and they all have different agendas: developer, utility companies, and different people, we [landscape architects] are, and need to be, the strongest advocates for the care and preservation of trees" (Anonymous, interview).

Bruno Latour emphasizes that no thing, and for that matter not even humans, speaks on its own, but always through some thing else. Importantly, he suggests that "[l]ike all modernist myths, the aberrant opposition between mute nature and speaking facts was aimed at making the speech of scientists indisputable" (Latour, 2004: 68). The scientists here are the engineers and the foresters, and their laboratory is the city street. In the process of negotiating human relationships, pipes are also distinguished from roots. What pipes or roots say through the voices of their unelected spokespersons is inscribed onto the physical design of the street. Simultaneously, the physical character of trees and pipes also prescribes the scope of the relationship between their respective professionals, providing a material framework for their practices. In this sense, "not only are humans as material as the material they mold, but humans themselves are molded . . . by the 'dead' matter with which they are surrounded" (Pels et al., 2002: 101). In other words, rather than solely being defined by processes of human signification, things may themselves illuminate their human and social context (Appadurai, 1986: 5).

Tree Recalcitrance

For it to work, the project of human governance must take the material nature of things into account. Yet things do not always comply with their social enrollment as such. Certain scholars have explained this phenomenon through a sense of "the world kicking back" (Whatmore, 2002: 5). Eileen Curran from Vancouver tells me, for example, that although the City wanted twenty-foot laneways to service the backs of the houses, "volunteer trees" kept popping up in those laneways (interview). Trees also die, despite the intentions of the distinguished experts that planted them (Curran, interview). Other experts point out that although they are carefully distanced from one another, trees still get infected by pests and mess up orderly city streets by dropping their fruits on cars and their leaves on raked sidewalks (Simon, interview).

The most common complaint about street trees "kicking back" is the unpredictable behavior of their roots. "[T]he roots have [their] own resistance to the situation, they change," Simon Stuart of Toronto tells me in an interview. Trees, and roots in particular, resist regulation by humans. But is this proposed tree resistance an anthropocentric figure of speech or an actual act of volition? While such a claim to consciousness by nonhumans may at first sound outrageous, it actually corresponds with certain human instincts: who has never experienced a vague sense that some things fail on purpose? This is especially true when nature is involved (Mill, 1998: 3, 5–6). Bruno Latour's work is helpful in this context. Latour suggests stepping out of the subject/object divide into a world of actancy (Latour, 2004: 75). He proposes the term "recalcitrance" to capture the subversiveness of nonhuman actions: "Anyone who believes that nonhumans are defined by strict obedience to the laws of causality must never have followed the slow development of a laboratory experiment. Anyone who believes, conversely, that humans are defined at the outset by freedom must never have appreciated the ease with which they keep silent and obey" (Latour, 2004: 81).

Dig-Safe

Most of the interviewees refer to the underground world as a condensed space of chaos and messiness. "The underground space is jam-packed," says Brookline's city engineer Condon in our interview. In order to manage the street's underground and coordinate between the various entities that operate in this space, the American legislator came up with the unison language of Dig-Safe. Applied across North America, Dig-Safe regulates underground construction by imposing a rigid form of communication between various city utilities. Elaborate regulations apply to Dig-Safe across the United States. For example, chapter 82 of Massachusetts' General Law requires a process of "premarking" the pavement with white paint before any excavations can be made in public or private rights of way (Mass. Gen. Laws ch. 82, § 40A). The "premarking" is followed by a "marking" stage, which identifies "the location of an underground facility by placing marks on the surface above and parallel to the center line of the facility" (220 C.M.R. § 99.02 2008), adding that "Within 72 hours . . . every company shall mark the location of an underground facility by applying a visible fluid, such as paint, on the ground above the facility" (220 C.M.R. § 99.05 2008). The colors of the marking are also specified, defining: "(1) red—electric power lines, cables, conduit or light cables; (2) yellow—gas, oil, petroleum, steam or other gaseous materials; (3) orange—communications cables or conduit, alarm or signal lines; (4) blue—water, irrigation and slurry lines; (5) green—sewer and drain lines; (6) white—premark of proposed excavation" (220 C.M.R. § 99.02 2008).

Underground matters are translated into on-the-ground representations through coded colors and straight lines. The Dig-Safe procedure reduces the language of communication to its crudest form: pipe locations are indicated by arrows and pipe intersections are marked by diamond shapes (220 C.M.R. § 99.02 2008). The complexity and depth of the underground world is flattened, literally, when projected and inscribed onto the concrete. Brown- and gray-colored pipes are translated into red, blue, and green arrows, while depth and width, as well as other compositions of this space, are mostly ignored. "I know, it looks great," remarks Boston's landscape architect, concluding cynically: "[T]hose people think they're sidewalk artists" (Anonymous, interview). Regardless, the simple arrow and color (de)signs are understood by all utility workers, facilitating complex mitigations and vocabulary adjustments without requiring personal interactions. The hieroglyphic language of Dig-Safe presents itself as unitary and neutral.

Figure 9.3 Dig-Safe, Ithaca, NY, USA.

Source: Irus Braverman, 2013.

But something is strikingly missing from the Dig-Safe picture: trees. No color in the Dig-Safe manual is assigned to map tree roots, and no arrows are marked on the pavement to represent their underground location. Moreover, the relevant legal norms blatantly ignore tree presence underground. How can one explain such blindness by the law? Brookline's tree warden suggests that trees are different from utilities in that their roots correspond with their aboveground location, so that anyone would know not to dig under the tree's "drip-line" (line of canopy; Brady, interview). Put differently, the tree's presence aboveground perceivably speaks for itself, rendering unnecessary the process of translation. The situation, however, is not that simple. Even among themselves, foresters contest the mirror reading of the tree's underground through its aboveground representation. For example, Toronto's urban forest specialist claims that roots reach at least three times the drip-line measure (Simon, interview).

If the root's location is not easy to ascertain without proper mapping, then why not utilize the Dig-Safe procedure to mark tree roots alongside pipes? Boston's urban forester MariClaire McCartan explains that unlike utilities, "the roots will grow wherever they can, and [only when] we pick up the concrete [will] we know where exactly the roots are" (interview). Hence, while pipes are "mappable" (however inaccurate this mapping might be) roots are deemed unpredictable and thereby unfit for regulatory mapping. Legal norms seem to take the trouble of regulating only what can be regulated by its nature. In this sense, legal norms and practices indeed take physical matters into account. Consequently, whereas they are tightly managed aboveground, trees are left to their own devices underground. This dual form of governance can again be explained through highlighting the importance of visibility to law. When aboveground, trees serve as a spectacle for the governance of nonhuman nature. Underground, however, the project of governance is much less relevant and tree control therefore becomes less important.

Conclusion

This chapter examined the project of urban governance from an unfamiliar angle: city street trees. Focusing on three spatial technologies—the grid, the grate, and Dig-Safe—the essay highlighted the importance of physical matter to the project of governance. It demonstrated that human ordering of physical things into the exclusive categories of either society or nature affects the level of their regulation. Aboveground, the tree's categorization as a thing of nature makes it more susceptible to human governance, in turn facilitating the city's domination of nature. At the same time, under the ground the tree's categorization as a thing of nature makes it less prone to human governance. The key to understanding this difference is in law's extreme bias toward visibility (Braverman, 2011).

The chapter suggested that alongside their binary categorization into living/nonliving and into nature/human, the visibility of things to humans affects the nature of their governance. Paying careful attention to visibility, it becomes apparent that the differentiation of street strata into above, on, and under ground is paramount to the regulation of city space. The construction of the natural tree grid aboveground represents a tight project of governance. The grate is somewhat less visible to humans and thus also less important as a project of governing nature; it thus represents a softer instance of management than that of the grid: one that negotiates between the tree bark and the human need for a flat surface. Finally, the underground management established through the Dig-Safe procedure takes only nonhuman things into account while ignoring trees altogether. This demonstrates that, especially in the case of the human regulation of nature, the legal bias towards visibility very much defines the extent of the human governance of things.

By exploring the tree's similarities and differences in relation to other things, the chapter also distinguished the particular thingness of the tree within what Latour calls the Parliament of Things (Latour, 2004: 227). Specifically, it suggested that the tree's thingness is unique in that it embodies a set of binary constructions. As aboveground street furniture, the tree has become an object of rigid regulations that reduce it to a product of detailed, calculable distances within a "lollypop" street order. The process of treescaping the modern grid utilizes both "lollypop" order and natural disorder to reinforce, and at the same time soften, the mechanical features of urban governance. Simultaneously, the tree's "living image" also subjects it to other forms of representation and regulation. Those become especially relevant in the city's underground space.

Furthermore, the tree's seemingly symmetrical physical existence above- and underground both reinforces and challenges the bifurcated stratification of urban space. On the one hand, tree management is split according to these socio-material structures, applying strict regulations over its aboveground dimensions while ignoring its underground features, as demonstrated through the Dig-Safe procedure. On the other hand, professional spokespersons are assigned to represent the entire tree, juxtaposing it with other things that are represented by engineers: namely, pipes and other utility lines. As it oscillates between objectivity and subjectivity, living and nonliving, human and nonhuman, orderly and disorderly, city and nature, the street tree's "thingness" determines the level of its human governance.

Importantly, the chapter also questioned the monopoly of the perception according to which humans govern things in general, and trees in particular. Weaving together Foucauldian perspectives on governmentality and Actor Network Theory's emphasis on actancy, this chapter challenged assumptions about nonhuman agency. The numerous ways in which trees "kick-back" exemplify how things might unexpectedly act against their human governors. The chapter also demonstrated how static hierarchies imposed upon things may bounce back at humans, asserting dominance, rivalries, and schemes of unification between humans according to the things they represent. In this respect, trees also govern, or at least act upon, humans.

Acknowledgments

A longer version of this chapter was published in 2008 in the *Tulane Environmental Law Journal* under the title 'Governing Certain Things: The Regulation of Street Trees in Four North American Cities.' This shorter version is published here with permission.

Notes

1 To sample different grate designs, see, for example, Ironsmith, Tree Grate Info, www.ironsmith.cc/TREE-GRATES-ABOUT.hmt (accessed 4 June 2013).

2 *See also* Ironsmith, ADA, www.ironsmith.cc/ADA.htm (accessed 4 June 2013) ("We have elected to make all of our grates with slot openings 1/2" or less because we believe it offers better all round pedestrian safety and comfort"). Engineers also give thought to how disabled individuals will be affected by the grates. *See id.;* U.S. Access Bd., Ground and Floor Surfaces Technical Bulletin, www.access-board.gov/guidelines-and-standards/buildings-and-sites/about-the-ada-standards/background/proposed-ada-and-aba-accessibility-guidelines/part-iii-technical-requirements (accessed 22 March 2014).

References

Akrich, M. (1992) 'The De-Scription of Technical Objects', in W.E. Bijker and J. Law (eds) *Shaping Technology/Building Society*, The MIT Press, Cambridge, MA, p. 205.

Appadurai, A. (ed) (1986) *The Social Life of Things: Commodities in Cultural Perspective*, Cambridge University Press, Cambridge, UK.

Braverman, I. (2011) 'Hidden in Plain View: Legal Geography from a Visual Perspective', *Journal of Law, Culture, and the Humanities,* 7(2): 173–186.

Brown, B. (2001) 'Thing Theory', *Critical Inquiry*, 28(1): 1–22.

Callon, M. (1986) 'Some Elements of a Sociology of Translation: Domestication of the Scallops and the Fisherman of St Brieuc Bay', in J. Law (ed) *Power, Action and Belief: A New Sociology of Knowledge?*, Routledge Kegan and Paul, New York, NY, pp. 196–222.

Callon, M. and Law, J. (1995) 'Agency and the Hybrid Collectif', *South Atlantic Quarterly*, 94: 481–507.

Foucault, M. (1970) *The Order of Things: An Archeology of the Human Sciences*, Vintage, New York, NY.

Foucault, M. (1991) 'Governmentality', in G. Burchell, C. Gordon and P. Miller (eds) *The Foucault Effect: Studies in Governmentality*, The University of Chicago Press, Chicago, IL, pp. 87–104.

Harrison, R. P. (1992) *Forests: The Shadow of Civilization*, University of Chicago Press. Chicago, IL.

Johnson, J. (1998) 'Mixing Humans and Nonhumans Together: The Sociology of a Door-Closer', *Social Problems*, 35(3): 298–310.

Latour, B. (1986) 'Visualization and Cognition: Thinking with Eyes and Hands', in H. Kuklick and E. Long (eds) *Knowledge and Society: Studies in the Sociology of Culture Past and Present*, Jai Press Inc., UK.

Latour, B. (1991) 'Technology is Society Made Durable', in J. Law (ed) *A Sociology of Monsters: Essays on Power, Technology and Domination*, Routledge, New York, NY, pp. 101–131.

Latour, B. (1993) *We Have Never Been Modern*. Harvard University Press, Cambridge, MA.

Latour, B. (1997) 'A Few Steps Toward an Anthropology of the Iconoclastic Gesture', *Science In Context*, 10(1): 63–83.

Latour, B. (2004) *Politics of Nature: How to Bring the Sciences into Democracy*, Harvard University Press, Cambridge, MA.

Mill, J. S. (1998) *Three Essays on Religion: Nature, the Utility of Religion, Theism*, Prometheus Books, Amherst, NY.

Miller, R. W. (1997) *Urban Forestry: Planning and Managing Urban Greenspaces*, Prentice Hall, Eaglewood Cliffs, NJ.

Mitchell, W. J. T. (2001) 'Romanticism and the Life of Things: Fossils, Totems, and Images', *Critical Inquiry*, 28(1): 167–184.

Pels, D., Hetherington, K. and Vandenberghe, F. (2002) 'The Status of the Object: Performances, Mediations, and Techniques', *Theory, Culture & Society*, 19: 1–21.

Pels, P. (1998) 'The Spirit of Matter: On Fetish, Rarity, Fact, and Fancy', in P. Spyer (ed) *Border Fetishisms: Material Objects in Unstable Spaces*, Routledge, New York, NY.

Rose, N., O'Malley, P. and Valverde, M. (2006) 'Governmentality', *Ann. Rev. L. & Soc. Sci.*, 2: 83–104.

Schama, S. (1995) *Landscape and Memory*, Harper Perrenial, London, UK.

Sennett, R. (1970) *The Uses of Disorder: Personal Identity and City Life*, W. W. Norton, New York, NY.

Sennett, R. (1992) *The Conscience of the Eye: The Design and Social Life of Cities*, W. W. Norton, New York, NY.

Street Tree Guidelines, City of Vancouver, For the Public Realm, Year 2011 Revision. https://vancouver.ca/files/cov/StreetTreeGuidelines.pdf (accessed 22 March 2014).

Whatmore, S. (2002) *Hybrid Geographies: Natures Cultures Spaces*, Sage Publication, London, UK.

Interviews

Anonymous, Boston's landscape architect, in Boston, MA, 7 October 2005.

Thomas Brady, Brookline's tree warden, in Brookline, MA, 28 September 2005.

Sherri Brokopp, Director of Community Forest Partnership, Urban Ecology Institute, Boston, MA, 3 November 2005.

Ian Buchanan, Manager of Natural Resources and Forestry Services, York Region, Toronto, Canada, 8 August 2005.

Thomas Condon, Brookline's senior civil engineer, Brookline, MA, 28 September 2005.

Eileen Curran, Streets Administrative Branch, City of Vancouver Engineering Services, Vancouver, Canada, 30 June 2005.

MariClaire McCartan, urban forester, Boston Parks and Recreation, Boston, MA, 14 October 2005.

Garry Onysko, Vancouver City tree inspector, Vancouver, Canada, 29 June 2005.

Peter Simon, urban forestry specialist, Planning and Protection, North District, Toronto Parks and Recreation, Toronto, Canada, 18 July 2005.

Bill Stephens, arborist technician, Vancouver Park-Board, Vancouver, Canada, 26 June 2005.

Richard Ubbens, Toronto's city forester, Toronto, Canada, 27 May 2005.

Carol Weinbaum, tree activist from Toronto's Casa Loma neighborhood, Toronto, Canada, 5 July 2005.

Laws and Statutes

220 C.M.R. § 99.02 (2008).

Guidelines for Planting City Boulevards, 9 April, 2013, City of Vancouver Engineering Services, www.city.vancouver.bc.ca/ENGSVCS/streets/greenways/guidelines.htm (accessed 4 June 2013).

Mass. Gen. Laws ch. 82, § 40, § 40A (2005).

Street Restoration Manual 2008. City of Vancouver (on file with the author).
Streetscape Guidelines for Boston's Major Roads 1999. Boston Transportation Department, www.cityofboston.gov/transportation/accessboston/pdfs/streetscape_guidelines.pdf (accessed 4 June 2013).
Traffic Control Manual for Work on Roadways 1999. Engineering Branch, BC Ministry of Transportation and Highways (on file with author).
Urban Tree Soil to Safely Increase Rooting Volumes, US Patent No. 5,849,069; filed 23 April 1996; issued 15 December 1998.

10 Four Arboricultures of the Tokyo Metropolis

High and Low, West and East, from Edo to 2020

Jay Bolthouse

Introduction

Home to 35 million people, Tokyo is the world's largest metropolis. Images of the city—for example the often-photographed 'scramble' at Shibuya Crossing, the city's busiest and most iconic intersection—paint a picture of relentless urbanity. The only flash of green is the stripe of the Yamanote Line speeding by as it circles and circulates the city. Yet the archipelago of Japan is two-thirds forest and the Tokyo metropolis, like the nation it dominates, contains not only many trees but its own arboricultures. A short stroll from Shibuya Crossing is Meiji Shrine, a forest sanctuary insulated from the city by a dense evergreen canopy (Figure 10.1). To the east, at the very centre of the city, is an even larger forest, the inaccessible green recesses of the Imperial Palace, encircled by moats but brimming with foliage for the eyes. In Tokyo's mountainous west, cryptomeria plantations protect the city from floods and help to quench its thirst, all while emitting clouds of spring pollen that force many Tokyoites to envelope themselves in pollen-tight gear or stay indoors for days on end. Emerging from Tokyo Bay, on land reclaimed with the city's waste, is a forest island expected to send cool breezes inland to curb the vicious ecology of urban heat island effect, 'guerilla rainstorms' and flash flooding. Even at the commanding heights of the city, towering over the low skyline of the eastern working class districts, is a tree. Although not a real tree, Sky Tree, the popularly selected name for the city's new 634-metre tall communications tower, is not unrelated to the city's arboriculture. Like the plastic foliage that marks the arrival of spring and fall in the shopping arcades far below and the utility poles that supplant so many trees on the city's streets, Sky Tree is part and parcel of the political ecological processes shaping the city and its arboricultures.

The primary goal of this chapter is to survey four arboricultural regions of the Tokyo metropolis—namely the High and Low Cities and the western and eastern peripheries—from early modern Edo to Tokyo's vision for 2020. A second but equally important aim is to make a case for engaging political ecology with approaches more attuned to the place constituting dimensions of human–tree relations. Political ecological work on urban forests—as pioneered by the 'Milwaukee School' led by Heynen and Perkins—has called attention to the uneven distribution of urban trees (e.g. Heynen, 2006), the contradictions of greening

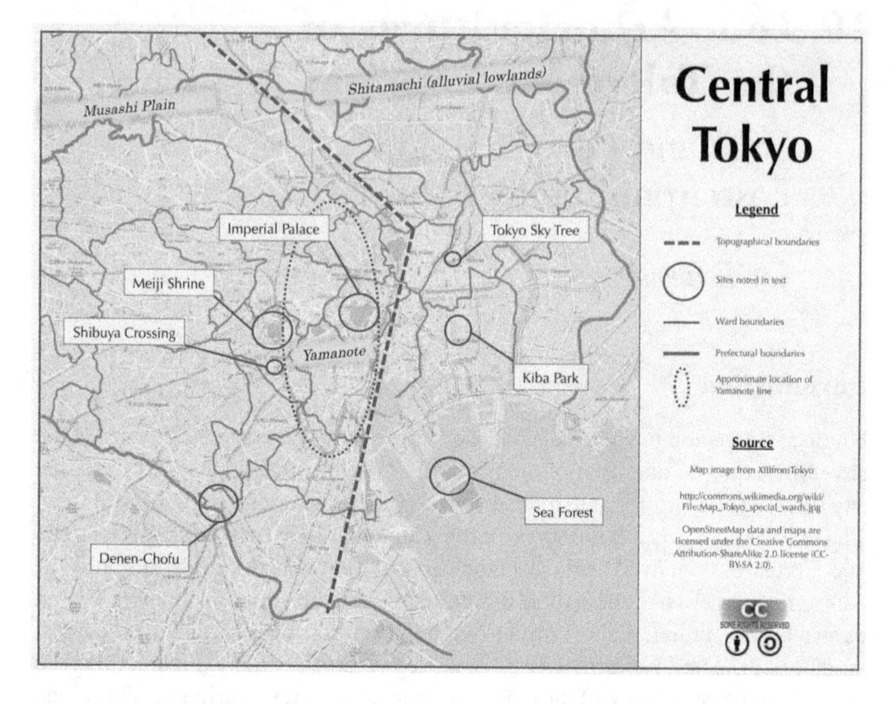

Figure 10.1 Map of central Tokyo, Japan, depicting sites mentioned in the text.

Source: http://commons.wikimedia.org/wiki/File:Map_Tokyo_special_wards.jpg - OpenStreet Map data and maps are licensed under the Creative Commons Attribution-ShareAlike 2.0 license (CC-BY-SA 2.0). Author: XIIIfromTokyo, 2012.

amidst neoliberalization (e.g. Heynen and Perkins, 2005) and the role of trees as mediators in securing the hegemony of market-oriented neoliberalizations through urban forest volunteering (e.g. Perkins, 2013). Yet while this work sheds profound light on the interconnections of arboreal ecologies and politico-metabolic processes, there remains a need for further consideration of the constitution of place-specific arboricultures, if we are to explain the political ecologies of urban trees across cities and cultures. Toward such ends, this chapter attempts to combine political ecology with the more sustained emphasis on people, place and trees found in the 'arboricultural' approach of Jones and Cloke (2002). Although limited to the UK, their work invites application further afield. And while Jones and Cloke (2002) pursue deeply engaged case studies of anthropo-arboreal dwelling at specific sites, their approach also offers a model for the briefer historical geographic surveys on offer in the short space of this chapter. In attempting to combine political ecological and arboricultural approaches, inspiration is also drawn from two pioneering studies: Banham's (1971) classic study of Los Angeles' 'four ecologies' and Walker's (1995) Banham-inspired reading of four of San Francisco's 'residential ecologies'. The fourfold accounts of these classics offer object lessons for a political ecology sensitive to place.

High City Arboriculture

> The city I am talking about (Tokyo) offers this precious paradox: it does possess a center, but this center is empty. The entire city turns around a site both forbidden and indifferent, a residence concealed beneath foliage . . .
>
> (Barthes, 1982: 30–2)

The Imperial Palace—Barthes' empty centre—is the central node in the arboriculture of Yamanote or the High City.[1] The term Yamanote (literally 'mountain hand') is derived from the finger-like ridges that extend eastward from the 'palm' of the Musashino Plain into the alluvial lowlands around Tokyo Bay. It was here that Tokugawa Ieyasu established his castle compound following unification of the realm in the early 1600s. To the east lay the Low City and its dense commercial and residential quarters. On the high ground alongside and to the west spread samurai quarters and daimyo villas. Dominated by neither commerce nor industry, but by elite residences, the High City of Edo was overarched by an expansive residential forest. Today, Yamanote is the heart of world city Tokyo. Although its forest cover is greatly reduced, a remnant archipelago of island forests float against an urban sea, and an arboriculture of prestige remains a defining characteristic of the district.

The Imperial Palace is an empty centre brimming with trees (Figure 10.2a–b). Indeed, within the 23 wards of central Tokyo, the Imperial Palace is by far the most tree-full site. As Nakashima suggests (1999), the Imperial Palace is the central node in an archipelago of large forest islands dotting the High City, each connected to the Imperial Family. The relationship between empty centre and full forest is, of course, not arbitrary. The full foliage of the arboreal veil is integral to the architecture of emptiness. It is the evergreen broadleaf trees that transform the centre into a deep-green recess of unseen and unknown proportions. While the concept of nothingness or emptiness in the Kyoto School of philosophy bears connections to the emperor system (e.g. Heisig, 2001), there is here an immediate connection between emperor and green empty centre. As Nakashima (1999) shows, an often-unrecognized symbolic relation exists between the emperor and the various greening campaigns that have shaped the mountain slopes and urban fabric of modern Japan. Neutral and innocent, the planting of trees in greening campaigns became linked to the emperor through his symbolic presence at greening events. The green islands of the High City reinforce this link between nature, nation and nurturing imperial figure.

The most-frequently visited island in Tokyo's green archipelago is Meiji Shrine. The most popular destination in Japan for New Year's Day shrine visits, this forest sanctuary draws in 3 million people annually on that day alone (see Havens, 2011). Located between the major hubs of Shinjuku and Shibuya, and immediately adjacent to trendy Harajuku Station, the shrine is engulfed by the city (Figure 10.2c). Originally non-forested public land, the site was selected in 1914 as an ideal site for deifying Emperor Meiji and Empress Shoken through the creation of a forested shrine sanctuary as well as an outer detached garden (*gaien*), a site later

Figure 10.2 Four images of Tokyo's High City arboricultures. *Top left* (a) Imperial Palace surrounded by moat; *top right* (b) the 'grey' approach to the 'green' empty centre; *bottom left* (c) island-like Meiji Shrine engulfed by the city; *bottom right* (d) the grey and green blur of a passing Yamanote train.

Source: Jay Bolthouse, 2013.

dedicated to national and international sporting events. Planners opted for an evergreen deciduous forest that would grow into a 'natural' forest as quickly as possible, concealing artifice beneath evergreen canopy (Nakashima, 1999). Of great interest is the simultaneously national, imperial and voluntary nature of the project. A total of 12,000 trees were donated from throughout the empire, including from the colonies of Sakhalin, Korea, Manchuria and Taiwan, and subsequently planted by 11,000 youth group members as part of patriotic national service (Nakashima, 1999; Havens, 2011). From its opening in 1921, Meiji Shrine became a key site in the city's life and development. The outer gardens became a site for a national sports' day in the 1920s, and the shrine area was later the centre for the ill-fated Olympics of 1940 and the actual centre of the Tokyo Olympics of 1964 (Tasgold, 2010). For the 2020 summer Olympic Games, the Meiji Shrine area is part of the 'heritage zone' to be crowned with a futuristic new Olympic stadium.

As the visitor to Tokyo quickly learns, Yamanote names the central artery of the city's circulation: the Yamanote Line. The Line circles central Tokyo and circulates the city, defining its nodal geography of multiple dispersed centres. Its grey-bodied, green-striped cars are an icon of the city, and also provide a useful metaphor for the arboriculture of the High City. On the grey side is the densely urbanized ground of the central city. On the green side are the large urban forest islands that dot the area. The empty centre and its satellites stand in deep-green contrast to the grey hardscape of the surrounding city. Indeed, when speaking of Tokyo's urban forest, one must invert Barthes' (1982) notion of an empty centre. While the empty centre of the Imperial Palace is full of trees, it is the commercial and residential areas of the central city that are arboreally empty. The 'Green Tokyo 10-year Project', the city's 10-year greening plan released in 2006, aims to bridge the green–grey divide by planting school playgrounds with grass, greening rooftops and planting trees along roads and waterways. Its most ambitious target is to double the city's street tree population from an estimated half-a-million to a million in a decade (Tokyo Metropolitan Government, 2007). Street trees are deemed necessary for fixing the city's heat island woes and improving its green image. However, while procuring adequate street tree habitat is difficult anywhere, it is particularly difficult in Tokyo. Owing to the ubiquitous utility poles on the city's streets, existing street trees are heavily pruned and burying wires to make way for new trees is cost prohibitive. Moreover, the very notion that shade is an amenity is at odds with a mode of urban ecological dwelling that demands sunshine for drying laundry and airing futons. As shown by Roth (2009), any new trees planted on Tokyo's streets are likely to be seen by local residents as a 'garbage problem' and not an amenity. Thus the metabolism of the grey–green divide of the high city—and its vicious ecology of urban heat island effect, 'guerilla rainstorms' and flash flooding—is not only political-ecologic, but also rooted in the arboriculture of the city and its inhabitants' modes of dwelling.

Low City Arboriculture

Below and east of Yamanote is the Low City. Like the High City, the arboriculture of the Low City can be partitioned into two zones: two arboricultures. In the north is shitamachi (literally 'below-town'), dominated by a dense residential–commercial ecology offering poor habitat for trees. In contrast, in the south, on land reclaimed from the bay, a more spacious environment provides openings for urban afforestation. If the older districts challenge tree provision, the new districts are challenged to produce an arboriculture that will help Tokyo maintain its position in the global hierarchy of cities.

In Edo the Low City was shitamachi, a densely inhabited maze of residence, commerce and industry, and a water city defined by its rivers, canals and the water-based transportation that circulated and connected its inhabitants. Today, many of the canals have been filled and shitamachi completely transformed. As Waley (2002) shows, the geography of the Low City has become inexact, diffusing across the lowlands east and north of the city while also retreating into an inner landscape of the geographical imagination. Shitamachi provided poor habitat for large trees. What did develop in this intensely inhabited district was a culture of displaying potted plants and bonsai, and this miniature arboriculture continues to provide the tight streets of the Low City with charming green relief (Figure 10.3a–b). Although large trees were found in peripheral gardens east of the Sumida River, from the turn of the twentieth century this garden periphery was transformed into an industrial and blue-collar residential zone (Waley, 2010). After the earthquake of 1923 and the fire bombings of the war, urban planning efforts widened streets allowing trees and parks to be inserted into the dense urban fabric (see Sorensen, 2002). Many of the defunct canals of the former water city were transformed into tree-lined waterways or greenways (Figure 10.3c). The largest of the post-industrial greenspaces east of the river is Kiba Park, a site of great interest here not only for its many trees but also for its displaced arboricultural history. Kiba (literally, 'tree-place') is named for lumberyards built here during the Edo era. Composed of a grid of canals, Kiba harboured flows of lumber transported from around the country, and later the world, to the city of Edo and its successor (see Totman, 1989). Yet as the city continually expanded bayward in the post-war period, the city's lumberyard was relocated bayside. Former canals were transformed into a green oasis in the older Low City, but little of the area's arboricultural history survived

Figure 10.3 Four images of Tokyo's Low City arboricultures. *Top left* (a) an intensely cultivated narrow alley in shitamachi; *top right* (b) bonsai and other small plants cultivated along the front of a residence in shitamachi; *bottom left* (c) former lumberyard canal converted into urban greenway in Kiba Park; *bottom right* (d) a view of Sea Forest from the recently constructed Tokyo Gate Bridge.

Source: Jay Bolthouse, 2013.

greening (Figure 10.3c). And evidence of the city's global forest consumption habit was pushed to the edges of the bay—out of sight and mind.

The bay has always been a receptacle not only for the city's surplus waste, but also its surplus of dreams and speculative visions (Pernice, 2006). Land reclamation was already at work in Edo, but large volumes of waste continue to push the city bayward. While skyrocketing land values put reclaimed lands in high demand before the real estate bubble burst in the early 1990s, weakened demand now allows for extensive land uses such as parks and greening. The most high-profile designs for developing surplus land are found in Tokyo's plans for the 2020 Olympic venues. While unsuccessfully wooing the International Olympic Committee since 1988, pressure has continually mounted to match the Olympic spectacles of its East Asian (i.e. Beijing) and world city (i.e. London) rivals (Kelly, 2010). Tokyo has managed to capture the 2020 Games and reinforce its world city status. Although the 'Heritage Zone' of the High City is slotted for the main stadium, the 'Bay Zone' is to host the majority of events and the largest development project: the Athlete's Village, a 44-acre and nearly billion-dollar development (Adelman and Chu, 2013). One Olympic venue—Sea Forest (*umi-no-mori*)—encapsulates the arboriculture of the city's watery edge (Figure 10.3d). Far from the city's inhabited districts, an 88-hectare island was reclaimed with over 12 million tons of garbage buried bayside from 1973 to 1987. In 2005, a proposal was made to turn this island, unromantically named the Inner Central Breakwater Reclamation Area, into a forest (Bureau of Port and Harbor TMG, 2005). Upon the plan's acceptance in 2007, the island was renamed Sea Forest and a 30-year plan hatched for developing a forest park (Umi-no-Mori, 2008). Sea Forest is the flagship initiative of the 'Green Tokyo 10-year Project' and by far the most popular donation destination for its 'Green Tokyo Fundraising Campaign', a fundraising scheme encouraging citizens and corporations to contribute monetarily to their preferred greening project (TMG, 2007; Midori no Tokyo Bokin, 2010). Sea Forest is portrayed as an example of sustainable recycling drawn from Japan's alleged tradition of living in harmony with nature. Yet even looking past the questionable notion of landfill as 'recycling', it is clear that there were few possible land uses for the island and that forest planting was, first and foremost, a natural fix for soil erosion on the windy site (Bureau of Port and Harbor TMG, 2005). Moreover, while the island is framed as an arboreal fix for Tokyo's heat island problem, the eco-technical claim that Sea Forest will send cool forest-conditioned breezes to the city is not backed up with data.

The city frames Sea Forest, and many of its greening projects, as representative of a new model of park and greenery provision in which government, corporations and non-governmental organizations (NPOs) work hand-in-hand (Umi-no-Mori, 2008). However, these mobilizations are equally representative of the revitalization of the volunteer, donation-based project to build Meiji Shrine and the post-war afforestation campaigns. From its inception Sea Forest promoters requested individual donations of 1,000 yen, aiming for, and achieving, a total of 800 million yen in only 3 years. Likewise, tree planting has been a volunteer effort, often using acorns raised by local schoolchildren. While it is tempting to read the project as neoliberal, we must first concede that Sea Forest is rooted in a certain arboriculture:

a culture of arboreal service manifested in Meiji Shrine and national afforestation campaigns and reoriented today in the direction of urban sustainability, Olympic and global ambitions.

The Arboriculture of the Western Periphery

Westward from the skyscraper districts of Shinjuku lies the Musashino Plain.[2] The plain is bounded by the Tama River on its south, the Ara River on its north and by the rising mountains that mark the transition to the central mountains of Honshu on its west. The Yamanote district is the far eastern portion of Musashino and the boundary between them is, and always has been, unclear. A simple rule of thumb would be to place Musashino west of the Yamanote loop line, but clearly its romantically infused geography is somewhere beyond the city.

Historically people first settled along narrow valleys carved into the upland. The upland itself was used only for hunting and foraging, since minimal water and poor soils made the uplands difficult for agriculture.[3] However, along with rapid growth of Edo, large areas of Musashino were pioneered by settlers from the city (see Robertson, 1991). These settlements consisted of a line of residences along a central axis and a long strip of narrow fields extending from each homestead. At the far rear of each farmstead a plot of coppice woodlands was established to supply fuel and fertilizer (Figure 10.4a). Surrounding non-agricultural areas were also converted to intensive coppice management to provide the growing city with fuel for home and industry (Yokohari and Bolthouse, 2011a).

From the turn of the twentieth century, Musashino became a highly desirable site for suburban development. Rapid expansion of train lines provided the means, but the discovery of Musashino as a suburban topos was propelled by literature (Yiu, 2006; Sand, 2009). In 1898 Kunikida Doppo published *Musashino*, a work of realist fiction that praised the aesthetic beauty of the area and set off a 'Musashino boom' (Kunikida, D. (2001 [1898]). Most important here is that Kunikida reserved highest praise for the arboriculture of the area. His intention to subvert the 'famous place' (*meisho*) logic of classical poetry (a tradition in which the pine held sway) led Kunikida to assert the realist, romantic and vernacular beauty of the area's coppiced oaks. Woodlands already tied to the city's metabolic flows became a symbolic resource, a topos on the edge of the city available for leisure, self-discovery and suburban colonization (Sand, 2009). Appended to the Musashino boom was the garden city concept. The same year Kunikida penned *Musashino*, Ebenezer Howard published his treatise on garden cities. In 1907, the powerful Home Ministry introduced this model to Japan as *denen-toshi* (countryside city), but without Howard's Fabian Socialism or utopian orientation (Watanabe, 1992). Denen Chofu, the first Japanese manifestation of the garden city, remains the most desirable suburban Tokyo address today. Here we will use Denen Chofu to formulize an axiom for Tokyo's residential arboricultures. Although seen as representative of Musashino's leafy suburbs, Denen Chofu has a canopy that is far denser than even the swanky developments nearby, making this site less representative and more of an exception. The exceptional nature of Denen Chofu's tree canopy is indicative of the near-total absence of the Anglo-American suburb. Even in the exceptional arboriculture of Denen Chofu, tree canopy does not dominate rooftops. And nowhere in Tokyo's centre or periphery will we find a 'residential ecology' similar to Walker's (1995) appraisal of Berkeley as a city ensconced in a wood (Figure 10.4b–c).

Turning to the arboriculture of Musahino today returns us to the topos of coppice woodlands. A century after Kunikida's discovery, the coppice woods of Musashino have been rediscovered, this time through the nostalgic and 'vanishing' idiom of *satoyama* ('village-forest'). Ivy (1995) describes how things traditional are lodged between presence and absence, provoking efforts to restore and recover that which is not gone but forever suspended at a point of 'vanishing'. Framing coppice woodlands as *satoyama* places them squarely in the category of vanishing (Bolthouse, 2013). Following decades of urbanization and the complete abandonment of woodland management, *satoyama* woodlands did indeed partially disappear. However, countless volunteer groups have formed to pursue conservation and restoration of vanishing woods (Takeuchi *et al.*, 2003). Central to this rediscovery was the animated film *My Neighbor Totoro*, a nostalgic depiction of the Sayama Hills area of

Figure 10.4 Four images of the arboricultures of Tokyo's western periphery. *Top left* (a) long, narrow agricultural field and coppice woods of Edo-era reclamation at Santome Shinden; *top right* (b) a gingko-lined street in Denen Chofu; *bottom left* (c) a residential street lacking trees in Denen Chofu; *bottom right* (d) sign commemorating the first forest plot purchased by the Totoro Fund.

Source: Jay Bolthouse, 2013.

Musashino in the 1950s before Japan's rapid post-war development (Yokohari and Bolthouse, 2011a). As Wegner (2010) suggests, the film was partially motivated by director Miyazaki Hayao's desire to draw public attention to the fact that traditional landscapes still existed, even on the periphery of cities. In other words that *satoyama* were vanishing, disappearing but still recoverable. In this effort the film achieved incredibly. Vanishing *satoyama* woods and rural landscapes have been rediscovered throughout Japan and the most prominent *satoyama* conservation efforts are those aiming to save the wooded stage of *My Neighbor Totoro* in western Musashino.

The Totoro Fund was established in 1990 with the explicit purpose of saving the *satoyama* woods of Totoro's Hometown (*Totoro-no-Furusato*) in the Sayama Hills (Totoro no Furusato Kikin, 2013). Although part of the broader *satoyama* revival, Totoro Fund is exceptional in that it is highly organized and receives donations from throughout Japan and abroad and has pursued the outright purchase of endangered forests in the area. This is an example of the way in which Musashino, like neighbouring Yamanote, is able to command prestige as a famed landscape. From the stately imperial forests of the High City, to the leafy garden suburbs of Musashino and the enchanted haunts of Totoro, the arboriculture of western Tokyo is rooted in a rich cultural–economic capital (Figure 10.4).

The Arboriculture of the Eastern Periphery

On Tokyo's eastern periphery, covering all of northern Chiba Prefecture minus its lowland edges, lies the Shimousa Plain, counterpart to the western Musashino. Shimousa Plain shares many similarities with Musashino.[4] Both were dialectically integrated with the development of Edo as resources supply zones. Both are home to numerous bedroom communities that today supply Tokyo with its daily influx of labouring bodies. However, the different historical geographies shaping west and east contribute to differently articulated arboricultures. If the arboriculture of the western periphery bears connections with the prestige of the High City, then the eastern periphery overlaps with the blue-collar nature of the Low City.

As on the western periphery, initial settlement of Shimousa Plain was limited to its edges and margins, leaving the upland largely undeveloped. Historically the plains were used as pasturelands for raising horses by powerful warlords. With unification of the realm, the Shogunate claimed direct control of the area and established two large pasturelands, Kogane Maki in the west and Sakura Maki in the east (Miyamoto and Yokohari, 2010). Already tied to the power ecology of Shogunate rule, the Shimousa Plateau and its pasturelands were further integrated with metropolitan developments in the Edo era as supply zones for food and fuel. Powerful local families were given large land grants as incentive to afforest and augment the supply of wood-hungry Edo (Miyamoto and Yokohari, 2010). With the Meiji Restoration, the cash-starved Meiji government ceded remaining pasturelands to jobless former samurai as a means of raising revenue and ensuring livelihoods for a disgruntled and politically dangerous class. The result was to parcelize and remove from the public domain a large portion of Tokyo's eastern periphery. The western Kogane pastureland can be insightfully compared with an analogous site in the margins of another world city. Although nominally pasture, its semi-forested condition and ownership by the Shogunate makes Kogane comparable to European royal forests and its location, roughly 20 km northeast of Edo-Tokyo makes it particularly comparable with Epping Forest, a former royal forest located a similar distance northeast of London. In the mid-nineteenth century, Epping Forest was targeted for enclosure by local lords seeking to develop farmland and housing. In response, a coalition of commoners and middle class professionals, led by the Commons Preservation Society, emerged to conserve the remaining commons of Epping Forest as a recreational space for the entire metropolis, eventually culminating in the Epping Forest Act of 1878 (Eversley, 1910). While tides had shifted from enclosure to preservation of remaining commons around imperial and world hegemonic London, Japan's belated and unequal entry into global political economics left the modernizing country and its leaders with the mission of catching up. Kogane Maki became a strategic site for modernization and revenue accumulation. Land was parcelled into small agricultural plots to provide both a political safety valve and an economic foothold into the world economy.

In great contrast to Epping Forest, the former grasslands and wood pasture of Kogane Maki are today a dense intermixture of urban–rural land uses. However, while the greenscapes of the Shimousa Plain were not seen as a civic commons in the nineteenth century, parts of the area are today being re-invented and re-assembled as commons. Although this region cannot command the cultural-economic capital necessary to purchase woodlands, shifts toward urban-oriented forest policies have enabled a group of seniors called 'Coppice Club' to re-assemble over 100 hectares of former common forests into a single management plan, as described in greater detail elsewhere (Bolthouse, 2013). The group has removed illegally disposed refuse and cleared away invading bamboo undergrowth (Figure 10.5a–b). They seek to open the woodlands up as new local commons and are pursuing the production of firewood and logs for mushroom cultivation. At another site, an agricultural plot located immediately in front of a suburban train station adjacent to Shimousa Plain (Figure 10.5c–d), urban citizens are transforming space through agricultural activities. Until a few years ago the adjacent farm household, a family with deep roots at this site, managed this large, highly visible plot. However, similar to many areas in Tokyo's peripheries, this plot has become difficult for the ageing farm household to manage. Gradually over the course of the last few years, it has been divided into numerous small parcels for the citizens of the neighbourhood to grow their own vegetables and flowers. Likewise, the *mikan* (*Citrus unshiu*) trees on the slope leading to the uplands are now leased annually by families in the neighbourhood and collected before the New Year's holiday, a custom somewhat analogous to the annual trip to the woods to procure a Christmas tree.

The senior citizen volunteers of Coppice Club are re-assembling a *satoyama* woodland commons. Likewise, the transference of the productive greenspaces to amateur farmers is creating an agricultural commons at the very centre of the neighbourhood. These ecological practices are not, however, unproblematic or apolitical. In ageing and slow-growth Japan, the abandonment of forest and

Figure 10.5 Four images of the arboricultures of Tokyo's eastern periphery. *Top left* (a) a 'before' image of Coppice Club members battling bamboo grass along a roadside on Tokyo's western periphery; *top right* (b) an 'after' image of the same site with a fence built by Coppice Club members with bamboo harvested from the forest site; *bottom left* (c) citizen-based agriculture in Happy Valley; *bottom right* (d) the same agricultural plot viewed from adjacent train platform.

Source: Jay Bolthouse, 2013.

agricultural land, as well as the difficulty of managing urban greenspaces and emergent vacant plots, creates growing pressure for citizens to spring into action (Yokohari and Bolthouse, 2011b). Nowhere is this shift more evident than the 'Citizen Management of the National Land' (*kokudo no kokumin teki keiei*) promotion efforts of the National Spatial Plan (Ministry of Land, Infrastructure, Transport and Tourism, 2013). Because the National Spatial Plan, since its inception in the post-war period, has been synonymous with central government-led remaking of the archipelago with heavy industry and infrastructure (McCormack, 2001), its promotion of citizen management of the national territory is nothing short of remarkable. These national-level, volunteer, management-promotion efforts, and their local prevalence on Tokyo's eastern periphery, perfectly illustrate the interplay of consent and coercion so prominent in an era of neoliberalizing greenspace management. But they are also representative of the recurrent arboriculture of service identified throughout this chapter.

Conclusion

This chapter surveyed the arboricultures of four regions of the Tokyo metropolis. These four arboricultures are in no way exhaustive of the region's arboreal ecologies. Nor are they as mutually exclusive as the neat divisions outlined in this chapter imply. Despite the enormity of the subject and the brevity of the surveys, what this chapter demonstrates is that diverse arboricultures populate the abstract notion of 'Tokyo's urban forest'. While providing a first cut at the arboreal ecologies of the world's largest metropolis, the chapter also pursued an engagement between political ecological and arboricultural approaches to urban trees. In Tokyo, the mundane but ecological daily habit of hanging laundry and drying futons in the sun has important consequences for the arboriculture of the city and the metabolism of the urban forest. Likewise, while this chapter has repeatedly called attention to volunteering practices that are analogous to the neoliberalization of urban greenspace management elsewhere, urban forest volunteering in Japan today is also part of an arboriculture of tree-service with deep roots. Politico-metabolic processes shape urban forests and trees. Yet the arboreal architectures of our surroundings are simultaneously constitutive of people and place, habits and habitus. Likewise, while trees and forests structure our ways of being in the world, foregrounding daily experience at the expense of global political economy leaves us with an impoverished perspective on the politics of local sites as well as the tasks ahead. There is a need to combine political ecology and arboriculture, and to construct a political ecology attuned to place.

Notes

1 On the High and Low Cities the best accounts remain Seidensticker (1985) and Jinnai (1995).
2 Rendering the Japanese term *daichi* in English is problematic. Although terms such as plateau, terrace or upland are often used, 'plain' appears to be the most common expression. For an explanation of the underlying geology in question, see Takeuchi *et al.* (2003).
3 On the natural history and cultural landscapes of the Kanto Plain see Takeuchi *et al.* (2003).
4 On Tokyo's peripheries (*kougai*) see, among others, Higuchi (2000).

References

Adelman, J. and Chu, K. (2013) 'Olympic front-runner Tokyo plans biggest project in 42 Years', *Bloomberg*, [online] July 24, www.bloomberg.com/news/2013-07-23/olympics-front-runner-tokyo-plans-biggest-project-in-42-years.html (accessed 13 August 2013).

Banham, R. (1971) *Los Angeles: The Architecture of Four Ecologies*, Harper and Row, New York City.

Barthes, R. (1982) *Empire of Signs*, Hill and Wang, New York City.

Bolthouse, J. (2013) 'End of tradition, reworking of custom: re-assembling satoyama woodlands on Tokyo's urban fringe', in I. Rotherham (ed) *Cultural Severance and the Environment*, Springer, Dordrecht.

Bureau of Port and Harbor Tokyo Metropolitan Government (2005) '*Chuoubouhateiuchigawa umi-no-mori (kashou) kousou: umi wo ikashi, mori wo tsukuri, hito wo sodateriru*', Tokyo Metropolitan Government, Tokyo, www.kouwan.metro.tokyo.jp/jigyo/plan/uminomori/honpen.html (accessed 12 August 2013).

Eversley, B. (1910) *Commons, Forests and Footpaths*, Cassell, London.

Havens, T.R.H. (2011) *Parkscapes: Green Spaces in Modern Japan*, University of Hawaii Press, Honolulu.

Heisig, J.W. (2001) *Philosophers of Nothingness: An Essay on the Kyoto School*, University of Hawaii Press, Honolulu.

Heynen N. (2006) 'Green urban political ecologies: toward a better understanding of inner-city environmental change', *Environment and Planning A*, 38(2): 499–516.

Heynen N. and Perkins, H. (2005) 'Scalar dialectics in green: urban private property and the contradictions of the neoliberalization of nature', *Capitalism Nature Socialism*, 16(1): 99–113.

Higuchi, T. (2000) *Kogai no Fukei—Edo kara Tokyo e*, Kyouiku Shuppan, Tokyo.

Ivy, M. (1995) *Discourses of the Vanishing: Modernity, Phantasm, Japan*, The University of Chicago Press, Chicago, IL.

Jinnai, H. (1995) *Tokyo: A Spatial Anthropology*, University of California Press, Berkeley, CA.

Jones, O. and Cloke, P. (2002) *Tree Cultures: The Place of Trees and Trees in Their Place*, Berg, Oxford.

Kelly, W.W. (2010) 'Asia pride, China fear, Tokyo anxiety: Japan looks back at Beijing 2008 and forward to London 2012 and Tokyo 2016', *The International Journal of the History of Sport*, 27(14–15): 2428–2439.

Kunikida, D. (2001 [1898]) 'Musashino', in Y. Tsubochi (ed) *Meijinobungaku 22: Kunikida Doppo*, Chikuma Shobo, Tokyo.

McCormack, G. (2001) *The Emptiness of Japanese Affluence*, M.E. Sharpe, New York City.

Midori no Tokyo Bokin (2010) '*Midorinotokyo-bokin Jikkou-iinkai*', www.midorinotokyo-bokin.jp (accessed 12 August 2013).

Ministry of Land, Infrastructure, Transport and Tourism (2013) '*Kokudo no kokumin teki keiei ni tsuite*', MLIT, www.mlit.go.jp/kokudoseisaku/kokudokeikaku_tk3_000014.html (accessed 12 August 2013).

Miyamoto, M. and Yokohari, M. (2010). 'Relationships between land grant processes of government-ruled pasturelands and transformation processes of land use after the modernization in Shimousa Plateau, Japan', *Journal of The Japanese Institute of Landscape Architecture*, 73(5): 631–636.

Nakashima, K. (1999) 'Representing nature and nation: national-land afforestation campaign and the production of forest in the 1960s–1970s Japan', in T. Mizuuchi (ed) *Nation, Region and the Politics of Geography in East Asia*, Osaka City University, Osaka.

Perkins, H. (2013) 'Consent to Neoliberal Hegemony through Coercive Urban Environmental Governance', *International Journal of Urban and Regional Research*, 37(1): 311–327.

Pernice, R. (2006) 'The transformation of Tokyo during the 1950s and early 1960s. Projects between city planning and urban utopia', *Journal of Asian Architecture and Building Engineering*, 5(2): 253–260.

Robertson, J. (1991) *Native and Newcomer: Making and Remaking a Japanese City*, University of California Press, Berkeley, CA.

Roth, J.H. (2009) 'Off with their heads! Resolving the "garbage problem" of autumn leaves in Kawagoe, Japan', *The Asia-Pacific Journal*, 40-2-09.

Sand, J. (2009) 'Landscape of contradictions: the bourgeois mind and the colonization of Tokyo's suburbs', *Japanese Studies*, 29(2): 173–192.

Seidensticker, E.G. (1985) *Low City, High City: Tokyo from Edo to the Earthquake,* Donald E. Ellis, San Francisco, CA.

Sorensen, A. (2002) *The Making of Urban Japan: Cities and Planning from Edo to the Twenty First Century*, Routledge, Abingdon, UK.

Takeuchi, K., Brown, R.D., Washitani, I., Tsunekawa, A., and Yokohari, M. (eds) (2003) *Satoyama: The Traditional Rural Landscape of Japan*, Springer Japan, Tokyo.

Tasgold, C. (2010) 'Modernity, space and national representation at the Tokyo Olympics 1964', *Urban History*, 37(2): 289–300.

Tokyo Metropolitan Government (2007) *Basic Policies for the 10-year Project for Green Tokyo: Regenerating Tokyo's Abundant Greenery*, TMG, Tokyo.

Totman, C. (1989) *The Green Archipelago: Forestry in Pre-industrial Japan*, University of California Press, Berkeley, CA.

Totoro no Furusato Kikin (2013) '*Koueki Zaidan Houjin Totoro no Furusato Kikin*', Tokorozawa, Japan, www.totoro.or.jp (accessed 12 August 2013).

Umi-no-Mori (2008) Bureau of Port and Harbor Tokyo Metropolitan Government, www.uminomori.metro.tokyo.jp (accessed 12 August 2013).

Watanabe, S. (1992) 'The Japanese garden city' in S. Ward (ed) *The Garden City: Past, Present and Future*, Spon Press, Abingdon, UK.

Waley, P. (2002) 'Moving the margins of Tokyo', *Urban Studies*, 39(9): 1533–1550.

Waley, P. (2010) 'From flowers to factories: a peregrination through changing landscapes on the edge of Tokyo', *Japan Forum*, 22(3–4): 281–306.

Walker, R. (1995) 'Landscape and city life: four ecologies of residence in the San Francisco Bay area', *Ecumene*, 2(1): 33–64.

Wegner, P. (2010) '"An unfinished project that was also a missed opportunity": Utopia and alternate history in Hayao Miyazaki's *My Neighbor Totoro*', *ImageText Interdisciplinary Comic Studies,* www.english.ufl.edu/imagetext/archives/v5_2/wegner/ (accessed 12 August 2013).

Yiu, A. (2006) '"Beautiful Town": the discovery of the suburbs and the vision of the garden city in late Meiji and Taisho literature', *Japan Forum*, 18: 315–338.

Yokohari, M., Bolthouse, J. (2011a) 'Keep it alive, don't freeze it: a conceptual perspective on the conservation of continuously evolving satoyama landscapes', *Landscape and Ecological Engineering*, 7(2): 207–216.

Yokohari, M. and Bolthouse, J. (2011b) 'Planning for the slow lane: The need to restore working greenspaces in maturing contexts', *Landscape and Urban Planning*, 100(4): 421–424.

11 The Unruly Tree

Stories from the Archives

Joanna Dean

Introduction

We like to tell stories about city trees. The stories shape our thinking, but more materially they shape our management of the trees. The meanings we find in these stories influence the choices we make when we plant trees in the city, they alter the ways that we trim and control the trees, and, finally, they inform our decisions to fell them. This chapter explores the history of three persistent narratives about city trees: the narrative of service, the narrative of power, and the narrative of heritage. I argue that these narratives are profoundly humanist in their subordination of the tree to human needs, and I suggest that there is another narrative buried in the archives: the story of the unruly tree. Here I trace the histories of three troublesome species in the city of Ottawa: the Manitoba maple (*Acer negundo*), the ornamental crab apple (*Malus sp.*), and the Lombardy poplar (*Populus nigra*). These are active agential trees that get in the way, cost money and resist human intentions. My point in unearthing their stories is to suggest that, for our own sake, as well as the sake of the trees, we need to move beyond narrowly anthropocentric narratives and consider the unruly tree in the urban landscape.

Persistent Narratives

The three traditional narratives traced here have been remarkably persistent. The first of these, the narrative of service, emerged in the late nineteenth century and returned in new form in the twenty-first century. In this narrative, the tree is selflessly providing services to the human residents of the modern city. The most valued environmental benefit provided in the nineteenth century was shade. In eastern North America, city trees were selected for their ability to grow quickly and provide deep shade during the hot and humid summers. William Saunders, the director of Canada's Central Experimental Farm, had tested hundreds of trees in the Farm's arboretum, and the notes that he prepared for public lectures show that he recommended the planting of large forest trees along city streets. Like many experts, he favoured the American elm (*Ulmus americana*): 'It grows rapidly and soon furnishes excellent shade.' But the maple came a close second: 'The maples stand deservedly high in public estimation as shade trees. Although not so tall as the elm or so graceful in habit, they form fine rounded heads, with masses of foliage, which afford a dense shade.' He identified as suitable street trees the red maple (*Acer rubrum*), 'the most decorative', the sugar maple (*Acer saccharum*), 'hardy', and the silver maple (*Acer saccharinum*), although it was 'less graceful' and liable to breakage. He had tested a total of 124 maple cultivars, from as far away as Japan, Hong Kong, and the Caucasus, but most did not survive Ottawa's cold winters. The Norway maple (*Acer platanoides*) was the only non-native referred to in his notes, highly recommended for its 'dense shade.'[1] Bernhard Fernow, Dean of the University of Toronto's Faculty of Forestry, shared Saunders appreciation for the maples. In his influential book, *The Care of Trees* (1910), he favoured the Norway maple as the 'most serviceable maple', 'a perfect shade tree, free from all troubles', but he recommended the planting of several types, including the Manitoba maple: 'It is . . . very shady, a good street tree for narrow streets . . .' (Fernow, 1910: 267).[2]

The centrality of shade is suggested by the common use of the term 'shade tree' to describe city trees. The term encompassed more than the blocking of sunlight; it also suggested other environmental benefits. As Saunders explained, trees cleaned the air of carbonic acid gas.

> When we breathe and take into our lungs pure air, which is a mixture of oxygen diluted with nitrogen, the oxygen combines with the carbon in the impure blood which is sent to the lungs to be purified and the air we breathe out contains that mixture of oxygen and carbon in the form of carbonic acid gas. All plants, shrubs and trees take in the carbonic acid gas, appropriate the carbon to the building of their substance and return the oxygen to the atmosphere. In this way vegetation purifies the air for the use of the animal world.[3]

The narrative of service declined with the advent of air conditioning, but it has recently reappeared in environmental discourse. In addition to shade, we now talk of such environmental services as pollution mitigation and storm water run-off, and emphasize the benefits of carbon sequestration for climate change rather than the absorption of carbonic acid. We quantify the forest with software, such as i-Tree, developed by the United States Department of Agriculture Forest Service, that refracts the living canopy of the urban forest into quantifiable benefits and services, calculating tonnes of carbon and particulate matter removed. To convince the sceptics in city council, the tonnes are crunched into dollars saved (Walton et al., 2008).[4] The narrative of service is popular among urban planners and urban forest advocates (Kirkpatrick, 2013a).The narrative is given an environmental justice edge by political ecologists like Nik Heynen who examine the inequitable distribution of the services of the urban forest (Heynen, 2003; Heynen and Lindsay, 2003; Heynen and Perkins, 2005; Heynen, 2006; Dean, 2011; Pham et al., 2011). The software is new, and the language different, but the point is an old one: that city trees serve us well.

The second narrative is the narrative of power. As landscape historians have observed, long lines of identical trees, alike in age and in type, speak of the human control of nature, and of a grace born of power. Georges-Eugène Haussmann tamed Paris with a geometry of wide modern avenues lined with chestnut trees. The effect was to pacify a city with beauty. In North America, rows of street trees served to tame the wilderness: as muddy frontier roads were transformed into urban avenues, forests were brought into line (Jacobs et al., 2002; Campanella, 2003; Lawrence, 2006). When, for example, the Ottawa Improvement Commission was charged with beautifying Canada's capital they immediately set about planting long boulevards of trees. William Saunders advised them on tree selection, and American elm, silver maple, and sugar maple trees were closely planted along residential streets. The trees masked empty lots, blighted housing, and industrial sites. They softened the hard edges of the rough lumber town, bringing the illusion at least of balance and harmony (Dean, 2005).

Serried ranks of identical street trees are not as popular today as they once were and the narrative of control over nature has gone out of fashion. We prefer Erik Jorgensen's metaphor of the urban forest.[5] We like to imagine our trees in more natural groupings, and emphasize biodiversity rather than massed symmetry. The satellite image of the canopy has replaced the photograph of the shady streetscape. But echoes of the old narrative persist. Landscape historians explore the peculiar beauty of ranks of trees, and show us how these trees spoke, as clearly as the built environment, of wealth and power (Pugh, 1988; Jacobs, et al., 2002; Lawrence, 2006). Modernist landscape architects continue to employ geometries of trees and plants.

The third narrative is the narrative of heritage. Paul Aird's influential definition of the heritage tree, adopted by the Ontario Heritage Tree Program, makes a nod to aboriginal uses of trees, but he emphasizes the beauty of an individual specimen and its associations with human history:

> A notable specimen because of its size, form, shape, beauty, age, colour, rarity, genetic constitution, or other distinctive features.
>
> A living relic that displays evidence of cultural modification by Aboriginal or non-Aboriginal people, including strips of bark or knot-free wood removed, test hole cut to determine soundness, furrows cut to collect pitch or sap, or blazes to mark a trail.
>
> A prominent community landmark.
>
> A specimen associated with a historic person, place, event or period.
>
> A representative of a crop grown by ancestors and their successors that is at risk of disappearing from cultivation.
>
> A tree associated with local folklore, myths, legends, or traditions.
>
> (Aird, 2005)

Some heritage trees were deliberately planted to commemorate wars and disasters, and others took on their commemorative duties over time. In his study of the American elm, *Republic of Shade*, Thomas Campanella traces the legends behind 'witness' or 'monument' elm trees, and their role in anchoring a community in the past. He notes that the fast-growing American elms served well as heritage trees, because they seemed older than they were. 'They were looked upon as keepsakes from a prelapsarian age . . . But even if an elm was in truth no aged relic, it could very well appear to be—and that was usually good enough' (2003: 48).

One of the most memorable of heritage trees was the horse chestnut (*Aesculus hippocastanum*) that stood outside Anne Frank's attic refuge in Amsterdam during the Second World War. Anne Frank made reference to the chestnut only a few times in her famous diary, but her references have made the tree iconic. 'Nearly every morning I go to the attic to blow the stuffy air out of my lungs', she wrote on 23 February 1944. 'From my favorite spot, the floor, I look up at the blue sky, the bare chestnut tree on whose branches little raindrops shone, and at the seagulls and other birds gliding on the wind and looking like silver and all that moved and thrilled the two of us so much we could not speak [sic]' (Frank, 2003). She was betrayed to the Nazis a few months after writing this entry, and her tree outlived her, becoming over time a living symbol of the Holocaust. Anne Frank's tree began to fail in the 1990s, and when plans were made for its removal in 2006, an alliance of local residents, experts and the national Dutch Tree Foundation raised hundreds of thousands of euros to build an elaborate steel structure to support its limbs. In spite of their efforts, the tree suddenly and unexpectedly succumbed to a windstorm in 2010.

Anne Frank's tree lives on: a virtual tree is online, courtesy of the Anne Frank Museum, and we are invited to post our own virtual leaf on it. Wood from the tree was distributed to selected organizations, and clones have been sent to other sites of mourning around the world. Most significantly, a sapling has emerged from the roots of the original tree and is expected to grow to full size by 2044. As the Support Anne Frank Tree Foundation noted in 2012: 'The tree and the struggle to preserve it in the last two years has fulfilled an important task in an extraordinary manner: the reawakening of the world's collective memory of the Holocaust and a call for tolerance and mutual respect. The seedlings planted all over the world will continue to spread the message, a grand and dignified final stage in the life of this tree.'[6] Anne Frank's chestnut tree has taken on a global significance.

Counter-Narratives

So . . . three narratives. The tree as a selfless service provider. The row of trees as grace and power. The tree as a living symbol of the past. These are three persisting narratives that have given city trees meaning for over a century. What could be the problem? I will argue that each of these narratives reduces the complexity and dynamism of the living tree, and each puts it to some human purpose.

Let us start with the shade trees and the narrative of service. There is no question that trees provide important environmental benefits, and there is no question that the quantification and monetization of those benefits, through i-Tree and other software programs, is necessary to persuade municipalities to fund urban forest management. The ability to correlate canopy cover to income and ethnicity has proven that these benefits are unevenly distributed, and that the urban poor are not well served by the urban forest. But we need to be aware that the math might not always come out on the side of the trees. A recent article calculating the full costs of an arboreal carbon sink in New York City was almost apologetic in its conclusions (Kovacs et al., 2013). And we need to be wary of the reduction of the living three-dimensional tree to a flattened canopy, and the further reduction of that canopy to a series of services. To define a tree by its canopy is a narrowly human perspective; trees exist below ground, as well as above ground, and from a tree's perspective it is possible that the more important part of its existence is underground in the complex mass of roots, soil, rhizomes, minerals, and water.

Let us consider those street trees. There is nothing particularly original in noting the anthropocentrism of our stories of the street tree; their symmetry is clearly symbolic of human control and order. But the narrative of the street tree ignores the history of arboreal resistance and fails to take into account the tree's own transgressive power. Our control over city trees was, and remains, hard fought.

Finally, let us look again at Anne Frank's horse chestnut, that symbol of tolerance and respect. As a child, I too had a massive horse chestnut outside my bedroom window. I was lucky enough to swing from its branches and play outside in its shade. Perhaps for that reason what I remember was the materiality of that tree—the waxy feel of the white petals on those striking candelabras of bloom, the tough prickliness of the large leaves, the spikes on the seedcase, and the way each case peeled open to reveal a bright brown conker. I remember the bitter taste of the conkers and their satisfying hardness. My horse chestnut was no symbol; it was a densely and sensuously physical tree.

The Unruly Tree

So I would like to try out a new historical narrative—trees that cause trouble. This is not a narrative that tree lovers are comfortable hearing, but there are, once we are alert to them, many trees behaving badly in the historical record. There are ornamental trees that make a mess; street trees that refuse to stay in line, or grow too big, too fast, or in the wrong way; and service trees that emit allergenic pollen. These are the nuisance trees, the invasive trees, the weed trees, and the dangerous trees. There are many counter-narratives that might be told of the city tree. My stories come from Ottawa; they feature a native species, the Manitoba maple; a Canadian cultivar, the Almey crab apple; and an exotic, the Lombardy poplar.

Residents of Ottawa followed William Saunders' advice and in the late nineteenth century they planted large fast-growing forest trees along their streets. But Saunders had not reckoned with their rate of growth, and the trees quickly crowded the streets and each other. The trees buckled sidewalks, they tangled with utility wires, and they created too much shade. Members of the Ottawa Horticultural Society lobbied the municipal authorities to better manage the trees. In response to the growing trees, and the influential lobbyists, the city changed their regulations: saplings were to be planted 16 feet, then 20 feet, then 30 feet apart. Eventually, the city hired a tree supervisor, Richard Waugh, who brought the unruly forest under control. He supervised the removal of thousands of trees from the closely planted streets, and trimmed many more. The city also banned difficult trees, and their list of difficult trees kept growing. In 1890 fast-growing and brittle species were banned: silver poplar (*Populus alba*), Balm of Gilead or cotton tree (*Populus balsamifera*), and willow (*Salix*). By 1923 the list had grown to include fruit-, nut-, and cone-bearing trees as well: butternut (*Juglans cinera*), cherry (*Prunus sp.*), chestnut (*Castanea sp.*), Manitoba maple (*Acer negundo*), sassafras (*Lauraceae sp.*), walnut (*Juglans sp.*), poplar (*Populus sp.*) (all kinds), and cone-bearing evergreens (all kinds) (Dean, 2005).

Banned from the streets, the weed-like Manitoba maple took to the backyards, and sprouted along fencelines and neglected lots. Geospatial mapping shows the maple canopies expanding to fill the available space in the yards of inner-city neighbourhoods (Dean, 2011). The errant maple even found refuge on the untended slopes of Parliament Hill, the national backyard, where construction debris, landslides, and the decline of American elm trees left an ecological niche ready to be filled. In 1980, a government report read: 'The vegetation along the north slope on Parliament Hill reflects the disturbed nature of the slope. The overall composition of the slope is mixed and dominated by Manitoba maple and cultivated escapees such as Norway maple (*Acer platanoides*), Siberian maple (*Acer ginnala*), and European honeysuckle (*Lonicera periclymenum*).'[7] In 2012, a C$9.5 million landscaping project was proposed to remove the 'invasive' Manitoba maple and its exotic cousin, the Norway maple, from the slopes of Parliament Hill and bring in a more compliant species mix. The project was halted, mainly because of the cost, and Manitoba maples continue to dominate entire sections of the iconic parliamentary slopes.[8]

The second trouble maker is the crab apple. Misadventures with large forest trees led to a preference for small, manageable trees like the ornamental crab apple tree. In the 1880s William Saunders had imported the cold-hardy Siberian crab apple (*Malus baccata*) and crossed it with standard apples in an attempt to produce an apple tree hardy enough for the Prairies. Federal horticulturalists hybridized the Siberian crab apple with the brightly coloured Redvein crab apple tree (*Malus pumila var.*) from the Caucasus mountains, to produce the Rosybloom ornamental crab apples, which melded Siberian cold hardiness with the deep rose flowers, purplish foliage, and maroon fruit of the Redvein (Preston, 1944; Crawford, 1987; von Baeyyer, 1990). The result was a made-in-Canada ornamental tree: tough enough for the northern winters and pretty enough to compete with Washington DC's cherry trees in the spring. One cultivar, the Almey, was selected as Canada's centennial tree and planted across the country in school yards, suburbs, and parks in 1966 and 1967 (see Figure 11.1). The tree was so popular that a second cultivar, the Royalty, had to be included in the official plantings.[9] Commemorative crab apple tree planting was at its most intense in Ottawa, where the municipal government gave away 18,000 saplings to individual homeowners and the federal government planted many thousands more along parkways. The federal Centennial Commission suggested that crab apple 'plantings could result in Ottawa receiving a special identity similar to that assigned Washington DC at cherry blossom time

. . . they should prove impressive when in full bloom and by then will become a commemorative feature of Centennial Year.'[10] The pretty pink blossoms did brighten the Ottawa springs in coming years, but the crab apple proved to be a problematic planting, prone to disease, and the abundant red crab apples became a messy nuisance underfoot on city sidewalks. By the time the trees reached their full commemorative glory, opinions had shifted. Trevor Cole, curator of the

Figure 11.1 Child planting a tree in an Ottawa suburb to celebrate Canada's Centennial, 1967, Ottawa, ON.

Source: Library and Archives Canada, MIKAN no. 3198625.

Central Experimental Farm, warned against planting either cultivar: 'Many crab varieties that were popular in the past and are still available are highly susceptible to certain diseases. Here are some you should steer clear of: "Almey", "Hopa", "Klem's Improved Bechtel", "Makamik", "Radiant Royalty", "Snowcloud" and Malus ionesis "Plena".'[11]

The final trouble maker is the Lombardy poplar.We encounter the poplar through the eyes of Mabel McKibbon, who lived in the shadow of a row of Lombardy poplars planted by the Ottawa Improvement Commission. Lombardy poplars were popular landscape trees in the eighteenth century, and spread from Italy through Europe, reaching North America in 1784. They were widely planted in North America as sentinels or to cover unsightly gaps in the urban landscape. They are still touted for this purpose by unscrupulous nurseries, who fail to mention their bad habits (Fernow, 1910: 301; Wood, 2000).

In 1910, as part of a federal beautification scheme, the Ottawa Improvement Commission drained a swamp, landscaped it as a sunken formal garden, and planted Lombardy poplars around the circumference (Dean, 2004) (Figure 11.2). The circle of poplars was intended to provide a green backdrop to the manicured park; they also screened empty lots and other eyesores. Lombardy poplars are fast growing, estimated to grow from 18 inches to six feet per year, and known to gain their full height of 100 feet or more in only two or three decades. The southern exposure and the rich moist swampy soil of Ottawa's Central Park would have been conducive to rapid growth, and the trees quickly began to overshadow Mabel McKibbon's house. In 1932, she wrote to the powerful chair of the Commission, William Ezra Matthews: '[The poplars] have ruined all the grass and now are ruining the garden and keeping my house damp and cold.'[12] Matthews blamed the tree growing to the north of her house, and offered to meet with her.[13] She was apparently not mollified, and rallied support from her neighbours, for a few months later Matthews received a petition from 14 residents of Rosebery Avenue demanding that the row of poplars be thinned: 'Owing to their great height and close planting, they do much towards shutting out the light and free circulation of any breeze, and in addition, stand rather in the nature of a menace to the houses over which they tower on windy days.'[14]

Matthews dismissed their concerns, but only a few years later the Commission was obliged to send a crew in to repair the damage from a falling tree in McKibbon's garden.[15] She thanked him, writing, 'Would you kindly convey my thanks to Mr. Stuart, Mr. Bell and the rest of the men who so promptly repaired the damage done my garden recently', and concluding pointedly: 'One more request, would you please see that something is done to the rest of the poplar trees in the vicinity of my property? Funerals [underlined] and house repairs [underlined] are costly things.'[16] The following year she wrote again to complain about the poplars, once in August, and then again in November, addressing her letter to the new chair of the Commission, Frederick Erskine Bronson: 'After past summer's incident, I am so terrified every time there is a windstorm. You promised to have something done to them as soon as the new Driveway was finished. So please do something before the frost gets into those trees for they get so brittle and being so old are always a menace.'[17] He promised to send someone out.

Figure 11.2 Ottawa's Central Park looking east, 1920s. Lombardy poplar trees define the edges of Central Park. The towering presence of the fast-growing poplars could be oppressive for the neighbours of the park.

Source: Library and Archives Canada, MIKAN no. 3318899.

The trees were topped, and the record is silent for six years. But a blueprint in the archives, suggesting locations in the park for war-time housing, indicates that the poplar trees were 75 feet high in 1942 and we find Mabel writing once more in 1943 demanding the trees be trimmed.[18] The war provided a ready excuse for the secretary of the Commission who replied:

> The Commission will look into the matter and see if it will be possible to again cut off the upper 20 to 30 feet of these trees. The Commission's skilled men in these matters were drafted some time ago for more essential duties and unless it is possible to secure some capable substitutes it may prove necessary to defer this work until after the war.[19]

His response is one of the few references in the archival records to the skill of the men who managed urban trees. In September of 1945, the war now ended, McKibbon wrote again with the familiar refrain:

> Now the trees at the rear of my property on Rosebery Ave are still towering 20 feet higher than the house shutting out the sun, ruining my gardens and on

> a windy day a danger. For if ever they fall (are so brittle) [sic] would certainly squash my house. Their height keeps my house dark and this time of year have to have lights on [sic]. We bought and built here long before the trees were here just for the sake of having lots of sun for a garden. But it has all been ruined and it doesn't seem fair that these trees are allowed to grow so high and good reason to believe that their roots have grown in my cellar and if any damage happens from them there will hold past Commission accountable ... Hope something will be done soon.[20]

They promised to look into it. A plaintive note in the file in December 1945, ('we are so shut in, just like an over grown hedge. Do wish you would do something about them') the last we hear from McKibbon, suggests that nothing had been done at that date.[21] But a 1946 aerial photograph reveals that she may have eventually won her war with the Commission and its trees. The photograph shows a ring of massive poplars ringing the western end of the park.[22] There is a conspicuous gap to the west of 32 Rosebery Avenue. It is possible that the Commission decided to remove one tree to let some afternoon light into Mabel's garden.

It was, at any rate, the return of the Commission's tree trimmers after the war that seems to have resolved the issue. Arborists appear only occasionally in the historical record, but it is they who contend with unruly trees; indeed, arboriculture might be defined as the practice of managing and hiding arboreal unruliness. As Paul Montpellier, Vancouver's city arborist, said: 'With trees you have to maintain them for people to see them as an amenity, so that they see them as a good thing . . . trees can be a huge pain in the ass' (Braverman, 2008).

Why tell stories of troublesome trees? Why focus on the Manitoba maple, the Rosybloom crab apple, and the Lombardy poplar when there are such majestic trees as the chestnut, the silver maple, and the American elm? And why stress unruliness over such admirable themes as service, power, and memory? These counter-narratives point to something—whether we understand it as wildness, alterity, or agency—that has disappeared in traditional narratives. Whether it is the Manitoba maple invading the slopes of Parliament Hill, the ornamental crab apple littering the sidewalks, or the Lombardy poplar threatening adjacent houses, the unruly tree is a tree with a purpose of its own.

There are two reasons that we need such trees in our urban environments, and that we need stories about such trees. I will start with the pragmatic reason: those who love trees need to come to terms with the Mabel McKibbons among us. Irus Braverman reminds us that many urban residents dislike or fear trees; she quotes Garry Onysco, a tree inspector with 16 years' experience in Vancouver, who points out: 'half the people love trees and half the people either hate trees, or don't care' (Braverman, 2008: 99). Braverman argues that many of tree haters are immigrants or minorities or working class. She argues that their views are silenced and even stigmatized by those urban forest advocates who have come to see the urban forest as a moral as well as an instrumental good that should be encouraged, even imposed, upon the wider population. Other researchers who have entertained the possibility of arboriphobia have also found significant portions of the population that do not love trees, and do not seem inclined to change their attitudes (Fraser and Kenney, 2000; Kirkpatrick et al., 2012, 2013a, 2013b). Braverman reverses the environmental justice analysis to argue that the absence of trees among African-Americans and other low-income groups is not necessarily an environmental injustice, for such groups may not want the expense of maintaining trees. Heynen has pointed out that the canopy cover in low-income neighbourhoods in Milwaukee is often made up of weed trees, like the Manitoba maples in Ottawa, that are liabilities rather than assets (Heynen, 2006). Braverman points finally to the coercive power of arboreal regulations; she emphasizes the regulation of marginal and criminal populations, but the story of Mabel McKibbon reminds us that even the middle class are subject to this power. Telling the story of the troublesome tree acknowledges the tensions inherent in sharing space with other living entities, like trees, and the social conflicts that arise around urban forest management.

There is also a more philosophical reason for thinking about the unruly tree. Posthumanists like Cary Wolfe are calling for us to 'rethink our taken for granted human modes of experience including the normal perceptual modes and affective states of *Homo sapiens* itself, by recontextualizing them in terms of the entire sensorium of other living beings' (Wolfe, 2010). His argument prompts us to acknowledge the existence of the arboreal sensoriums that share our urban spaces. Others, like Michael Marder, are pressing further. In *Plant Thinking: A Philosophy of Vegetal Life* (2013) Marder calls into question our understanding of thought, to suggest a non-cognitive, non-imagistic, and non-ideational mode of thinking practised by plants (and also, he suggests, by humans). Science offers less speculative routes past the limitations of the human sensorium. In *Tree: A Life Story* (2004) David Suzuki and Wayne Grady tell the story of a Douglas fir over 550 years. They use science to understand the subterranean and ecologically interconnected existence of the fir (*Abies sp.*), and their book points, in its profoundly tree-centred perspective and emphasis on the living materiality of the fir, towards other ways of seeing the tree. Their work comes closest to representing the dense physicality of my childhood chestnut.

The narratives told here of the Manitoba maple, the ornamental crab apple, and the Lombardy poplar are still anthropocentric: it is only from our perspective that trees cause trouble. From their perspective they are simply pursuing their own purposes, finding water in drainage systems, spreading a canopy to capture the sun, and dropping their progeny on our sidewalks. A recognition of the trouble that they cause is, however, a first step in recognizing that other purposes, besides the human ones, exist, even in the modern city.

Notes

1 William Saunders, 'Forest Lectures', Saunders Family Papers, vol. 4393, Library and Archives Canada (LAC), Record Group (RG) 17, vol. 13.

2 Fernow dedicated *The Care of Trees* to William Saunders, 'a venerable friend . . . who has devoted a lifetime in advancing knowledge of tree growth in Canada, both in its useful and ornamental aspects'. Fernow makes reference to Saunders' *Catalogue of Trees*

and Shrubs (1899), and uses his findings for hardiness in his extensive 'List of Trees for Shade and Ornament' (p. 200).

3 William Saunders, 'Forest Lectures', Saunders Family Papers, vol. 4393, LAC, RG 17, vol. 13.

4 The i-Tree software was developed by the USDA Forest Service. See 'What is i-Tree', www.itreetools.org (accessed 5 November 2013). For a Canadian application, see the report on Toronto's urban forest, *Every Tree Counts: A Portrait of Toronto's Urban Forest*, (Toronto: City of Toronto Parks, n.d.) available on the i-Tree site. The urban forestry organization LEAF (Local Enhancement and Appreciation of Forests) in Toronto, has created an Ontario Residential Tree Benefits Calculator for visitors at www.yourleaf.org/estimator (accessed 5 November 2013). Similar estimators have been created in the United States by large arboriculture firms, Casey Trees and Davey Tree Expert Company.

5 Erik Jorgensen taught at the University of Toronto's Faculty of Forestry. His role in the genesis of the term urban forest is described in Dean (2008).

6 Support Anne Frank Tree Foundation website at www.support-annefranktree.nl/node/206 (accessed 4 November 2013). For the virtual tree, see the Anne Frank Tree website at www.annefrank.org/annefranktree (accessed 4 November 2013).

7 'Rehabilitation of North Slopes Parliament Hill, Ottawa, Vegetation and Soil Considerations, Phase 1'. Report to Public Works Canada by Ecological Services for Planning, Ltd., October 1980, p. 8.

8 See Dean Beeby, 'Feds Scrap Parliament Hill tree project, citing expense', *The Star*, Thursday October 20, 2011, www.thestar.com/news/canada/2011/10/20/feds_scrap_parliament_hill_tree_project_citing_expense.html (accessed 2 November 2013).

9 The Almey cultivar was originally named Morden Rosybloom 452, and subsequently named for the Canadian Pacific Railway horticulturalist, J. R. Almey.

10 National Capital Area Program, Beautification of Road Approaches, p. 2. LAC, RG 69, vol. 508, Centennial Commission.

11 Trevor Cole, 'Growing Crabapples', *Canadian Gardening*, undated, www.canadiangardening.com/gardens/fruit-and-vegetable-gardening/growing-crabapples/a/1407 (accessed 1 November 2013).

12 Mrs M. McKibbon, 32 Rosebery Avenue to W. E. Matthews, Chair of the Ottawa Improvement Commission (OIC), July 15, 1932. LAC, RG 34, Vol. 233, File 115 (1,2) Central Park.

13 W. E. Matthews to M. McKibbon, July 23, 1932. Matthews was the president of Matthews and Laing Meat Packing Company. He had lived a few blocks from Rosebery Avenue, having invested in land when the Commission constructed the neighbouring Clemow Avenue. He lived at 221 Clemow Avenue until the early 1920s.

14 Residents of 35, 37, 43, 22, 24, 37, 39, 51, 53, 59, 47, 45, 32, 36 Rosebery Avenue to W. E. Matthews, November 5, 1932. LAC RG 34, Vol. 233, File 115 (1,2) Central Park.

15 W. E. Matthews to F. M. Journeaux, Nov 19, 1932. LAC, RG 34, Vol. 233, File 115 (1,2) Central Park. 'We have received a communication signed by 14 residents of Rosebery regarding the trees on the north side of Central Park. We do not know who sent the petition but knowing you personally desire to say that when Spring comes we will go in to the matter thoroughly but in the meantime the trees will not do any harm.'

16 M. McKibbon, 32 Rosebery Avenue to W. E. Matthews, July 16, 1932. LAC, RG 34, Vol. 233, File 115 (1,2) Central Park. A reply from the Commission secretary on July 17, 1932 acknowledged receipt of her letter, 'expressing appreciation for the attention recently given by the Commission to repairing damage caused to your garden by the falling of a tree'. The response pointedly did not address her demands.

17 M. McKibbon, 32 Rosebery Avenue, to F. E. D. Bronson, Chair of the OIC, August 31, 1936, and November 23, 1936. LAC, RG 34, Vol. 233, File 115 (1,2) Central Park. A pencilled notation on her irate August letter (she notes that the previous year the tree had missed her house by a few feet) indicates that she is Mrs G. W. McKibbon. Her

November letter opens: 'The winter will soon be here, rendering those trees in Central Park at the bottom of my property more dangerous than ever.' The Commission secretary replies, on November 25, 1936, to say that arrangements have been made to trim the tree.

18 M. McKibbon to F. E. D. Bronson, November 1, 1943. LAC, RG 34, Vol. 233, File 115 (1,2) Central Park. The blueprint—for proposed military buildings—is dated September 5, 1942 and is in the same file.

19 H. R. Cram, Secretary of the OIC, to M. McKibbon, November 22, 1943. LAC, RG 34, Vol. 233, File 115 (1,2) Central Park.

20 M. McKibbon to F. E. D. Bronson, Chair of the OIC, September 24, 1945. LAC, RG 34, Vol. 233, File 115 (1,2) Central Park.

21 M. McKibbon to F. E. D. Bronson, December 7, 1945. LAC, RG 34, Vol. 233, File 115 (1,2) Central Park.

22 National Air Photo Library Reel A10371, Number 208, 1946.

References

Aird, P. (2005) *Forestry Chronicle*, 81(4): 593.

Braverman, I. (2008) 'Everybody Loves Trees: Policing American Cities Through Street Trees' *Duke Environmental Law and Policy Forum*, 19: 81–118.

Campanella, T. J. (2003) *Republic of Shade: New England and the American Elm*, Yale University Press, New Haven, CO.

Crawford, P. (1987) 'The Horticultural Odyssey of Isabella Preston', *Canadian Horticultural History, an Interdisciplinary Journal*, 1(3).

Dean, J. (2004) 'Ottawa's Central Park: Esthetic Forestry vs Ornamental Gardens', *News of Forest History*, 36–37, Proceedings of the International IUFRO Conference 'Woodlands–Cultural Heritage' 3–5 May 2004, Vienna, Austria, pp. 21–30.

Dean, J. (2005) '"Said tree is a veritable nuisance": Ottawa's Street Trees, 1869–1939', *Urban History Review: Revue d'histoire urbaine*, XXXIV(1): 46–57.

Dean, J. (2008) 'Seeing Trees, Thinking Forests: Urban Forestry at the University of Toronto in the 1960s', in *Method and Meaning in Canadian Environmental History*, A. MacEachern and W. Turkel (eds) Thomson Nelson, Toronto, ON.

Dean, J. (2011) 'The Social Production of a Canadian Urban Forest', in *Environmental and Social Justice in the City: Historical Perspectives*, R. Rodger and G. Massard-Guilbaud (eds) White Horse Press, Isle of Harris.

Every Tree Counts: A Portrait of Toronto's Urban Forest, (Toronto: City of Toronto Parks, n.d.) www.itreetools.org/resources/reports/Toronto_Every_Tree_Counts.pdf (accessed 16 June 2014).

Fernow, B.E. (1910) *The Care of Trees in Lawn, Street and Park, with a List of Trees and Shrubs for Decorative Use*, Henry Holt and Company, New York, NY.

Frank, A. (2003) *The Revised Critical Edition of The Diary of Anne Frank*, David Barnouw and Gerrold van der Stroom (eds) Doubleday, New York, NY.

Fraser, E. D. and Kenney, A. (2000) 'Cultural Background and Landscape History as Factors Affecting Perceptions', *Journal of Arboriculture*, 26: 106–112.

Heynen, N. (2003) 'The Scalar Production of Injustice within the Urban Forest', *Antipode*, 35(5): 980–998.

Heynen, N. (2006) 'Green Urban Political Ecologies: Toward a Better Understanding of Inner City Environmental Change', *Environment and Planning A*, 38(3): 499–516.

Heynen, N. and Lindsay, G. (2003) 'Correlates of Urban Forest Canopy Cover: Implications for Local Public Works', *Public Works Management and Policy*, 8(1): 33–47.

Heynen, N. and Perkins, H. (2005) 'Scalar Dialectics in Green: Urban Private Property and the Contradictions of the Neo-liberalization of Nature', *Capitalism, Nature, Socialism*, 16(1): 99–113.

Jacobs, A. B., MacDonald, E., and Rofé, Y. (2002) *The Boulevard Book: History, Evolution, Design of Multiway Boulevards*, MIT Press, Cambridge, MA.

Kirkpatrick, J. D., Davison A., and Daniels, G. D. (2012) 'Resident Attitudes towards Trees Influence the Planting and Removal of Different Types of Trees in Eastern Australian Cities', *Landscape and Urban Planning*, 107: 147–158.

Kirkpatrick, J. D., Davison A., and Daniels, G. D. (2013a) 'How Tree Professionals View Trees and Conflicts about Trees in Australia's Urban Forest', *Landscape and Urban Planning*, 119: 124–130.

Kirkpatrick, J. D., Davison, A., and Daniels, G. D. (2013b) 'Sinners, Scapegoats or Fashion Victims', *Geoforum*, 48: 165–176.

Kovacs, K., Haight, R. G., Jung, S., Locke, D. H., and O'Neil-Dunne, J. (2013) 'The Marginal Cost of Carbon Abatement from Planting Street Trees in New York City', *Ecological Economics*, 95(1): 1–10.

Lawrence, H. W. (2006) *City Trees: A Historical Geography from the Renaissance through the Nineteenth Century*, University of Virginia Press, Charlottesville, VI.

Marder, M. (2013) 'What is Plant-Thinking', *Klesis: Revue Philosophique*, 25: 124–143.

Pham, T.-T.-H., Apparicio, P., Séguin, A.-M., and Gagon, M. (2011) 'Mapping the Greenscape and Environmental Equity in Montreal: An Application of Remote Sensing and GIS', *Lecture Notes in Geoinformation and Cartography—Mapping Environmental Issues in the City: Arts and Cartography Cross Perspectives*, S. Caquard, L. Vaughan, and W. Cartwright (eds) Springer, pp. 30–48.

Preston, I. (1944) 'Rosybloom Crabapples for Northern Gardens', *The Journal of the New York Botanical Garden*, 45(536): 169–192.

Pugh, S. (1988) *Garden, Nature, Language*, Manchester University Press, Manchester, UK.

'Rehabilitation of North Slopes Parliament Hill, Ottawa, Vegetation and Soil Considerations, Phase 1'. Report to Public Works Canada by Ecological Services for Planning, Ltd, October 1980: 8.

Saunders, W. and Macoun W. T (1899) Catalogue of Trees and Shrubs, Bulletin 2. Central Experimental Farm, Ottawa, ON.

Suzuki, D. and Grady, W. (2004) *Tree: A Life Story*, Greystone Books, Vancouver, BC.

Talarchuk, G. M. (1997) 'The Urban Forest of New Orleans: An Exploratory Analysis of Relationships', *Urban Geography*, 18(6): 65–86.

Von Baeyer, E. (1990) 'Isabella Preston', in *Despite the Odds: Essays on Canadian Women and Science*, M. Ainley (ed), Vehicule Press, Montreal, QC.

Walton, J. T., Nowak, D. J., and Greenfeld, E, J. (2008) 'Assessing Urban Forest Canopy Cover', *Arboriculture and Urban Forestry*, 34(6): 334–40.

Wolfe, C. (2010) *What is Posthumanism?* University of Minnesota, Minneapolis, MI.

Wood, G. D. (2000) 'A Most Dangerous Tree: The Lombardy Poplar in Landscape Gardening', *The Arnoldia*, 54(1): 24–30.

12 Seeking Citizenship

The Norway Maple (*Acer platanoides*) in Canada

Brendon M. H. Larson

Introduction

I stood transfixed by the chiaroscuro of leaves and sky-gaps overhead. Deep in Peter's Woods, about an hour east of Toronto, Ontario, I felt like I had a glimpse of how this part of southern Canada used to be. The sugar maples (*Acer saccharum*) towering overhead had been here for centuries and some were nearly a metre in diameter, huge for this region. I was touring the province in the summer of 1998 documenting the province's heritage woodlands, the best we had to show after a long history of despoliation. Peter's Woods was exemplary, regaling all the features expected of an old-growth forest: hillocks where large sugar maples had once stood, now toppled by windstorms; scarlet tanagers calling from high overhead in the branches; parasitic ichneumon wasps drilling their ovipositors into the wood of fallen logs to lay their eggs on horntail larvae deep within; and diverse herbs—trilliums, jack-in-the-pulpits, hepaticas—thriving in the rich soil and deep shade. I felt blessed to partake of this quintessential experience of being a Canadian: the grandeur of a native sugar maple forest (Figure 12.1).

When people around the world think of Canada, there are certain things that come to mind: snow, igloos, hockey, and perhaps Canadians' endearing tendency to apologize for faults they didn't even commit. Yet for Canadians, the ultimate symbols of our shared identity revolve around the leaf of the sugar maple. What could be more Canadian? Sugar maple trees exude that sweet elixir, maple syrup, which is highlighted by 'Canadian Maple' and 'Maple Dip' donuts in our storied Tim Horton's donut-and-coffee chain. A sugar maple leaf is the centrepiece of our easily

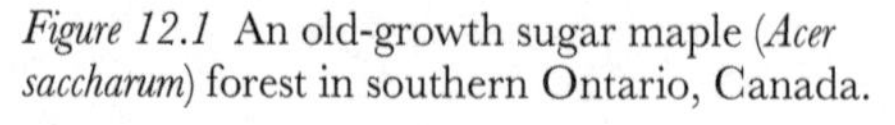

Figure 12.1 An old-growth sugar maple (*Acer saccharum*) forest in southern Ontario, Canada.

Source: B. Larson.

recognized red and white flag, unfurled by politicians across the country and until recently worn with pride by backpackers (including American ones) around the world to exemplify our 'cool' identity (e.g. Carstairs, 2006). A similar leaf is emblazoned on the jersey of our national hockey teams and on the bottles of a popular Canadian beer, Molson Canadian.

Imagine Canadians' surprise, then, when an interloper appeared. The front of the disputed new polymer Canadian $20 banknote sports a single, large maple leaf, further entrenching maple in our collective consciousness (Figure 12.2). Unfortunately, it's the wrong maple (Mahoney, 2013). Rather than one of Canada's ten native maple species, the bigleaf maple (*Acer macrophyllum*) or vine maple (*Acer circinatum*) of extreme southwestern British Columbia, the Manitoba maple (*Acer negundo*) of the prairies (the only Canadian maple with multi-part leaves, which look like an ash), or the aforementioned sugar maple of eastern Canada, it's a Norway maple (*Acer platanoides*).[1] Our symbol was under threat, not least because Norway maple is not only a species of urban areas but also one of those pesky non-native, invasive species. As one American blogger put it, 'The problem is that maple leaves are Canada's *thing*. Like how some nations' thing is communism, or being the world economic leader, or producing generation after generation of beautiful super-models. Canada's thing is that there are leaves there' (Weaver, 2013).

The issue catapulted into the media. The senior botanist at the Atlantic Canada Conservation Data Centre led the charge: 'It's a species that's invasive in eastern Canada and is displacing some of our native species, and it's probably not an appropriate species to be putting on our native currency' (UPI, 2013). Comments on blogs shared dismay that the Canadian government could commit such a gaffe (though others asked, 'who cares?'), one soon propagated in millions of pockets

Figure 12.2 Canada's $20 polymer banknote, which was released November 7, 2012. The actual banknote is pale greenish in colour. The single large leaf in the upper left corner of the banknote is thought to be a Norway maple (*Acer platanoides*); the band of leaves on the right side are more typical of 'Canadian' maple leaves (e.g. the one on the flag).

Source: This image is used with the permission of the Bank of Canada.

across the country like the inexorable spread of so many Norway maple seedlings.

In this chapter, I wish to examine this discussion through a political ecology lens. There is a history—both ecological and human—to how the sugar maple became Canadians' national symbol, just as there is a history to how Norway maple has been evaluated—and both have political dimensions.[2] In particular, the dichotomy between the two species rests on a distinction between native and non-native species, which itself depends to a large extent on that undying imbroglio of much environmental thought, the idea of wilderness. The idea of wilderness only makes sense relationally, as an ideal contrasted with urban places where people are markedly present. Thus, we are faced with the irony of considering sugar maple more 'Canadian' than Norway maple even though the latter species is more urban and thus occurs in greater proximity to where Canadians live. In this light, it would be striking if a species like Norway maple can never quite belong in Canada, even though it was probably here before Canada became a country, we introduced it intentionally, and then we dispersed it widely to serve an important role in urban areas. In fact, as I detail below, some people give great value to urban Norway maple trees. After examining these ideas, I will argue for granting Canadian citizenship to Norway maple; it is unable to speak, but by its own agency—in establishing, spreading, taking its place in our neighbourhoods, and now making it onto our currency—it clearly merits (if not demands) such citizenship.

A Brief History of Sugar Maple and Norway Maple in Canada

Why is sugar maple considered representative of Canada? The maple leaf has been symbolically associated with Canada in war and peace since the mid-nineteenth century, yet its significance rose dramatically when it became the centre of the national flag of Canada on 15 February 1965 (followed by its official proclamation as national arboreal emblem on 25 April 1996 (Canadian Heritage, 2011)). Even setting aside the question of the overall design and colours of the flag, the story of how and why the maple leaf was selected for the flag is a complex weave of history, landscape, and politics that led to the highly contentious Great Flag Debate of 1964 and culminated with Prime Minister Lester Pearson's eloquent argument for its 'heraldic propriety' against John Diefenbaker's atypically lax response (Matheson, 1980). Until this time, Canadians had either flown the Union Jack (the flag of the Commonwealth) or the 'Red Ensign' (which had a small Union Jack in its corner) depending on context, but the search was on for a single flag that was distinctively Canadian. One must keep in mind the critical importance of such national symbolism for a young country, with one commentator rejecting the beaver, for example, because 'I would like an animal that gets his nose off the ground a little farther' (Matheson, 1980: 116), and another questioning the choice of the red autumnal colour of maple leaves (as opposed to vibrant green) because it would symbolize 'decline, decay and approaching death' (Matheson, 1980: 19).

The Prime Minister argued that 'The Maple Leaf itself has been accepted as a Canadian symbol since long before Confederation. It is deep in our history and in our traditions. Contemporary records all through the nineteenth century are full of references to it' (Matheson, 1980: 83). For example, during the Prince of Wales' visit to the country in 1860, 'all citizens of Upper Canada were asked to wear sprigs of maple leaves as the emblem of their land so that . . . they would be known to the world as Canadians' (Matheson, 1980: 83). He also detailed its emblematic importance for Canadians in times of war, not least that 'the official badges of the three armed services all include maple leaves and have for a hundred years' (Matheson, 1980: 83). It appears that 'in practical terms it was [World War I] itself that fixed in the public mind that Canada's badge was the maple leaf' (Matheson, 1980: 7), so shortly thereafter, in 1921, it became the Arms of Canada by Royal Proclamation.

It is worth noting that Pearson was promoting the emblem of three conjoined maple leaves (and a different flag design overall) because it was in use in most of the contexts he mentioned and had the stamp of approval of British royalty going back to 1868. The final design of the flag, in red and white and with a single maple leaf, was ultimately inspired by the flag of the Royal Military College and the recognition that a single leaf was both parsimonious and endorsed by many recent uses, perhaps most exemplary and poignant being 'the single maple leaf carved in stone at the head of each overseas grave for Canada's fallen dead [in World War II]' (Matheson, 1980: 124).

The single leaf also had the advantage of symbolizing unity, whereas three leaves would beg the question of what each one meant (perhaps the Trinity, the three branches of the military, or something else). The political context here was discontent in Quebec, so Pearson offered 'a healing image, the maple leaf, living evidence of the land universally beloved. The symbol spoke to all Canadians, not of empire nor of powers or principalities, but of soil' (Matheson, 1980: 138). Some considered this a concession to French-speaking Canada, though this view was perhaps marked by unfortunate 'anti-French Canadian prejudices' (Matheson, 1980: 89).

Matheson reports that he ultimately chose sugar maple because 'not only did it have a handsome leaf but also this tree had been familiar to the Indians, the Habitants, and the Upper Empire Loyalists, for whom it had produced furniture, food and fuel. Most important, the leaf was visually familiar to all Canadians' (Matheson, 1980: 177). To the extent that the maple leaves used for prior Canadian symbols resembled sugar maple, this may have been true, but one could hardly justify selecting sugar maple based on its distribution. Sugar maple is restricted to eastern Canada, occurring south of 50° latitude in southern Ontario (with a tiny pocket in northwestern Ontario), southern Quebec, and the Maritime provinces. It is found much more widely in the United States, where it is the state tree of four states (New York, Vermont, West Virginia, and Wisconsin; Wikipedia 2013). Yet its selection does reflect the dominance of central Canadian interests in the country's politics (at least until recently) and its distribution coincides with where the majority of Canadians live, including not only two of our three largest metropolitan areas, Toronto and Montreal, but the Canadian capital, Ottawa. Undoubtedly, as any Canadian botanist would indicate, it is difficult to find a 'native' species of anything (except humans—and the species associated with them) that occurs from British Columbia's temperate rainforest east to Maritime krummholz, and from the desert of the Okanagan, shortgrass steppe of southern Saskatchewan and 'Carolinian' forest of southern Ontario north to the Artic Circle. Regardless, sugar maple doesn't really have much to do with anyone who lives anywhere in the northern or western 'hinterlands' or within Canada's urban areas.

In contrast, Norway maple is the most widespread of European maples, found across that continent, and was introduced to North America in the mid-eighteenth century. Rather than being a species of 'wild lands', the Norway maple is concentrated in urban areas. It has been widely planted as an urban street tree, especially after the loss of so many elm trees to Dutch elm disease in the 1970s. The Norway maple is popular for several reasons, including 'a vigorous early growth rate, desirable form and size, the capacity to withstand many urban impacts (e.g. pavement, moderate levels of pollution, dusts, and dry soils) and the abilities to transplant well, grow on a wide variety of soils, and withstand ice and snow damage better than other maples' (Nowak and Rowntree, 1990: 291). More recently, it has been promoted for the production of 'fast shade' and protection from UV radiation. Yet when the Norway maple escapes into nearby natural areas and parks, it tends to spread, and can eventually grow quite densely, contributing to the demise of other herbs and trees (see below). The concern about Norway maple is that it is found across more and more of the country; it is considered an invasive species because it is non-native, it spreads and it has ecological impacts and/or socio-economic costs (for example, to control it). In short, it appears that sugar maple is to nature and paper banknotes as Norway maple is to culture and plastic banknotes, as well as all the forces seeking to depredate Canadian nature.

Revisiting the Concepts of Non-Native and Invasive

Nonetheless, there are some uncomfortable ironies here in how we tend to think about non-native species. Species like Norway maple may sometimes impact human preferences, yet so did many of the native species that the settlers eliminated—or nearly—because they were in the way of their expansive spread a century or two ago (and which is ongoing). There is also a perverse logic here in the context of the history of indigenous peoples in North America. Sugar maple is considered Canadian because it was here before European settlement, but we have often been less respectful of the other beings who lived here then, the native peoples (and species). And if your response is to note how much damage invasive plants cause, keep in mind the warning words of many tribal elders when white settlers first appeared. Relative to the impact that European immigrants have wrought on the land, invasive species have had relatively minor effects.

And the whole idea of nativeness is fraught with challenges (e.g. Larson, 2007; Chew and Hamilton, 2011; Head, 2012). Species do not belong in any particular, static place: only 10,000 years ago (a blink in geological time), for example, there was no sugar maple anywhere in Canada because the land was under several kilometres of ice. Further, we are by no means consistent in our judgements about native and non-native species. Briefly consider how we look upon a much more prevalent non-native species of urban areas: Kentucky bluegrass (and related grass species) that we encourage to invade our lawns. As Michael Pollan explains in *Second Nature: A Gardener's Education*, the lawn 'looks sort of natural—it's green; it grows—but in fact it represents a subjugation of the forest as utter and complete as a parking lot. Every species is forcibly excluded from the landscape but one, and this is forbidden to grow longer than the owner's little finger. A lawn is nature under totalitarian rule' (Pollan, 1991: 48). Many of the species now considered invasive were once desired: witness the introduction of garlic mustard for salad and purple loosestrife for colour in a garden.

We admittedly accept some non-native species, as nearly any urban garden will reveal, but of course the problem people have with Norway maple—like other invasive species—is the problems it causes when it 'escapes' from a controlled urban context. It is widely planted in cities, so its wind-borne seeds allow it to disperse far into nearby woodlands, where its shade-tolerant seedlings can survive in the understory for long periods, sometimes growing to form the canopy. Norway maple tends to produce more leaves than sugar maple (but it less efficiently utilizes water and thus does less well in drier sites), its canopy leafs out early and can thus inhibit the growth of native species, and its seedlings out-compete native ones (e.g. Morrison and Mauck, 2007; Galbraith-Kent and Handel, 2008; Paquette et al., 2012). A report by Environment Canada in 1993 included it among a list of 32 'minor invasive aliens' of upland habitats—though not among the four 'principal' or five 'moderate' ones (White et al., 1993). It was mostly a problem in natural areas near urban areas, for example in ravines around Toronto.

More recently, when several Canadian and French ecologists compared the growth of Norway maple in Canada with its growth in Europe, they found that its density was quite similar on the two continents and that in Canada its density 'was not related to native tree species density.' It had also 'invaded only 9 percent of the forests sampled in southern Ontario, Canada.' They concluded that it 'is not aggressively invasive in southern Ontario' (Lamarque et al., 2012: 803). This is not to say that on a local scale it doesn't spread. It's also possible that, like other slow-growing trees, we're still in a 'lag' phase between the establishment of trees and the later expansive spread after the trees reach reproductive age (e.g. Wangen and Webster, 2006). Either way, it seems safe to say that the Norway maple is here to stay.

The Case for Norway Maple's Citizenship

If the Norway maple is here to stay (and perhaps even behaving) then we at least need to consider accepting its presence, effectively granting it citizenship.[3] An interesting thing to note about how biologists conceptualize non-native species, though, is that it appears they can never attain citizenship. As historian Peter Coates (2006) explains, two principles of human citizenship are pertinent here: citizenship based on one's place of birth (*jus soli*, 'right of soil') or on the citizenship of one's parents (*jus sanguinis*, 'right of blood'). By the former principle, a 'naturalized' non-native species should effectively become a Canadian citizen (after the first generation); however this is not the case, so even after multiple generations the latter principle cannot apply either. The species apparently retains bad blood, so it can never quite belong despite the important roles it plays (see below).

Yet ultimately it is people and societies that grant citizenship, and there are several indications that people at some level accept Norway maple. I would wager that before this news broke (and even now) most Canadians couldn't have distinguished Norway maple from sugar maple (Figure 12.3). The contrast between the shape of their leaves is certainly subtle (and quite variable) and thus it is not used as a definitive characteristic for the two species in Gleason and Cronquist's (1991) authoritative *Manual of Vascular Plants of Northeastern United States and Adjacent Canada* (nor in other manuals, e.g. Voss and Reznicek, 2012). The small print therein notes that the leaves of Norway maple are 'much like those of [sugar maple]' (1991: 352), though often having a few extra lobes. To distinguish them, it's better to turn to more categorical differences: Norway maple has petals (sugar maple does not), bigger samaras with widely diverging 'wings' (as opposed to parallel ones), and its petiole-stem contains a milky juice (whereas sugar maple's is clear), not to mention recognizably different bark.

For a related, technical reason, one could argue that the Norway maple is not a threat to Canadian symbolism because the design chosen for the leaf on the flag had more to do with physics than botanical detail. Matheson reports that there are 'some twenty-three' points on a sugar maple leaf, but the 'stylized design finally chosen has eleven points' (Matheson, 1980: 178). He chose this design because it was 'the best available model of the sugar maple leaf for display upon a flag surface under moderate and mean conditions of wind . . . It was selected after I had studied its performance under varying velocities in the National Research Laboratory Wind Tunnel (p.178).' This information accords with Canadian Heritage's view: 'It is the generic maple species that is being proclaimed as Canada's arboreal emblem . . . At least one of the ten species grows naturally in every province' (Canadian Heritage, 2011). So much for the botanical realism of the new $20 banknote, the

Figure 12.3 The contrasting leaves and samaras of sugar maple (two leaves to the left) and Norway maple (three leaves to the right).

Source: B. Larson.

implication being that an image that looks a little more like a Norway maple should not be interpreted as too much of a threat to our identity.

Norway maple may not have the same vibe as the statuesque sugar maple, yet there are more significant indications that at least some urbanites have accepted it. As part of a fundraiser a few years ago, a Toronto non-profit organization, Local Enhancement & Appreciation of Forests (LEAF), asked 16 'respected Torontonians' to select their favourite tree (LEAF, 2010). About two-thirds of them chose a non-native species. Not one chose the sugar maple, but two Canadian Broadcasting Corporation (CBC) radio hosts—Jian Ghomeshi (of CBC's *Q*) and Matt Galloway (of CBC Toronto's *Metro Morning*)—chose its European cousin, the Norway maple.

Ghomeshi describes a very intimate relationship with his Norway maple tree, perhaps even as his muse: 'This tree has been loyal to me. It has never failed to bring solace or perspective in the madness of my world. The living beauty and muscular body of this tree—MY tree—never tires. I have leaned on this tree, literally. Many times I have written down ideas and played my guitar in the shadow of this tree (LEAF, 2010). Elsewhere, he wrote, 'In the middle of a busy urban centre this tree has given me a getaway; a nod to life outside human-created concrete jungles . . . I cherish this tree' (Dempsey, 2010).

In contrast, Galloway lauds the value of a Norway maple for a local community: 'There are a lot of great trees in this city, but my first choice would be one that I've spent a lot of time under; the giant tree on the north side of Christie Pits, by the baseball diamond. Aesthetically, it's lovely, but it's also really important from a community perspective. Sundays in Christie Pits watching baseball wouldn't be the same without the tree to lounge under' (LEAF, 2010).

Isn't the problem here simply one of education? After all, these are celebrities rather than biologists. Yet even education campaigns would be challenged to undo a lifetime of experience with particular Norway maple trees. A large percentage of Canadians live in Toronto (and other cities in eastern Canada) and many of them have grown up experiencing Norway maple rather than sugar maple along local city streets and in parks (Figure 12.4). Norway maples have been widely planted there because of their environmental tolerance. According to the *York Region Urban Forest Study* (TRCA, 2012), about 4 percent of the canopy across the city of Richmond Hill, Ontario (near Toronto) is Norway maple, rising to 7 percent in the nearby city of Markham (vs about 13 percent for sugar maple in both regions). However, these statistics are a bit misleading because Norway maple is the most abundant species in residential areas, with sugar maple essentially absent there yet more common overall because it is prevalent in more marginal, agricultural and commercial-industrial sites (e.g. TRCA, 2013). Across the City of Toronto, in fact, Norway maple 'ranks first in leaf area coverage, providing 15 percent of citywide leaf area (152km^2)' (Millward and Sabir, 2011: 180).

In short, people grow to love what they experience. Canada is a multicultural country too, one composed of citizens from many different countries, so why shouldn't the same apply to tree species? Shouldn't we be more accepting of this diversity, of the possibility, for example, that some immigrant Europeans might feel more at home here in Canada with these Norway maples in their

Figure 12.4 Forty-year old Norway maple (*Acer platanoides*) trees along a residential street in Guelph, Ontario, Canada.

Source: B. Larson.

neighbourhood? The analogy between diverse peoples and diverse trees is a strained one, though one often made. In Golden Gate Park, San Francisco, conservationists proposed to remove all the non-native eucalyptus trees (from Australia), yet this proposal was ultimately rejected because so many people had come to not only associate them with the park, but to adopt them as a symbol of the California landscape (Coates, 2006). As a city supervisor argued, 'How many of us are "invasive exotics" who have taken root in the San Francisco soil, have thrived and flourished here, and now contribute to the wonderful mix that constitutes present-day San Francisco?' (Todd, 2002). Some scholars further argue that anti-nativism is essentially xenophobia (see Simberloff, 2003 for discussion; Katz, 2013). Regardless, Norway maple is entrenched on the city streets of many towns in eastern Canada, and to many people those little extra lobes may not be enough to erase the association with the Canadian nation.

It's not as if the banknotes had not gone through pre-testing. The Bank of Canada hired a market research firm to conduct focus groups in six Canadian cities in 2009 about potential themes for the new plastic banknotes. The focus groups identified 'diversity, inclusiveness, acceptance of others/multiculturalism' as key preferences. Their proposed images were then tested, some of the most popular being 'children of different ethnic backgrounds building a snowman; faces of individuals from different cultures celebrating Canada Day; an image of a hand of many colours; and children of different ethnic backgrounds playing hockey'. However, none of these was accepted, the bank instead choosing more traditional images that 'lack reference to Canada's diversity of ethnicity, culture and colour' (Beeby, 2013). This choice led to controversy in the case of the draft $100 banknote: later focus groups interpreted the image of a woman researcher on it to be Asian, thereby presenting 'a stereotype of Asians excelling in technology and/or the sciences', so the bank replaced the image with one of 'neutral ethnicity', leading to charges of racism and ongoing 'prejudice against visible minorities' (Beeby, 2012). Yet through all of this process, to my knowledge the focus groups did not raise issues about the Norway maple leaf: it appears to have been the only 'non-neutral' image that eluded the process.

More deeply, in a multicultural context the old association between Canada and wilderness may be wearing thin. European studies have shown, for example, that 'immigrants hardly ever visit non-urban green areas' (Buijs et al., 2009: 113), mainly because they perceive and relate to nature differently, that is, more functionally and with less emphasis on non-urban areas. Furthermore, historians have demolished the illusion that North America was a 'pristine wilderness' prior to the arrival of Europeans. In a classic analysis, William Cronon (1996) argues that these views are outdated, blithely purged of the indigenous peoples that were eliminated during European colonization, and in fact harmful for conservation. Sugar maple is in part a synecdoche for wilderness (or nature, more generally) at least relative to an urbanite like Norway maple, with conservation thus traditionally prioritizing the former.

Yet as Cronon points out, conservation of the future needs to engage with the places where we actually live and work and play (among many other authors reiterating and updating this argument, see Francis et al., 2012; Newman and Dale, 2013).[4] Urban forests are important, not least because the species there form an important connection to 'nature' for urban residents, almost regardless of the particular species in question. Like many other invasive species, Norway maples are to a large extent just good at associating with humans; for example, ecological studies have emphasized how dependent its spread is upon 'roads and trails' that 'facilitate longer distance dispersals than would be expected given the biology of the species' (Wangen and Webster, 2006: 258). This is just another way in which this is not a story of invasive species versus humans, but of invasive species intertwined with humans (see Robbins, 2004). The alternative view is antithetical to modern ecological and evolutionary science to the extent that it considers nature to be static and humans to be outside of nature. Instead, these 'successful' species are doing what species have always done and will always do: spreading themselves, and often as a consequence of choices we've made.

In addition to these questions of cultural preference and symbolism, there are also very good economic reasons to be more accepting of the Norway maple and its contribution to Canada. In short, these urban Norway maples play an important role in several ecosystem services, including improved air quality and carbon sequestration and reduced energy use. Their value, particularly when we also include experiences such as Ghomeshi's and Galloway's, is largely unquantifiable, though there have been attempts. Millward and Sabir (2011), for example, set out to quantify the value of trees in Allan Gardens, a 'historic public park in downtown Toronto' that has larger trees than is typical of Toronto. They found 309 trees of 43 species, which had annual benefits of $26,326 USD, but especially interesting was their finding that 'the flow of benefits from Allan Gardens' urban forest is heavily dependent upon Norway maple, a finding mirrored across much of Toronto's urban forest. Norway maple provides the greatest overall annual benefits ($4846 total; $113 per tree) and as a species contributes 17.5 percent of the environmental value and 20 percent of the aesthetic value provided by trees in the park' (2011: 177). As we've seen, it is widespread in many urban areas, so these benefits would be greatly compounded.

Conclusion

I have argued that Norway maple is now as Canadian as sugar maple, to the extent that a tree (or any animal, for that matter) can ever be considered to have a nationality. Species are not tied to nation-states; more radically, neither are they even truly tied to particular biogeographic locations on the earth's surface. Sugar maple's association with Canada is a temporary evolutionary-ecological stage because species never stay in one place for very long (on evolutionary or longer timelines), but flow with the tide of history and change, regardless of whether human opinions and values adapt apace. Is a 'Canadian' species one that is mostly found in the northeastern United States, but which is undoubtedly important in the constitution of eastern Canadian forests (sugar maple)? Or is it one that co-occurs with many Canadians in multicultural urban areas (Norway maple)? This is not ultimately an ecological or botanical issue, but a question for Canadians everywhere.

There is a good case to be made for the importance of both species to Canada's national symbolism. Norway maple could not reasonably replace sugar maple as a symbol of Canada (not least because it has not been planted in urban areas across the country), yet both fit within the bounds of the stylized leaf Canadians see on their flag. And the move to recognize the agency of Norway maple, and its place in the lives of many Canadians, could help to break down recalcitrant barriers between nature and culture.

There are several ways that this story could unfold. On the one hand, many urban Norway maple trees are old and prone to 'failure' in windstorms, and when they fall some people will advocate for replacing them with native alternative species. It also makes sense to be wary of planting Norway maple near local areas where people prefer extant forests. The Regional Municipality of York, for example, is developing recommendations for tree plantings along roads, where the conditions are harsh and thus only very hardy species survive. Silver Maple (*Acer saccharinum*)—but not sugar maple—is recommended as one of the top species, with Norway maple as a more minor contributor to future plantings, though with a caveat that it should not be planted near natural areas (Ian Buchanan, Regional Municipality of York, pers. comm.).

Yet there's another evolving dimension to this story, which gets quite interesting. By late summer, the leaves of many Norway maple trees in eastern Canada are covered by black 'tar spots'. One of the common explanations for the spread of non-native species in new regions, like the Norway maple in Ontario, is that they have escaped their 'natural enemies', which have been left behind in their native range. This reason has been given as an explanation for the success of Norway maple because it can grow more quickly than sugar maple, though recent ecological studies have shown that the two species experience similar rates of herbivory and disease (Morrison and Mauck, 2007). But the tar spot disease is a recent arrival from Norway maple populations in Europe. Thus, it is perhaps unsurprising that ecological studies in urban forest on Mount Royal, Montreal found that recent outbreaks of the disease led to a 'very sharp decline in sapling and tree growth, together with high mortality of Norway maple saplings' (Lapointe and Brisson, 2011: 63). This process may well, in time, shift the balance and reduce the invasiveness of Norway maple.

Yet future changes may bode even worse for sugar maple. With warming temperatures, the range of sugar maple will gradually shift further north. In fact, by about 2050, models predict that the climate in Polar Bear Provincial Park—created for the conservation of Polar Bears—will support sugar maple (McKenney et al., 2010), though of course the loamy soil it requires will not be there in the tundra. On the one hand, the shift of sugar maple's range further into Canada might make it a more appropriate representative of the country, but on the other hand, it might be unable to do so because of concurrent changes (including higher frequencies of fire) and might even become extirpated or at least less healthy in many regions. The predictions for the maple syrup industry are dire (Reynolds, 2010).

I am not sure if Norway maple will fare any better under climate change than sugar maple, though there is some evidence that it may be better able to take advantage of an extended fall growing season (Paquette et al., 2012). At least in the short term it nonetheless provides many urban people with a local maple tree and symbolically, a little piece of nature. Canada is not just about the landscapes and people that were here in the past, but about the mix of landscapes and people that are here now. We have modified most of the landscape now, so it is important to learn how to live in these places, whether urban or agricultural.

I look out at a Norway maple in my front yard in Guelph, Ontario every day, like most of my neighbours, and for good reason: the tree was widely planted in the 1960s because it does well in the urban environment. When I first arrived, a local plant nativist visited and told me to cut it down. It is undoubtedly difficult to grow most plants beneath it, and each spring we have to eliminate its dense offspring. But I couldn't cut it down: it's a good-size tree, not particularly beautiful, yet its leaves rustle in the breeze, it shades the house, and it has been here much longer than I have—and its species has occurred in Canada since before Confederation.

Notes

1. Canada's other maple species have leaves with lobes that appear superficially similar to a sugar maple (*Acer saccharum*), but occur in dissimilar habitats among other differences. Black (*Acer nigrum*), red (*Acer rubrum*), and silver maple (*Acer saccharinum*) are trees, and mountain (*Acer spicatum*) and striped maple (*Acer pensylvanicum*) are shrubs, which have a largely eastern distribution in Canada more or less like sugar maple. Douglas Maple (*Acer glabrum*) occurs through much of southern British Columbia, including Vancouver Island.
2. There is a burgeoning literature concerning the political ecology of invasion biology, which I do not have space to engage here. However, excellent case studies pertinent for thinking about urban trees and/or national identity include Foster and Sandberg (2004), Head and Muir (2004), Carruthers and Robin (2010), and Lien and Davison (2012).

3 Partly in the interest of space, I play relatively loosely with the personification of invasive species necessary to grant them citizenship. However, biologists' framing of them as 'invasive' species arguably set the precedent (e.g. Larson, 2008; though for varied discussion also see Simberloff, 2003, and references therein; Head and Muir, 2004; Coates, 2006).
4 Numerous ecologists have also come to this realization, e.g. Rosenzweig (2003).

Acknowledgements

I appreciate helpful comments from Anders Sandberg and Catriona Sandilands that helped to improve significantly an earlier version of this chapter.

References

Beeby, D. (2012) 'Bank of Canada bans image of Asian woman from $100 bill', http://business.financialpost.com/2012/08/17/bank-of-canada-bans-image-of-asian-woman-from-100-bill (accessed 24 July 2013).

Beeby, D. (2013) 'Gays, turban-wearing RCMP officers nixed as images on Canada's new bills: report', http://business.financialpost.com/2013/02/11/gays-turban-wearing-rcmp-officers-nixed-as-images-on-canadas-new-bills-report (accessed 24 July 2013).

Buijs, A. E., Elands, B. H. M. and Langers, F. (2009) 'No wilderness for immigrants: Cultural differences in images of nature and landscape preferences', *Landscape and Urban Planning*, 91: 113–123.

Canadian Heritage (2011) 'The maple tree', www.pch.gc.ca/pgm/ceem-cced/symbl/o2-eng.cfm (accessed 22 July 2013).

Carruthers, J. and Robin, L. (2010) 'Taxonomic imperialism in the battles for *Acacia*: Identity and science in South Africa and Australia', *Transactions of the Royal Society of South Africa*, 65: 48–64.

Carstairs, C. (2006) '"Roots" nationalism: Branding English Canada cool in the 1980s and 1990s', *Social History/Histoire Sociale*, 39: 235–255.

Chew, M. K. and Hamilton, A. L. (2011) 'The rise and fall of biotic nativeness: A historical perspective', in D. M. Richardson (ed) *Fifty Years of Invasion Ecology: The Legacy of Charles Elton*, Blackwell, Oxford.

Coates, P. (2006) *American Perceptions of Immigrant and Invasive Species: Strangers on the Land*, University of California Press, Berkeley, CA.

Cronon, W. (1996) 'The trouble with wilderness; or, getting back to the wrong nature', *Environmental History*, 1: 7–28.

Dempsey, A. (2010) 'Toronto's favourite leafy giants', www.thestar.com/news/gta/2010/07/13/torontos_favourite_leafy_giants.html (accessed 24 July 2013).

Foster, J. and Sandberg, L. A. (2004) 'Friends or foe? Invasive species and public green space in Toronto', *Geographical Review*, 94: 178–198.

Francis, R. A., Lorimer, J. and Raco, M. (2012) 'Urban ecosystems as "natural" homes for biogeographical boundary crossings', *Transactions of the Institute of British Geographers*, 37: 183–190.

Galbraith-Kent, S. L. and Handel, S. N. (2008) 'Invasive *Acer platanoides* inhibits native sapling growth in forest understorey communities', *Journal of Ecology*, 96: 293–302.

Gleason, H. A. and Cronquist, A. (1991) *Manual of Vascular Plants of Northeastern United States and Adjacent Canada*, Second Edition, New York Botanical Garden, New York City.

Head, L. (2012) 'Decentering 1788: Beyond biotic nativeness', *Geographical Research*, 50: 166–178.

Head, L. and Muir, P. (2004) 'Nativeness, invasiveness, and nation in Australian plants', *Geographical Review*, 94: 199–217.

Katz, E. (2013) 'The Nazi comparison in the debate over restoration: nativism and domination', *Environmental Values* (forthcoming).

Lamarque, L. J., Delzon, S., Sloan, M. H. and Lortie, C. J. (2012) 'Biogeographical contrasts to assess local and regional patterns of invasion: a case study with two reciprocally introduced exotic maple trees', *Ecography*, 35: 803–810.

Lapointe, M. and Brisson, J. (2011) 'Tar spot disease on Norway maple in North America: Quantifying the impacts of a reunion between an invasive tree species and its adventive natural enemy in an urban forest', *Ecoscience*, 18: 63–69.

Larson, B. M. H. (2007) 'Who's invading what? Systems thinking about invasive species', *Canadian Journal of Plant Sciences*, 87: 993–999.

Larson, B. M. H. (2008) 'Entangled biological, cultural, and linguistic origins of the war on invasive species', in R. Frank, R. Dirven, T. Ziemke and E. Bernárdez (eds) *Body, Language and Mind. Volume 2: Sociocultural Situatedness*, Mouton de Gruyter, New York City.

LEAF (2010) 'Toronto's favourite trees', www.yourleaf.org/torontos-favourite-trees (accessed 24 July 2013).

Lien, M. E. and Davison, A. (2012) 'Roots, rupture and remembrance: The Tasmanian lives of the Monterey Pine', *Journal of Material Culture*, 15: 233–253.

McKenney, D. W., Pedlar, J. H., Lawrence, K., Gray, P. A., Colombo, S. J. and Crins, W. J. (2010) *Current and Projected Future Climatic Conditions for Ecoregions and Selected Natural Heritage Areas in Ontario*, Queen's Printer for Ontario.

Mahoney, J. (2013) '$20 bill's maple leaf isn't Canadian, botanists say', www.theglobeandmail.com/news/national/20-bills-maple-leaf-isnt-canadian-botanists-say/article7519375/?cmpid=rss1 (accessed 26 July 2013).

Matheson, J. R. (1980) *Canada's Flag: A Search for a Country*, G. K. Hall and Co., Boston, MA.

Millward, A. A. and Sabir, S. (2011) 'Benefits of a forested urban park: What is the value of Allan Gardens to the city of Toronto, Canada?', *Landscape and Urban Planning*, 100: 177–188.

Morrison, J. A. and Mauck, K. (2007) 'Experimental field comparison of native and non-native maple seedlings: Natural enemies, ecophysiology, growth and survival', *Journal of Ecology*, 95: 1036–1049.

Newman, L. and Dale, A. (2013) 'Celebrating the mundane: Nature and the built environment', *Environmental Values*, 22: 401–413.

Nowak, D. J. and Rowntree, R. A. (1990) 'History and range of Norway Maple', *Journal of Arboriculture*, 16: 291–296.

Paquette, A., Fontaine, B., Berninger, F., Dubois, K., Lechowicz, M. J., Messier, C., Posada, J. M., Valladares, F. and Brisson, J. (2012) 'Norway maple displays greater seasonal growth and phenotypic plasticity to light than native sugar maple', *Tree Physiology*, 32: 1339–1347.

Pollan, M. (1991) *Second Nature: A Gardener's Education*, Grove Press, New York City.

Reynolds, J. (2010) 'Will maple syrup disappear?', *Canadian Geographic*, www.canadiangeographic.ca/magazine/oct10/discovery.asp (accessed 25 July 2013).

Robbins, P. (2004) 'Comparing invasive networks: Cultural and political biographies of invasive species', *Geographical Review*, 94: 139–156.

Rosenzweig, M. (2003) *Win-Win Ecology: How the Earth's Species Can Survive in the Midst of Human Enterprise*, Oxford University Press, Oxford.

Simberloff D. (2003) 'Confronting invasive species: a form of xenophobia?', *Biological Invasions*, 5: 179–192.

Todd, K. (2002) 'A new language is needed to win the day for native species', http://grist.org/article/correct (accessed 24 July 2013).

TRCA (2012) *York Region Urban Forest Study: Summary Report*, Toronto and Region Conservation Authority, Downsview, ON.

TRCA (2013) *Town of Markham Urban Forest Study: Technical Report*, Toronto and Region Conservation Authority, Downsview, ON.

UPI (2013) 'Canadian $20 bears Norwegian leaf', www.upi.com/Odd_News/2013/01/18/Canadian-20-bears-Norwegian-leaf/UPI-85071358536959 (accessed 22 July 2013).

Voss, E. G. and Reznicek, A. A. (2012) *Field Manual of Michigan Flora*, University of Michigan Press, Ann Arbor, MI.

Wangen, S. R. and Webster, C. R. (2006) 'Potential for multiple lag phases during biotic invasions: Reconstructing an invasion of the exotic tree *Acer platanoides*', *Journal of Applied Ecology*, 43: 258–268.

Weaver, C. (2013) 'Canada proudly prints wrong maple leaf on its currency because Canada can't do anything right', http://gawker.com/5978787 (accessed 26 July 2013).

White D. J., Haber, E. and Keddy, C. (1993) *Invasive Plants of Natural Habitats in Canada: An Integrated Review of Wetland and Upland Species and Legislation Governing their Control*, Environment Canada, Ottawa.

Wikipedia (2013) 'List of U.S. state trees', http://en.wikipedia.org/wiki/List_of_U.S._state_trees (accessed 22 July 2013).

13 Queering the Urban Forest

Invasions, Mutualisms, and Eco-Political Creativity with the Tree of Heaven (*Ailanthus altissima*)

Darren Patrick

Introduction

How would a tree build the city? Such a question requires us to think in a rather unconventional way about urban planning and urban ecologies, not to mention the actual or potential agency of non-human actors and forces in both of those areas. Trees do not apparently urbanize, at least not in any immediately recognizable way, unless you liken climax forests to the dense aggregations of skyward buildings found in the downtown cores of today's global cities. And so my question is more provocative than literal, especially because I ask it with a particular tree in mind: *Ailanthus altissima*.

Present on every continent except Antarctica, *A. altissima* is more commonly known as Tree of Heaven. While I use that name throughout this chapter, it is just one of many names that have proliferated before and after the tree's arrival in North America in 1784 (Shah, 1997). Other monikers include: 'ghetto palm,'[1] on account of its apparent affinity for economically impoverished areas; the 'cum blossom,'[2] for the distinctive smell emitted by flowering male trees in early summer; and 'the tree of hell,'[3] owing to its frequent demonization as an invasive, 'non-native,' unruly, and excessively successful species. Despite its negative or weedy reputation, it is important to note that Tree of Heaven is not only demonized. As Kowarik and Säumel (2007) point out, Tree of Heaven also has a long history of more affirmative and intentional uses including as traditional/folk Chinese medicine, as an ornamental tree in Europe and elsewhere, and as a buffer against the effects of erosion, to name a few notable examples (214–215).

Keeping this diversity of attitudes, uses, and material histories of Tree of Heaven in mind, in this chapter I take the tree as both an object of analysis and an opportunity for targeted intervention into some concepts and practices that circulate in urban political ecology and urban forestry. Through three empirical cases, I open some grounded conceptual questions regarding the ethics and politics of our entanglement with trees both in actual cities and in a broader context of global urbanization. Each case serves to open possibilities for critically examining practices and discourses of both academic scholarship in urban political ecology and in our everyday engagements with the urban forest, in a broad sense.

The first case interprets the eradication of Tree of Heaven from New York City's High Line in terms of invasiveness and gentrification. The second case takes

Detroit's Tree of Heaven Woodshop as an opportunity to discuss mutualism and alliance. The third case extends Stefan Tiron's and Vadim Tiganas' eco-political art project 'Ghetto Palm.'

Theoretical and Intellectual Context

My rather broad opening question can be situated in context of my current research project on queering urban ecologies.[4] As a critical and theoretical gesture, queering has roots in critiques and resistances to normative genders (male/female) and sexualities (hetero/homo) and the social, legal, and conceptual binaries that uphold and define them. Importantly, queer theory also critiques the dominance of reproduction and traditional family in mainstream visions of nature and culture alike. As in this chapter, claims to 'queering' in scholarship and theory have, for some time now, extended beyond human sexuality and gender. With this in mind, it is important to recall that 'queer,' as both a noun and verb, would have no meaning without lived experiences of marginality and oppression based on failure to conform to dominant modes of expressing and embodying gender and sexuality.

For me and many others, the power and provocation of queering in theory cannot be separated from ongoing risks and struggles of queer in practice. Queer most often functions as a disruption to the smooth functioning of oppressive and hierarchical discourses and practices. Nevertheless, its institutionalization *as theory* increasingly risks becoming a hegemonic force in its own right. This risk invites queers to acknowledge the roots of our theory and practice in critical thought and action emerging from lived experiences of sexism, racism, ableism, classism, and colonization. The variety of intersectional lived experiences makes concepts and practices of mutualism and alliance particularly important, even as theoretical 'queering' and 'queerwork' maintain their roots in critiques of sexuality and gender. So, whether through living indeterminacy and fluidity of gender and sexuality or by making demands for recognition and respect for queer lives and bodies, the gesture of queering speaks of the urgent necessity to create worlds which differ from received understandings of 'the way things should be.' This chapter is one space in which to imagine a queerer world.

Coming to ecology, this is perhaps why I am most interested in weedy or, following the groundbreaking work of Herbert Sukopp (2003; 2008a [2002]; 2008b [1973]), ruderal ecologies.[5] These are the fuzzy 'fragments, edges, and matrices' which queer geographer, artist, and landscape theorist Gordon Brent Ingram (2010) has so intriguingly interpreted in the context of Vancouver and its cruisey Stanley Park. I particularly appreciate Ingram's attention to ecological detail as he links homoeroticism to biophysical landscapes (261–263). In many respects, Ingram's queering of landscape ecology builds on earlier explorations (Bell and Valentine, 1995; Ingram et al., 1997; Browne et al., 2007; Knopp, 2007) of relationships between queer theory/studies and geography.

A more recent example of the mutual influence of urban ecology and geography is Matthew Gandy's (2012) investigation into queer ecologies in the context of London's Abney Park Cemetery. I claim kinship with Gandy's mode of investigation, especially his question, "What are the political implications of queering urban nature?" (2012: 737). In his effort to address this question, Gandy speaks to both urban ecologists and queer geographers interested in 'unruly spaces' (2012: 740–741). He seeks "possibilities for political constellations that might come into being as an alternative to the contemporary presence of utilitarian or historicist approaches to urban nature" but concludes that "At present . . . such instances are more coincidental than coordinated" (2012: 741–742). At its core, Gandy's intervention aims to intensify the "political dimensions of urban ecology" to see if they resonate with "the cultural and material complexities of urban space" (2012: 737).

Like Gandy, I am attracted to the idea of alliance (2012: 737–740), and, in the second case presented below, I experiment with linking human/non-human alliances to a biological concept of mutualism. My wager is that paying both political and ecological attention to intentional mutualisms between humans and non-humans might help spaces to become something other than 'waste,' 'brownfields,' or 'development parcels,' let alone the greenwashed urban renewal projects of tomorrow.

Here, it is important to emphasize, as Gandy (2013) does, that the geographic field of urban political ecology still has some distance to travel in terms of its engagement with the scientific insights into ecologies of urban wastelands. While this chapter does not enact an exhaustive integration of "key ecological insights into the distinctiveness of urban space" with "historically grounded studies of capitalist urbanization," which Gandy helpfully suggests as a trajectory for research, I do pay close attention to the interplay of metaphor, interpretation, and material reality in both urban and ecological modes (2013: 1304). In this sense, Tree of Heaven is one point of entry into thinking and writing the diverse mutualisms of these strands of thinking.

Collectively, much of the work cited above continues to inspire both queers *in* geography and the queering *of* geography. In this chapter, my desire to queer urban political ecology takes the form of asking us to think with and through Tree of Heaven as a 'queer figure' for both urban political ecology and urban forestry. On my reading, the tree's queerness has to do as much with its disruption of normal or mainstream modes of thinking and doing urban ecology and forestry as it does with some more fine-grained details related to its frequently 'unruly' character and capacities. Thus, in a mode related to, yet distinct from, Gandy's, this chapter might be read as an attempt to address some ecological dimensions of queerness with the help of a particular species of tree. By placing an emphasis on queerness as both method and material, I read and write with Tree of Heaven in order to (re)imagine both the actual histories and speculative futures of sites of intense urban ecological and spatial transformation.

Wherefore 'Queering the Urban Forest'?

Before diving into the cases, I'd like to offer four axes of significance for my gesture of queering the urban forest: poetic, philosophical, political, and ethical. My intention is to give greater definition to the multiple senses of 'queering' that operate throughout my analysis.

Poetically, the tree is a figure that embodies meaning and carries symbolic power. Its ways of being different and doing differently suggest a vocabulary which challenges normative, neoliberal, and heterosexualized urbanization and city building. The poetics should not lead to a preoccupation with the tree as a metaphor alone, though it can certainly be read as such quite productively (Adamson, 2012). Instead, it should suggest that the tree, like all material, ruptures our ability to understand the actual world in strictly mechanical, positivist, or managerial terms.

The philosophical dimension of this chapter takes a cue from Michael Marder (2011), who, drawing on philosophers Gilles Deleuze and Félix Guattari, points out that 'Plants are the weeds of metaphysics: devalued, unwanted in its carefully cultivated garden, yet growing in between classical metaphysical categories of the thing, the animal, the human . . . and quietly gaining the upper hand over that which is cherished, tamed, 'useful'" (2011: 487). So, in the spaces and moments in which we find a particular plant—whether demonized as a weed or exploited as a resource—we can be sure we have found some conceptually rich ground in which to start digging. On my reading, Tree of Heaven is a materially existing concept that often thrives in biologically and economically impoverished—but richly historical—ground.

The tree's globalization points to openings for thinking about the politics of Tree of Heaven on multiple scales. Here we might think about both the instrumental and non-instrumental ways in which Tree of Heaven travels and proliferates, as well as the ways in which it is eradicated and erased. On the one hand, human actions and decisions such as planting the tree create intentional and instrumental relationships. On the other hand, the tree's success as a weed is, at least in part, a result of its capacity for spreading into new territories by, for example, traveling on transportation networks. The latter raises questions of the unintentional or, from a human perspective, non-instrumental alliances that it forges with us. Indeed, the very disagreement around the tree's 'value' as a resource versus its 'threat' as an invasive or unwanted presence troubles ideas of intentionality. So, one aspect of the tree's political significance points to questions of its management for human ends.

A second political opening occurs on a micro-biological level. Tree of Heaven produces a substance called ailanthone, an allelopathic[6] secondary metabolite which, while "hav[ing] no apparent role in life processes or plant structure," (Heisey, 1997: 28) is found both to inhibit the growth of certain other plant species via distribution through the soil and to affect certain herbivorous insects which ingest the plant (Kowarik and Säumel, 2007: 227–228). As a biological capacity, allelopathy is one way in which the Tree of Heaven interacts with other beings as it establishes relationships in and to its environment. Because of its frequently inhibitory function, allelopathy is one justification for demonizing Tree of Heaven as an enemy/threat to biodiversity.

Further, paying attention to non-reproductive processes such as allelopathy encourages us to expand, even to queer, our understanding of the tree's successional presence by pushing beyond a narrow, and arguably heteronormative, framework of reproduction. I bring it up here to raise questions of the terms on which beings, human and otherwise, are themselves recognized as having a 'right' to inhabit, use, or proliferate in particular spaces, especially when their modes of interacting are deemed 'threatening' in the politically ambiguous milieu of urban biodiversity (Gandy, 2013: 1305). Whether on micro or macro scales, the tree's many entanglements become both politically and ethically sensible when we begin to ask questions about which territories it is 'permitted' to inhabit and on whose terms.

Ethically, as a writer, researcher, and urban dweller, I myself am entangled with the tree. It grows in my backyard. It is a co-producer of the literal and symbolic atmosphere of my immediate environment. It is charismatic (Lorimer, 2007) and implicated in waves of ecological succession that help define the post-industrial, 'wasted,' or excessive character of the spaces in which, for reasons both willful and unintentional, I often find myself conducting research. Drawing on Sarah Whatmore (2002) and Val Plumwood (1993; 2002) among others, Paul Cloke and Owain Jones (2003) point to the ways in which a grounded 'ethical mindfulness' in our encounters with trees "will represent a complex mix of 'flying by instinct' and the 'bringing-to-life' of wider ethical codes" by virtue of which "questions of how to act will depend on the context of the encounter" (2003: 201).

I take this call to wider ethical codes seriously insofar as it offers a way to challenge and even dismantle dominant ethical frameworks, including greenwashed versions of urban sustainability, which continue to foreground human reproduction and the reproduction of preferred non-humans, as modes 'protecting the environment.' One plant-related example involves the emphasis in park planning on so-called 'native' plants over 'non-native' or, worse, 'invasive' species, such as Tree of Heaven. These bio- and ecological discourses and the practices they inform are far from neutral. Indeed, they are marked by an insistent patriarchal and racialized nativity, nationalism, heteronormativity, and colonialism (see O'Brien, 2006 and Peretti, 1998). The ethical dimension of my inquiry concerns vegetal adaptations to anthropogenic environments, perhaps especially instances in which specific plants defy, exploit, or benefit from human schemes. I take such instances as an opportunity to pose an ethical question of how 'we,' as a human collectivity, might both live with and learn from 'their' vegetal insistence.

Tree of Heaven is a speculative link in a longer chain of queer and ecological theory and practice. This chapter deploys queer reading and writing of Tree of Heaven. A queer reading looks not only at how the tree reproduces but also at its dis/location, to the symbolic and material crisis of its overabundance, and to its space-making tactics. A queer writing of Tree of Heaven does not speak of success and failures in anything like absolute terms. Instead, it is a gesture of mutualist alliance with the tree. I write as a way of asking where it takes us and where we are taken if we follow it.

What kinds of spaces must we walk through to find it? Where are these spaces in the city? Who and what uses those spaces? Who and what disappears when Tree of Heaven disappears? The tree tells stories and reveals patterns of displacement, destruction, renewal, and dwelling. It helps us to locate concepts and practices that bring new creativity to bear in dealing with the violent displacements and volatile crises of urban capitalism. Lurking around freshly sprouting stands of Tree of

Heaven is just one way to ask how anything survives, let alone thrives, in the midst of those crises and displacements.

High Line: Invasion and Gentrification

Originally constructed between 1929 and 1934, New York City's High Line removed rail traffic from the streets of Manhattan's West Side, promoting both public safety and industrial efficiency for the factories served by those trains. By 1980, Manhattan was in the throes of deindustrialization; many of the factories served by the elevated rail line had been shuttered or abandoned. Some former industrial and commercial buildings had been converted into residential buildings and co-ops for 'pioneer' gentrifiers who, for reasons both economic and cultural, sought out edgier, less-established areas of the city in which to live (see Zukin, 1982). Still other spaces became opportunistic locations for artistic production as artists who had squatted and claimed abandoned lofts in SoHo, further downtown, started migrating northward. With its proximity to the West Side piers, which, by the 1980s were a well-established cruising zone and social space for frequently racialized and often homeless queer and trans* youth,[7] the High Line and surrounding area also served as a zone for queer sex and socialization, both essential conditions for survival.

The pace and scale of gentrification of downtown Manhattan neighborhoods has varied. Neil Smith (1996) identifies the West Village, which is the southernmost neighborhood through which the High Line runs, as having been gentrified by the 1950s (1996: 96). But this did not necessarily include a second-wave of gentrifiers—gay men and, to a lesser extent, lesbians—who not only began settling in the West Village but also in Chelsea, the neighborhood in which the greater portion of the High Line's 1.45-mile-long (2.33 km) stretch is concentrated. In the largely industrial West Chelsea, the blocks nearest the High Line would eventually become home to art galleries and museums attracted to plentiful space and able to take advantage of the commercial zoning with minimal variances.

The latest wave of gentrification in Chelsea, ushered in by the redevelopment of the High Line, has relied on a complex political and planning process which ensured the maintenance of the galleries while allowing property owners with parcels under the High Line to sell their development rights within newly rezoned areas in a development corridor along the High Line. The underlying zoning and planning aspects of the redevelopment leveraged the increasing attractiveness of the neighborhood to high-flying developers and aspiring (or actual) bourgeois urban dwellers, smoothing the path for their invasion of the primed neighborhoods around the soon-to-be-park.

And what about Tree of Heaven (Figure 13.1)? With persistence befitting a weed, the tree erupted into my archive as I was researching the history of the redevelopment of the High Line. Opened to the public in 2008, the self-consciously sustainable urban park, with its 'mostly native plant' landscape has received significant praise as it has diffused as a model for post-industrial urban greening. Urban park advocates, architects, city planners, and ordinary citizens alike often

Figure 13.1 *Ailanthus*. 30th Street and 11th Ave, Manhattan, NY, USA, 2002 (removed 2002).

Source: Image courtesy Benjamin Swett. Source: Swett, B. (2013) *New York City of Trees*, Quantuck Lane Press, New York City, NY.

hold up the High Line as a prime example of the revivifying potential of 'designed waste spaces.' Only more recently has the project begun to face popular criticism on account of its intensification of ongoing gentrification in the neighborhoods surrounding the park. In this context, Joshua David and Robert Hammond, the co-founders of Friends of the High Line, the organization that spearheaded the redevelopment effort, have deployed a number of different discourses to account for their apparent complicity—or at least their own ambivalence—in the hyper-gentrifying landscape which the park helped to consolidate.

David and Hammond, both white gay men, routinely discussed their attraction to the abandoned High Line in terms that blended its weedy character with its reputation as a gay cruising space. These two aspects of the site's aesthetic and material configuration prior to redevelopment would eventually become important precedents for the designed landscape. Among the seventy-five plants that grew on the abandoned High Line, Tree of Heaven stood as one of the most charismatic inhabitants (New York City Department of City Planning, 2005: Appendix C). Owing to its ability to thrive in nutrient-poor soils and its prolific capacity for growth—up to 6 feet in a single season (Fryer, 2010)—the High Line's stands of Tree of Heaven rose high above many of its more ground-loving plant inhabitants. Nevertheless, as Friends of the High Line, in collaboration with the landscape architect James Corner and garden designer Piet Oudolf, proceeded to institutionalize and instrumentalize the redevelopment of the space in the broad service of real estate capital, Tree of Heaven disappeared from the list of botanical inhabitants.

Taking a cue from Anne Rademacher's (2011) research on the human settlement of riparian zones in the context of rapid urban expansion in Kathmandu, we might pay close attention to the ways in which the eco-logic of the High Line served as a political/ideological buffer against real estate speculation and exploitation of an 'underutilized' zone of urban nature. Rademacher offers a version of urban ecology as "a fragmented and contested, yet engaged, universal" which "reminds us of the absence of neutral spaces from which to imagine or forge environmental change" and, instead, demands that we pay close attention to the political dimensions of any 'call' to urban ecological awareness or action (2011: 183). While the context and location of Rademacher's ethnographically anchored observation thwarts any direct comparison, we might draw inspiration from her observation in noting that the High Line redevelopment relied both implicitly and explicitly on a logic of ecological improvement in order to depoliticize critical aspects of the project's impacts on both the existing landscape of the space and the neighborhood in which that landscape was embedded.

One notable impact of the project was the manner in which it reconfigured longstanding community demands for small-scale affordable housing as it virtually guaranteed intensified gentrification. Despite the fact that local residents and long-time members of the Community Board consistently advocated for affordable housing as part of lower-density developments, the ideology of ecological and economic value was leveraged against these concerns.

And so, with the arrival of the redesigned High Line, scores of luxury residential units also arrived, growing at an alarming rate around the edges of the carefully manicured promenade, inviting an invasion all their own. The inhibitory effects of gentrification on spontaneous inhabitation, queer cruising, and unmanaged successional emergence can provocatively be read alongside Tree of Heaven's allelopathic capacity. In this sense, we can see that the High Line is no innocent 'ecological value-added' in the midst of inevitable gentrification and redevelopment. Instead, it performs a naturalization of gentrification. That its erasures included both specific human political imaginaries and various non-human worlds should, in this context, come as no surprise.

Detroit Tree of Heaven Woodshop: Mutualism and Alliance

In 2005, artists Mitchell Cope, Ingo Vetter, and Annette Weisser formed the Detroit Tree of Heaven Woodshop. The group was drawn both to Detroit and to Tree of Heaven owing to a shared interest in the environmental and social impacts of deindustrialization. Of their approach to the tree, they write:

> Commonly disregarded as a weed better to be extinguished, we look at [Tree of Heaven] as a post-industrial resource, and take advantage of its ubiquity. Since the tree is growing very fast, the wood is of poor quality by conventional standards but processable if correctly cured. We actually like the unpredictability of the material and work with it instead of against it.
>
> (Detroit Tree of Heaven Woodshop, n.d., n.p.)

Combined with their view of the tree as a prime example of "anarchic invasiveness," the artists' approach to working with, rather than against or in spite of, the tree speaks to practical and political potentials of a mutualist alliance. Boucher et al. (1982) trace the scientific and epistemological origins of the concept of mutualism, arguing that its importance for figures such as the anarchist Peter Kropotkin and the Quaker pacifist Warder C. Allee led it to be understudied by ecologists in the "20th century because of its association with left-wing politics" (1982: 318). In addition to explicating various forms of mutualism (symbiotic versus non-symbiotic, direct versus indirect) the authors describe the distribution of the phenomenon itself, noting that ecological mutualisms appear in both stable and marginal (i.e. disturbed) milieus (1982: 328). Tree of Heaven's predominance in disturbed urban ecologies suggests that, beyond its allelopathic capacity (a phenomenon that is distinguished from mutualism in that it is not uniformly beneficial to both organisms involved in the relation), it might also be implicated in a variety of mutualist relations.

Nevertheless, ecological and biological science, as the authors point out, often avoids the more explicitly political dimensions of such relations. Stepping across the divide between human and non-human, the Detroit Tree of Heaven Woodshop inaugurates a mutualist alliance which metabolizes the physical tree while being careful to produce discourses that do not demonize the tree for its deft responsiveness and adaptation to anthropogenic tendencies. They began their relationship with the tree from a position of intentional political awareness of the conditions which support its proliferation. From this position, they engaged the tree as a transformative material, turning it into museum benches and picture frames for a traveling exhibition detailing the processes and politics of shrinking cities, including Detroit. Eventually, the collective began planting the trees intentionally (Figure 13.2), a gesture of good faith and mutualist intention in light of their consumption of the tree itself.

The Woodshop's cultivation of a relationship with the tree raises important points for an ethics and politics of human/non-human urban ecological mutualism.

Figure 13.2 Ailanthus altissima on Detroit Tree of Heaven Woodshop's tree farm as pictured for an exhibition at the SMART Museum of Art (Chicago, IL, USA). Speaking about the farm, the Woodshop collective writes, "Afforest your Neighborhood. Detroit Tree of Heaven Woodshop has established its first Tree of Heaven Farm on a vacant Detroit city lot for future harvest. We planted seedlings in beds of car tires. The tires protect the young trees while they are growing but also determine their lifetime to a size when the trunks are suitable for processing. We assume this period of growth to be approx. 40 years. Within this timespan we will maintain the plantation and keep the lot free of any kind of real estate speculation or building activity. The plantation has been realized with the support of the SMART Museum of Art, University of Chicago and a documentation will be on display in the 'Heartland' exhibition" (n.p.). Image courtesy of the artists.

Source: Detroit Tree of Heaven Woodshop, available online at www.treeofheavenwoodshop.com/exhibitions2/smart-museum-of-art/ (accessed 17 November 2013).

Namely, an intentional approach to metabolizing the tree creates localized relationships which support both human and non-human actors. The humans do discursive and symbolic work on behalf of the tree while the tree provides material to that effort and insight into the history of abandonment and deindustrialization in the city. If, in the current global economic and ecological conjuncture, the production of disturbed urban ecologies in which Tree of Heaven thrives seems inevitable, then the Woodshop might be understood as a form of eco-art which resists the violences and erasures of those productions by helping the tree itself to become 'worthy' of political consideration. Tree of Heaven Woodshop's emphasis on both material and representational strategies of mutual diffusion suggests that a more sensuous and practical engagement with Tree of Heaven can push back against insistent demands for productivity, redevelopment, and 'improvement.' In this context, the tree is not merely a transitional object en route to neoliberal redevelopment and management strategies. It is, instead, an active component of a holistic symbolic and material network which can support efforts to forge intentional and responsible ecological and forestry practices in the forgotten and abandoned spaces of urban capitalism. The Woodshop's method is not devoid of ethical problems—the death of individual trees, for example—but it does embody a creative potential expressed in alternative valuations and modes of relating to urban ecologies.

'Ghetto Palm': Eco-political Creativity

One final example expresses an even more radical set of possibilities that issues forth from building the city (or at least our understanding of it) with and through Tree of Heaven. The artist-researchers Stefan Tiron and Vadim Tiganas (2010) introduce their "alien species urban ecology invasive biology ecocritical project" with a breathless recollection of Tree of Heaven's long history of entanglement with colonialism, nationalism, urbanization, gardening, and biology. Their trenchant analysis concludes by arguing that:

> Tree of Heaven is a migrant plant that simply didn't want to work for us any more, didn't actually play its prefigured role in human horticultural propaganda. It stands out as one of the most resistent [sic] organisms to this sort of propaganda and against the benign and interested gardeners and well-intentioned planters of the postindustrial world. As a species it is not just an accusatory reminder but also more importantly about [sic] what happens to introduced species when they are not anymore the favorites, the pets, the symbols of the system. Where do they fall out of grace, grow and reproduce as unwanted illegals or newly identified pests?
>
> (2010: n.p.)

Tiganas and Tiron take unhesitating artistic license as they debunk 'horticultural propaganda' that seeks to demonize Tree of Heaven without addressing the biopolitical and ethical dimensions of the tree's globalization. As a whole, the website documents a project whose scope ranges from eco-art to citizen-science to science-fiction, with each approach taking up an interesting position with respect to the policy document that prompted their effort in the first place. Working in Chişinău, Moldova, which they identify as 'the GREEN city par excellance,' Tiganas and Tiron first encountered Tree of Heaven in a report which identified it as a threat to biodiversity conservation (2010: n.p., emphasis in the original). With a sense of curiosity about the tree's presence in their own context, they set out to map the tree's distribution in Chişinău and, in the process, built a visual and experiential archive capable of speaking back to the scientific consensus on the 'threat' represented by "non native, alien species, invasive species, foreign species, introduced species, non indigenous or exotic" plants (2010: n.p.). In a critique that reaches all the way back to the Great Oxygenation Event (2.4 billion years ago) the artists develop a significant manifesto demonstrating the inherency of toxicity to evolutionary transformation, highlighting the importance of shedding notions of innocence and biological purity, and ascribing not just adaptive, but also intentional and resistant, strategic intention to the tree itself (Figure 13.3). Indeed, their narrative treats humans as far from centrally important in the emergence of Tree of Heaven as a globally successful plant. And yet, they understand the present era to be one in which humans are grappling to come to terms with increasingly complex, large-scale, temporally massive ecological processes that are responsive to humans, but in varied, often convoluted ways. How can we take responsibility for this responsiveness?

Figure 13.3 Tehnică pentru o viață (Technique for a life time). According to Vadim Tiganas and Stefan Tiron, this image of *A. altissima* in Arad, Romania "totally encapsulate[s] the great survivalist techniques of this urbanized species."

Source: http://ghettopalm.tumblr.com (accessed 17 November 2013). Photographer: @fra_pio.

While space prevents me from unfolding their project in greater detail, the opening statement alone provides ample support for some concluding remarks on the potential of a queer urban ecology as expressed through the particular figure of Tree of Heaven.

Conclusion

Throughout this chapter, I have invoked a variety of examples to illustrate how Tree of Heaven can aid in queering conceptions of the urban forest. In the first example, Tree of Heaven is connected to the erasure and redevelopment of an abandoned landscape. We should not see the tree's disappearance as incidental to either the narrative or the material processes of gay and green gentrification. Instead, by recalling its presence and noting its absence, we invite a form of queer memory and mourning for a space of successional ecological emergence and for counter-normative social relations. This case also demonstrated how 'the queer' can be enrolled in a process of its own destruction. Rather than seeing Tree of Heaven as an unruly inhabitant better displaced than embraced, the second example calls us to ask how, and on whose terms, we might ally with the tree in mutualist relation. Here, queerness directs us toward recognition and engagement with difference as a matter of both practice and principle. Here, the tree is 'a queer thing,' queer as noun; once deemed out of place, it is eventually recognized for the enormous amount of work it does to create that very place. In the last case, Tree of Heaven is not only recognized as 'a queer thing,' it also prompts the authors to rewrite a particular history of place from the perspective of the tree itself. The 'Ghetto Palm' project suggests Tree of Heaven as 'queer thing' or 'queer being' and 'queer doing,' or queer as verb. This leads us to understand not only what we can make the tree do, but also what is enabled in us if we begin to work with, rather than against, the tree's persistence.

The foregoing examples have also shown what can happen when Tree of Heaven is involved in practical human efforts to shape and interact with the urban forest. In the first example, the tree was one among many plants that supported an initially indeterminate, but eventually overdetermined, human desire to preserve and redevelop New York City's High Line. Accorded with little more than the symbolic power of 'greening,' the tree was rendered as a passive and expendable component of a superficially 're-wilded' urban landscape. Unlike the depoliticizing erasure of the tree from the manicured landscape of the High Line, a move that invokes the aesthetic and imaginative potential of weediness only to eradicate it materially, the Detroit Tree of Heaven Woodshop relies on an inverse approach. In the latter case, the tree is treated as an allied urban dweller, capable of expressing historical and political dimensions of Detroit's deindustrialization by virtue of its placement, growth, and botanical capacity. Interpreted as intentional mutualism, the Tree of Heaven Woodshop suggests an approach to the tree which blends cultural drives with natural processes in a novel eco-art project.

Perhaps more than the first two examples Tiganas and Tiron's 'Ghetto Palm' project pushes us to think of the tree in radical poetic, philosophical, ethical, and political terms. Even though the project does not directly invoke queerness, it deploys a biopolitical analysis with important precedents in Michel Foucault's (1990 [1978]) genealogical critiques of the emergence of sexuality as the axis along which the modern state manages and produces life, population, and territory. Tiganas and Tiron's inquiry, and my own attempt at following Tree of Heaven to various territories and contexts, suggests that ecology, along with sexuality, is an increasingly important axis of power and resistance.

If we accept Tiganas and Tiron's analysis and understand the reproduction of unwanted non-humans as a primary ideological and material justification for the imposition of ecological management, then I would argue that queering those relations is an indispensable part of any plan which seeks to change either the practices or the ideology of urban ecology and forestry. If we extend this agenda even further than transforming the frameworks of urban ecology, for instance to the very politics of city building and ethical dimensions of human-development strategies, then queering is what calls us to begin by denaturalizing our own ideas, our own strategies, our own territories. Starting with the ground on which we stand reveals that we have always already displaced someone or something. Attempts to 'naturalize' certain plants, animals, or people cannot ever be viewed innocently; they are laden with mystified and obscured agendas that often serve narrow political interests. Keeping this in mind, those of us with a passion for the urban forest might begin to understand that the challenge of working with that which is already here, especially if it resists our efforts, offers profound transformative possibilities not just for the forest itself, but also for our entanglements with it.

Notes

1 I first encountered this name in the incredible project 'Ghetto Palm,' by researcher-artists Stefan Tiron and Vadim Tiganas, discussed at the end of this chapter.

2 This particular nom de guerre, which I have heard colloquially—with surprising frequency—among gay friends in both Toronto and New York, has also received an indirect mention in the *Toronto Star* (Dempsey, 2010).

3 Tree of Hell is, again, a colloquial name. One particularly interesting instance in which the name appears is online (http://davesgarden.com/, n.d.) in a forum "for gardeners, by gardeners," where users are able to cast a vote of positive, neutral, or negative for specific species of plants and then to contextualize their vote. As of this writing, Tree of Heaven has a score of 14, 11, and 76, respectively. Some users voting down the tree describe it as, "a spawn of Satan himself"; "evil"; "An awful, awful plant that destroys native forests and can destroy your landscape . . . One of the most invasive species on earth, only good in China, terribly stinking." Even among this scathing and racialized vegetal vitriol, there are a few eloquent defenses of the tree, nearly all of which include references to its urban-ness, in particular its affinity for polluted ground and disturbed wayside ecologies, spaces which are nutritionally impoverished, industrially polluted, and distinctly anthropogenic. One user calls the Tree of Heaven, "A great urban tree," noting that "They even grow in Brooklyn."

4 Important scholarly references include texts in queer ecology (Giffney and Hird, 2008; Mortimer-Sandilands, 2008; Mortimer-Sandilands and Erickson, 2010a, 2010b) and urban political ecology (Heynen et al., 2006; Heynen, 2006; Bunce, 2009; Quastel, 2009; Foster, 2010).

5 According to the OED Online (2013b), 'ruderal' is an ecological term dating to 1865. Its first sense is adjectival: "of a plant: growing on waste ground or among rubbish, esp. as a pioneer . . . (also) designating a habitat of this type." Its second life is as a noun describing particular plants. In one recent use, ruderals are described as "plants of disturbed habitats or particularly stressful ones, where long term survival of individual plants is unlikely." It is precisely this implication that captures my interest and prompts questions like: What kinds of work—what creativity—is immanent to ruderals and ruderal ecologies?

6 Allelopathy (OED Online, 2013a) is derived from the ancient Greek. In Hellenistic Greek, the word means 'being subjects of mutual influence.' The English word combines the prefix 'allo-,' which is used to form "nouns and adjectives with the sense 'other, different(ly),'" with the German '-pathy(ie),' which has denotations of "being affected," or, more pathologically, of "diseases and disorders of a specific kind or affecting a specified part." In my ecological reading, the aspect of allelopathy which emphasizes the 'deleterious,' 'toxic,' and 'harmful' nature of these relations is prominent, yet it is not clear from the etymological information that this sense has ancient origins. As a prefix, 'allo-' is used in many scientific lexicons to inaugurate relations based on difference, isomorphy, and instability. Allelopathy seems to drift between homeopathic and toxic senses, depending on the bodies and relations implicated. This is particularly interesting to note in light of Tree of Heaven's use in Chinese medicine. For more on the medicinal and ecological benefits of Tree of Heaven see Scott (2010: 295–301).

7 As Sam Killerman helpfully explains, "Trans* is one word for a variety of identities that are incredibly diverse, but share one simple, common denominator: a trans* person is not your traditional cisgender wo/man. Beyond that, there is a lot of variation." (Killerman, S. 2012. "What does the asterisk in 'trans*' stand for?" Available online: http://itspronouncedmetrosexual.com/2012/05/what-does-the-asterisk-in-trans-stand-for/, accessed 31 March 2014.)

References

Adamson, J. (2012) '"Spiky Green Life": Environmental Justice Themes in Sapphire's PUSH', in E. McNeil, N. Lester, D. Fulton, and L. Myles (eds) *Sapphire's Literary*

Breakthrough: Erotic Literacies, Feminist Pedagogies, Environmental Justice Perspectives, Palgrave Macmillan, New York City.

Bell, D. and Valentine, G. (eds) (1995) *Mapping Desire: Geographies of Sexualities*, Routledge, London.

Boucher, D., James, S. and Keeler, K. (1982) 'The Ecology of Mutualism', *Annual Review of Ecology and Systematics*, 13: 315–347.

Browne, K., Lim, J. and Brown, G. (eds) (2007) *Geographies of Sexualities*, Ashgate, Aldershot, UK.

Bunce, S. (2009) 'Developing Sustainability: Sustainability Policy and Gentrification on Toronto's Waterfront', *Local Environment*, 14(7): 651–667.

Cloke, P. and Jones, O. (2003) 'Grounding Ethical Mindfulness for/in Nature: Trees in their Places', *Ethics, Place & Environment,* 6(3): 195–213.

Davesgarden.com (n.d.) 'PlantFiles: Tree of Heaven, Chinese Sumac, Stink Tree, *Ailanthus Altissima*', http://davesgarden.com/guides/pf/go/1699/ (accessed 24 November 2013).

Dempsey, A. (2010) 'The Heavenly Tree with a Seedy Side', *Toronto Star* [online], 15 July, http://www.thestar.com/news/gta/2010/07/15/the_heavenly_tree_with_a_seedy_side.html (accessed 4 April 2013).

Detroit Tree of Heaven Woodshop (n.d.), www.treeofheavenwoodshop.com (accessed 18 August 2013).

Foster, J. (2010) 'Off Track, in Nature: Constructing Ecology on Old Rail Lines in Paris and New York', *Nature and Culture,* 5(3): 316–337.

Foucault, M. (1990 [1978]) *The History of Sexuality: An Introduction*, Vol. 1, R. Hurly (trans), Vintage Books, New York City.

Fryer, J. (2010) '*Ailanthus altissima*'. In: *Fire Effects Information System*, [online]. US Department of Agriculture, Forest Service, Rocky Mountain Research Station, Fire Sciences Laboratory (Producer), www.fs.fed.us/database/feis/plants/tree/ailalt/all.html (accessed 31 March 2014).

Gandy, M. (2012) 'Queer Ecology: Nature, Sexuality, and Heterotopic Alliances', *Environment and Planning D: Society and Space*, 30(4): 727–747.

Gandy, M. (2013) 'Marginalia: Aesthetics, Ecology, and Urban Wastelands', *Annals of the Association of American Geographers,* 103(6): 1301–1316.

Giffney, N. and Hird, M.J. (eds) (2008) *Queering the Non/Human*, Ashgate, Aldershot, UK.

Heisey, R.M. (1997) 'Allelopathy and the Secret Life of *Ailanthus Altissima*', *Arnoldia,* 57(3): 28–36.

Heynen, N. (2006) 'The Political Ecology of Uneven Urban Green Space: The Impact of Political Economy on Race and Ethnicity in Producing Environmental Inequality in Milwaukee', *Urban Affairs Review,* 42(1): 3–25.

Heynen, N., Kaika, M. and Swyngedouw, E. (eds) (2006) *In the Nature of Cities: Urban Political Ecology and the Politics of Urban Metabolism*, Routledge, London.

Ingram, G.B. (2010) 'Fragments, Edges, Matricies: Retheorizing the Formation of a So-called Gay Ghetto through Queering Landscape Ecology', in C. Mortimer-Sandilands and B. Erickson (eds) *Queer Ecologies: Sex, Nature, Desire, Politics*, Indiana University Press, Bloomington, IN.

Ingram, G.B., Bouthillette, A.-M. and Retter, Y. (eds) (1997) *Queers in Space: Communities | Public Places | Sites of Resistance*, Bay Press, Seattle, WA.

Knopp, L. (2007) 'From Lesbian and Gay to Queer Geographies: Pasts, Prospects, and Possibilities', in K. Browne, J. Lim and G. Brown (eds) *Geographies of Sexualities: Theories, Practices, and Politics*, Ashgate, Aldershot, UK.

Kowarik, I. and Säumel, I. (2007) 'Biological Flora of Central Europe: *Ailanthus Altissima* (Mill.) Swingle', *Perspectives in Plant Ecology, Evolution and Systematics*, 8(4): 207–237.

Lorimer, J. (2007) 'Nonhuman Charisma', *Environment and Planning D: Society and Space*, 25(5): 911–932.

Marder, M. (2011) 'Vegetal Anti-metaphysics: Learning from Plants', *Continental Philosophy Review*, 44: 469–489.

Mortimer-Sandilands, C. (2008) 'Queering Ecocultural Studies', *Cultural Studies*, 22(3–4): 455–476.

Mortimer-Sandilands, C. and Erickson, B. (2010a) 'A Genealogy of Queer Ecologies', in C. Mortimer-Sandilands and B. Erickson (eds) *Queer Ecologies: Sex, Nature, Politics, Desire*, Indiana University Press, Bloomington, IN.

Mortimer-Sandilands, C. and Erickson, B. (eds) (2010b) *Queer Ecologies: Sex, Nature, Politics, Desire*, Indiana University Press, Bloomington, IN.

New York City Department of City Planning (2005) 'Final Environmental Impact Statement for the Special West Chelsea District Rezoning and High Line Open Space Project' [online], www.nyc.gov/html/dcp/html/westchelsea/westchelsea5.shtml (accessed 17 November 2013).

O'Brien, W. (2006) 'Exotic Invasions, Nativism, and Ecological Restoration: on the Persistence of a Contentious Debate', *Ethics, Place & Environment*, 9(1): 63–77.

OED Online (2013a) 'Allelopathy, n.', www.oed.com/view/Entry/241607?redirectedFrom=allelopathy (accessed 19 August 2013).

OED Online (2013b) 'Ruderal, adj. and n.', www.oed.com/view/Entry/276667?redirectedFrom=ruderal#eid (accessed 19 August 2013).

Peretti, J. (1998) 'Nativism and Nature: Rethinking Biological Invasion', *Environmental Values*, 7(2): 183–192.

Plumwood, V. (1993) *Feminism and the Mastery of Nature*, Routledge, London.

Plumwood, V. (2002) *Environmental Culture: The Ecological Crisis of Reason*, Routledge, London.

Quastel, N. (2009) 'Political Ecologies of Gentrification', *Urban Geography*, 30(7): 694–725.

Rademacher, A. (2011) *Reigning the River: Urban Ecologies and Political Transformation in Kathmandu*, Duke University Press, Durham, NC.

Scott, T.L. (2010) *Invasive Plant Medicine: The Ecological Benefits and Healing Abilities of Invasives*, Healing Arts Press, Rochester, VT.

Shah, B. (1997) 'The Checkered Career of *Ailanthus Altissima*', *Arnoldia*, 57(3): 21–27.

Smith, N. (1996) The New Urban Frontier: Gentrification and the Revanchist City, Routledge, New York City.

Sukopp, H. (2003) 'Flora and Vegetation Reflecting the Urban History of Berlin', *DIE ERDE*, 134(3): 295–316.

Sukopp, H. (2008a [2002]) 'On the Early History of Urban Ecology in Europe', in J. Marzluff, E. Shulenberger, W. Endlicher, M. Alberti, G. Bradley, C. Ryan, C. ZumBrunnen, U. Simon (eds) *Urban Ecology: An International Perspective on the Interaction Between Humans and Nature*, Springer, New York City.

Sukopp, H. (2008b [1973]) 'The City as a Subject for Ecological Research', in J. Marzluff, E. Shulenberger, W. Endlicher, M. Alberti, G. Bradley, C. Ryan, C. ZumBrunnen, U. Simon (eds) *Urban Ecology: An International Perspective on the Interaction between Humans and Nature*, Springer, New York City.

Tiganas, V. and Tiron, S. (2010) 'Context of the project, project members and bioterrorist aims', http://ghettopalm.tumblr.com/post/1162070473/context-of-the-project-project-members-and, (accessed 14 August 2013).

Whatmore, S. (2002) *Hybrid Geographies*, Sage Publications Limited, London.

Zukin, S. (1982) *Loft Living: Culture and Capital in Urban Change*, Johns Hopkins University Press, Baltimore, MD.

14 The Thin End of the Green Wedge

Berlin's Planned and Unplanned Urban Landscapes

Cynthia Imogen Hammond

Introduction

This chapter explores the cultural and ecological politics of the urban forest in Berlin. The focus of this exploration is not Berlin's iconic urban parks and forests such as Teufelsfenn, Jungfernheide, or Tiergarten, but rather the uneven trajectory of the "green wedge," an early-twentieth-century city planning tactic that surfaces in different ways in the decades following 1945 in Berlin. I use three texts to trace the uneven path of the green wedge. F. Brinkmann's "Modern City Planning" (1910) introduces the green wedge as a carefully inserted slice of "nature" in the city that fosters the German people's connection with the forests beyond. Herbert Sukopp et al.'s 1979 essay, "The Soil, Flora, and Vegetation of Berlin's Wastelands" and Jouni Häkli's 1996 paper, "Culture and Politics of Nature in the City: The Case of Berlin's 'Green Wedge'" help situate the ecology and politics of unplanned urban landscapes and forests in Berlin, pre- and post-unification, understanding how ruderal or self-seeded landscapes transformed Berlin, but were themselves subject to new forces in the form of urban revitalization.

To understand the recent cultural and ecological politics of the green wedge, I employ a case study: the green roof of the Canadian embassy in Berlin, designed in 2005 by Canada's foremost landscape architect, Cornelia Hahn Oberlander[1] (Figure 14.1). The embassy occupies a wedge-shaped portion of the historic, octagonal Leipziger Platz in central Berlin. Leipziger Platz was once part of the infamous "death strip," the militarized land enclosed between the two sides of the Berlin Wall between 1961 and 1989. But this site has deeper social, design, and landscape histories prior to the Cold War, which are themselves important agents in the unfolding of the embassy's present-day appearance. The link between my case study and the themes of this volume is the form that Hahn Oberlander chose: a fully functioning, miniature forest landscape. This chapter argues that urban landscapes, their ecology and their capacity to retain morphological links to key moments in the past, are crucial agents in the unfolding form of the city, and urban ecologies of the future.

The intent of this chapter is twofold. First, I aim to contribute to the growing literature on the non-human agencies of urban landscapes (Schama, 1995; Cronon, 1996; Grosz, 2008; Saint-Laurent, 2000; Sabloff, 2001) by drawing on a history of

Figure 14.1 Design for green roof, Canadian Embassy. 17 Leipziger Platz, Berlin, Germany (completed 2005). Cornelia Hahn Oberlander, landscape architect, in collaboration with KPMB Architects and HOK/Urbana Architects.

Source: Image courtesy of Buenck+Fehse, Berlin.

site-specific, secondary literature. The spirit of this literature is to critically identify and unpack the network of actors who play formative roles in the creation and evolution of such spaces. Relatedly, this chapter aims to make a critical intervention into the usual modes of writing architectural history (and the history of landscape architecture) by offering an alternative to the emphasis on human authorial intent as the main locus for meaning in the designed environment. My argument depends on the idea that urban ecologies have social, political, and cultural dimensions that must be explored historically in order to address the often tightly laced human and non-human agencies at play in urban landscape design.

The Green Wedge

> The love of the German citizens for nature, especially for the forest, is well known, and in nearly every province there exist so-called forest-protection leagues with thousands of members who arouse public interest to preserve their very precious forests.
>
> (Brinkmann, 1910: 26)

In 1910, a Berlin-based municipal engineer named Brinkmann observed the draw of nature, specifically forests, for German citizens. His brief article was not an exposition on the leagues of forest preservationists cited above, nor was it a study of the forests themselves. Instead, it informed readers about Germany's advances in modern urban land usage, and advocated in particular the "green wedge" as an alternative to utilitarian city planning and to ring-city models. "We intend to have these green areas located at intervals around the community," Brinkmann declared, "intersecting the latter like a wedge in order to facilitate traffic and general development of the city" (1910: 26).

Brinkmann does not cite Ebenezer Howard (1850–1928), the English city planner who generated the idea of the green wedge as part of his concept of the Garden City. It is clear however that Brinkmann, like many people interested in solving the early twentieth-century problems of congestion and density, had encountered Howard's 1902 book, *Garden Cities of To-Morrow*. Howard argued for a schematic reordering of the city into pie-shaped wedges, each providing equal access to institutions and greenspace. Each wedge, he theorized, could be successfully organized into multiple city-segments, gradually expanding to form concentric rings of urban development. In his gendered description of this plan he writes, "Human society and the beauty of nature are meant to be enjoyed together . . . [just as] man and woman . . . supplement each other, so should town and country" (1902: 17).

Howard's influential concept presupposed that all urban dwellers sought the same amenities (houses in which to raise families, decorative gardens, access to religion, escape from the city). Howard's intentions for virtually unlimited urban parkland landscapes, devoid of an ecological base seem as problematic today as his assumption that women would stay home, tending the baby. If Brinkmann had any criticisms of Howard's vision, we don't hear them, but in the interesting passage at the midpoint of Brinkmann's text, quoted above, we can see how the green wedge concept evoked, not a peaceful vision of heteronormative British countryside, but a powerful vision of the importance of the forest. Green wedges would be, in this view, the thin end of the wedge of nationalist spirit as it intersected with land attachment, or *heimat*.

Brinkmann's text summons what scholars have identified as the centrality of the forest for the Romantic concept of German identity and nationhood (Schama, 1995; Lekan, 2009; Wilson, 2012). The enfolding of forest/nature and city/culture is, however, an important lived nuance in this symbolization. According to historian Alexander Freund, nineteenth-century Berliners celebrated and commemorated their liberation from Napoleon by planting an oak tree, not in the forest, but in front of the Berlin courthouse, and by adorning their "houses and streets with oak leaves, branches, and wreaths" (Freund, 2012: 119). Historian Thomas Lekan describes how in 1896, "a German factory worker" attending a picnic on Chicago's North Side proclaimed, "Nothing thrills a German more than a festival in the woods . . . I forgot that I was so far, so distant from my homeland" (Lekan, 2005: 141). Even one well-placed urban tree, one well-designed green wedge, it would seem, could make direct links to the forests beyond the city, and the sense of identity associated with those forests. The human designer was the unquestioned agent who made these connections possible.

The flow of agency from human to non-human would prove rather more complex in the century that followed Brinkmann's confident assessment of "Germany city planning." Following the devastation of Berlin during the Second World War, the division of Berlin into East and West created a violent, irregular fissure through the city, and the country. These factors, combined with de-industrialization, contributed to a vast increase in the amount of unprogrammed terrain in the city: terrain that was not instrumentalized toward commerce, industry, housing, and so forth. Bombed buildings lay in rubble or gradually disintegrated; sometimes their ruins were hauled away, but tracts of land, particularly near the Berlin Wall become enormous landscapes of ruin. Yet these ruins and their neglect provided surprisingly good conditions for biological regeneration.

Weeds and Wastelands

> The definition of [certain] plants as "weeds" in the denigrating sense of the word, represents a viewpoint from a time when the production of biomass was the only goal. This evaluation does not consider the great protective importance of these plants in cities, settlements, and industrial areas.
>
> (Sukopp et al. 1979: 121)

In 1979, West Berlin scientists Herbert Sukopp, Hans-Peter Blume, and Wolfram Kunick researched the changes to the local urban ecosystem, assessing the condition of the city's accidental, "man-made" habitats one generation after the end of the Second World War. Historian of science Jens Lachmund has studied the work of the members of the Institute of Ecology at the Technical University of West Berlin. He observes that their special relationship to the city was shaped in part by the isolation of their geopolitical working conditions, in a walled, "frontier" city they could not leave (Lachmund, 2011: 204–205, 208–209). The rise of Cold War spatial divisions led to a new focus for these botanists—urban flora and fauna. Cut off from their normal fieldwork sites, the scientists had two types of urban ecology at hand: (1) areas within the city that had never seen any form of development, such as woodlands and wetlands; and (2) wastelands, large and small, that were the product of war. Political history therefore shaped what the researchers had access to, and how they interpreted it. Part of the context of their work was the West Berlin city government's wish to substantially develop the city's existing/surviving built fabric and infrastructure. The Institute of Ecology's findings were crucial to the growing desire to defend "Green Berlin" and its future against such development (Lachmund, 2011: 211).

Working with graduate students, the Institute's botanists discovered that while some urban wastelands remained subject to drought and were scarcely vegetated, other areas of rubble and fallen brick had degenerated, formed soil with fallen leaves and, due to their location in the city, were favorable to "hemerochores." Hemerochores are biological agents, so-called "pioneer" or non-indigenous plants that populate and transform a given environment as a direct or indirect result of human action, such as aerial bombing[2] (Sukopp et al., 1979: 118–219). A key factor in the hemerochores' transformation of Berlin was the wastelands' relative locations within East and West parts of the city. Because the division was not as clear in the early years as it would be after the construction of the Berlin Wall, confusion over jurisdiction or fear of confrontation left certain sites on the western side relatively free of human intervention.

Immediately adjacent to the line separating West from East, Potsdamer Platz was one such area. The bricks of the collapsed walls of the Potsdamer station were particularly conducive to new organic growth. Writing of their visit, 28 years after the bombing, the authors note with admiration that the former station has spawned complete "closed vegetation cover" (Sukopp et al., 1979: 118). The ecologists also discovered that there was a higher level of biodiversity in such ruderal areas compared to the rural zones surrounding Berlin (1979: 119). Black and white images illustrate this. One photograph shows brave young plants seeding the bleak plane in front of Berlin's Görlitz railway station, while another image depicts a lavishly forested railway embankment, covered with the Tree of Heaven. A large drawing provides a bio-mapping of one "derelict site in the center of West Berlin," and depicts the complex, self-seeded pattern created by over 25 species on one lot. The beauty, diversity, and sustainability of these landscapes prompted the authors to argue for their value, not only to Berlin's ecology but also for its urban future. "The prevailing plants of the future," Sukopp et al. write,

> will certainly be those which adjust themselves best to man-made sites . . . Just as man got stimulation for creating new rural landscapes by studying natural processes, so he can still when planning landscapes for urban conditions.
> (1979: 121–126)

The authors then list flora which can be relied upon to thrive in Berlin's climate—flattened meadow grass, hawkweeds, coltsfoot, birch trees, scots pine, tree of heaven, butterfly bush, traveller's joy, dog roses—plants that will continue to promote the "biotic communities" that the scientists observed on their fieldwork.

As if anticipating the changes soon to come, Sukopp et al. also caution that architectural development would be the death of these remarkable landscapes. Among the factors they identify as threats to urban biodiversity, lowering the ground-water level, leveling and accumulation of soil—all of which are normally undertaken in preparation for construction—are seen as key detriments to these landscapes. Lachmund observes that the Institute's work resulted in:

> a representation of the city as a hybrid environment in which spontaneous biological processes and various forms of human influence were closely

> interwoven. The urban ecosystem was conceived of as being natural and cultural at the same time.
>
> (2011: 214)

Sukopp et al.'s text, just one of many the Institute produced prior to reunification, underscores how the "weeds" in the "wastelands" of Berlin began to have an impact upon the political and planning decisions of their era. One of the most important outcomes of the Institute's work was the re-imagining of West Berlin as a series of biotope networks: small, interlinked habitats, or stable biological communities that provide support for multiple species and act as corridors, within the city, to other nodes in the network. The Institute's activities following the publication of this article, specifically, their "biotope-mapping" surveys of West Berlin (1979–84), did not only discursively reimagine the nature–culture binary; they also led to the state-sponsored West Berlin Species Protection and Landscape Program (Lachmund, 2011: 213). With their work on West Berlin, the Institute of Ecology challenged the way that nature and culture had been opposed, discursively, for much of modern history.

Bürgerinitiativen and the Green Wedge

The interest in unplanned urban landscapes during the 1970s extended to the New Social Movements in Germany, and Bürgerinitiativen (citizens' action committees), which were "concerned with municipal policies and the urban environment" (Rosol, 2010: 551). In 1996, Jouni Häkli explored the role of Bürgerinitiativen in the concept of urban nature and the shifting fortunes of massive ruderal landscapes in central Berlin. The Gleisdreieck, in an adjacent zone to Potsdamer Platz and Leipziger Platz, was a former train-marshalling yard that had been gradually scaled down during the post-1945 era, and then abandoned, much like the Südgelände, another rail yard located about 5 kilometers south. Hemerochores self-seeded these sites, which were then occupied by various animals and birds. Over time, citizens came to find the biodiversity, scale, and the potential of these landscapes remarkable. Häkli calls them "green wedges," islands of "spontaneous flora and fauna in the middle of rationally planned urban space, providing the city with an element of nature rarely found in European metropolises" (1996: 125). Just as Brinkmann saw the green wedge as a powerful connective link to a nature beyond the city, so too did the advocates of the Gleisdreieck and Südgelände view these urban wildernesses as unique places where nature had, despite human death and destruction, thrived.

The connection between Brinkmann and Häkli's use of "green wedge" quickly breaks down when the referent comes under scrutiny. Rather than summon a singular, essentialist notion of Germany identity, the urban landscapes that Häkli refers to as green wedges were valued, in the 1990s, for what they represented in terms of difference. Häkli remarks upon the impressive diversity of not only species but entire habitats in which hundreds of species exist side by side in the metropolitan region (1996: 129). Rather than see the value of this diversity as somehow separate from the urban conditions that underpinned it, Häkli writes of the Südgelände,

> The site's uniqueness stems from the many traces of its earlier uses. A careless wanderer may stumble across a rail cross-tie, old platforms, halls, wagons, railway bridges and tracks. The contrast between spontaneous vegetation and abandoned anthropogenic structures is fascinating, and haunting, because it highlights the finiteness of industrial culture.
>
> (1996: 129)

Within this view, it was not nature as extra-urban, pristine and separate that inspired the wastelands' advocates; it was instead the proliferation of nature (Grosz, 2008) within the city that moved them to act. In actor-network terms, thousands of flora and fauna, thousands of acres of rubble, thousands of bombs, had acted within the network of the "wounded city" of Berlin (Till, 2012), reshaping it and, over time, prompting some residents to action.

By the late 1980s, the discourse surrounding these sites was urgently concerned with their protection. The scientific approach, encapsulated in the critical notion of the urban biotope (culture and nature entwined) had influence, but advocates had to contend with the reality that "it was particularly difficult to create a positive feeling toward the bombed-out areas because many citizens associated them more with death and war than with the experience of nature" (Lachmund, 2011: 220). In 1987 the Förderverein Natur-park Südgelände formed to "save the green-wedge" from redevelopment as a new freight railway zone (Häkli, 1996: 130). The becoming-landscape of the wastelands exerted a powerful effect over the activists, who in one instance "dressed up as plants and animals when they invaded the office of the West Berlin office of planning and construction" to plead for the wildlife that would be harmed if the new railway plan took place (Lachmund, 2011: 218). The costumes, alongside other tactics, helped to raise awareness about, and the visibility of, the landscapes.

Efforts continued throughout the early 1990s to safeguard the Gleisdreieck and other ruderal sites, with varied results. Some became the dumping ground for the detritus from major post-reunification building projects (Häkli, 1996: 135), while other sites, particularly in Kreuzberg and the "Diplomatic Quarter," west of Potsdamer Platz could not be saved (Lachmund 2011: 219). Both the Gleisdreieck and the Südgelände would eventually gain protection as urban parklands (Südgelände in 2000, Gleisdreieck in 2013) although the ruderal wilderness the activists prized was not saved as such. Today, one can enjoy contemporary art alongside a coffee in the Südgelände café, while in the Gleisdreieck a portion of the ruderal flora is preserved for educational purposes, but visitors are primarily invited to enjoy newly programmed zones for skating, a rose garden, and "sporting activities" (VisitBerlin, 2013). While it is true that, as Lachmund asserts, Berlin "became naturalized, and nature became urbanized" during the pivotal years pre- and post-unification (2011: 222), it is also true that lingering ideas of the appropriateness of wilderness in the city shaped the final form of these spaces.

Häkli's thesis, that the "antipodal relationship between nature and city" thwarted the landscapes' advocates' efforts, remains relevant. Attempting to analyze the issue, Häkli identifies three views of nature in modern ideology:

> Nature's externalization is reflected in our conception of "virgin nature," a genuine and original nature untouched by humans (first nature), in our understanding of nature's cultivation and domestication in agriculture (second nature), and finally, in the symbolic and tamed nature of our urban parks and gardens (third nature) . . . the idea that cities and "nature" don't mix still prevails in wider society.
>
> (1996: 131)

William Cronon, writing in the same year as Häkli, makes a similar observation, that attempts to save "nature" are thwarted by "specific habits of thinking that flow from this complex cultural construction called wilderness" (1996: 81). As long as "true nature" or "real wilderness" are perceived as essentially separate from human culture, humans fail to understand the ways in which we are part of, and subject to, the natural. We are blind, too, to the ways in which the non-human plays constitutive roles in, and is central to, human culture. The bricks from Potsdamer station encouraged a thriving, miniature ecosystem to take root in the ruins of the Second World War. Seeing this ecosystem as mere "weeds" means, to quote Ian Laurie, that we remain "reluctant to forego [our] well-established role as the ascendant force" in the creation and development of the city (1979: xvi). The case of Berlin's planned and unplanned landscapes cumulatively suggest, however, an alternative.

The Canadian Embassy in Berlin: Architecture, History, and Politics

The decade following German reunification in 1990 was an optimistic, if fevered, time. Major corporations, eager to capitalize on the international focus on the city, rushed to claim their place while tourists chipped the former wall into pocketable pieces. Architecture and engineering firms flocked to the restored capital too, taking advantage of the first building boom in years. National embassies that had decamped to Munich during the Cold War returned to the city. New buildings were used to announce foreign presence in, and support for, what Holm and Kuhn (2011) refer to as the annexation of the East by the West. The gradual removal of the Berlin Wall and the clearing of the final, built traces of the Second World War and the Cold War created multiple opportunities for new construction within the former death strip. Potsdamer Platz became a high-tech, media- and brand-saturated spectacle within ten years (Lehrer, 2003: 390–396).

At the same time, some adjacent sectors such as Leipziger Platz were subject to a new "nationally coded simplicity" (Huyssen, 1997: 62). In these areas, building heights, program requirements, and fenestration options were strictly controlled by the controversially historicist vision of Berlin's head city planner, Hans Stimmann. Architectural critic Sebastian Schmaling questions the built-in "assumption that new design elements can sanitize and purify historically burdened places" (Schmaling, 2005–06: 27). Similarly, for cultural critic Andreas Huyssen, the post-reunification built culture of Berlin is such that "banal images of a national past [jostle against] equally banal images of a global future" along the path of the former wall (Huyssen, 1997: 62). This condition is perhaps most visible at the juncture of Leipziger Platz and Potsdamer Platz. Here, high-tech, logo-branded skyscrapers tower over the regulated forms of Leipziger Platz, where the size of window openings, the color of cladding materials, the shape of building footprints, programmes, and cornice heights were all determined before any architects were hired (Figure 14.2).

This historically loaded and architecturally various stretch of cityscape is the physical context of the Canadian embassy in Berlin. In 1998, the Department of Foreign Affairs and International Trade (DFAIT) chose this site, at some remove from other embassy districts, to be "at the hub of the resurgent capital" (Dörries, 2005: 2). "Both the location and the building provide a major public diplomacy opportunity," DFAIT asserted, "as a place to establish a powerful image of Canada to Berliners" (DFAIT, 1998: 36). Accordingly, DFAIT charged the embassy's architects with the task of creating a "strong and recognizable Canadian image" in the overall design and choice of building materials (DFAIT, 1998: 46). The architects, Kuwabara Payne McKenna Blumberg (KPMB), Gagnon Letellier Cyr, Smith Carter, and HOK/Urbana, respected this directive. The embassy is "faced with Tyndall limestone from Manitoba, with featured materials inside including Douglas-fir from British Columbia, black granite and maple from Quebec [and] Eramosa marble from Ontario" (Embassy of Canada in Berlin, 2011).

Figure 14.2 Canadian Embassy (middle-right) at Leipziger Platz, Berlin, with the towers of Potsdamer Platz to the left.

Source: Image courtesy of fotodesignberlin.

What the client, the building regulations, and the developers all essentially ignored was the site's cultural landscape. In the 1920s Leipziger Platz and Potsdamer Platz were, jointly, one of the most important destinations within Europe for entertainment, shopping, and pleasure. One of the most popular stores on Leipziger Platz was Wertheim's department store, part of a chain founded by the Jewish-born Berliner, Georg Wertheim (1857–1939). The store's enormous success may be one reason why the Nazis targeted Leipziger Platz. Journalist Alan Freeman recounts, "the Wertheims had several properties in the area . . . One of these properties was 17 Leipziger Platz" (1997: A1). This is today the address of the Canadian embassy in Berlin.

In 1938 Wertheim was pressured to relinquish 17 Leipziger Platz to the Nazis along with two adjacent properties for "rock-bottom prices." Propaganda Minister Paul Joseph Goebbels took an office at 17 Leipziger Platz. From here, the Nazis' program of anti-Semitic propaganda expanded (Freeman, 1997: A1). So too did the Nazis' extensive building program which, under architect Albert Speer's direction, was to ultimately include Leipziger Platz in a "megalomaniac north-south axis from the Great Hall in the north to Hitler's triumphal arch in the south." (Huyssen, 1997: 65) The axis did not materialize, but the Nazis' presence here made this site a target during air raids. The bombing damaged Wertheim's store and decimated 17 Leipziger Platz. After the War, the Communists expropriated and demolished all remaining traces of these buildings; their absence would mark the physical division between East and West Berlin.

Andreas Huyssen has observed how the "politics of willful forgetting" is a form of power that is strongly at work in the historicist or critical reconstructionist approach to rebuilding Berlin, post-1989 (1997: 60; see also Lehrer, 2003; Schmaling, 2005–06). At best this approach fails, and at worst it refuses, to contend with the complexities of sites such as the former Wertheim's store and the former death strip, leaving the work of commemoration to monolithic projects such as the Memorial to the Murdered Jews of Europe (Peter Eisenman, completed one year prior to the opening of the Canadian embassy in 2004). From the moment that DFAIT announced the embassy's future location, the choice was controversial, given its past. The Jewish Claims Conference (JCC), acting on behalf of the Karstadt AG Group, now the owner of the former Wertheim shares, made a land claim for the site of 17 Leipziger Platz in 1997. The JCC asserted that, as the land had been relinquished under duress in the 1930s, the resale to Canada was thus in question, and the German government should not be the beneficiary of the $28.5 million CAD profit (Bentley Mays, 2006). Yet the project continued and, as of this writing, the last empty lot—where the Wertheim's flagship store once stood—is almost rebuilt in a bizarre mix of corporate architecture and historicist deference. The city park once at the center of the Leipziger Platz octagon has not been forgotten. In its backwards leap over 80 years of painful history, the reconstruction of Leipziger Platz to pre-Nazi times rushes open-arms toward a green wedge of just the sort that Brinkmann would have approved: the original landscape design for Leipziger Platz, by Josef Lenné (1789–1866), was a tidy, geometric plot of green, with the odd tree for effect.

The Green Roof at the Canadian Embassy: Art and Landscape

In keeping with Berlin's policy of incorporating green roofs on all new construction, the competition mandate called for the incorporation of a self-sustaining, living roof for the embassy. Cornelia Hahn Oberlander was an obvious choice as collaborator. She is a globally recognized leader in landscape architecture, sustainable design, and green-roof construction (DFAIT, 1998: 45; see also Manus and Rochon, 2005; Cantor, 2008). As the designer of the public and indoor gardens at the National Gallery of Canada in Ottawa (1988), the landscape for Vancouver's Museum of Anthropology (1979), and over 100 public and private commissions, Hahn Oberlander was well prepared to deal with the complicated public–private partnership that controlled the new embassy project[3] (Therrien, 2005: 184).

Much has been made of Hahn Oberlander's personal history in relation to this roof. She was born in Germany, and left with her family because of Nazi persecution. Leaving her childhood garden behind in Berlin was, she acknowledges, a formative experience (Stinson, 2008; Hahn Oberlander 2010, 2012). It would be easy to view the Berlin roof project, her only commission in the city, as a kind of homecoming. When asked, however, Hahn Oberlander resists any suggestion that the rooftop landscape at the Canadian embassy is a personal or autobiographical project. Her interest, she maintains, was ecological and aesthetic (Hahn Oberlander, 2010). A closer look at the roof itself, as a landscape, will bear out the designer's stance.

Hahn Oberlander had two objectives when she set out to design the green roof of the Canadian embassy: (1) to create a compelling image of Canada and (2) to replace the footprint of the building on the roof. Every building uses up land and thus makes a footprint on the earth. Green roofs aim to offset this loss by transferring that footprint to the roof. Green roofs can have the purely pragmatic purpose of regulating temperature, or they can be decorative or recreational spaces. Hahn Oberlander's design concept for the new embassy was an abstracted river set in a miniature forest landscape. This concept refers directly to the Mackenzie River Delta in Canada's Northwest Territories.

The Mackenzie River, Canada's longest waterway, is less iconically "Canadian" than the Great Lakes district or the Rockies, but it is well known as a habitat to a variety of flora and fauna including Beluga whales, moose, black bears, snow geese, tundra swans, and many species of evergreen. Traditionally the home of the Aklavik, Inuvik, Tuktoyaktuk, and Dene nations, the Mackenzie River valley witnessed devastating colonial invasion (it was once known as "Disappointment River"), and near-genocide through the strategic exposure of these Nations to fatal viruses. In recent years, these nations have regained their population base and have begun to fight demands to exploit the region's oil fields (Notzke, 1995: 195–196).

Hahn Oberlander was familiar with the social, ecological, and geographic specificities of the Northwest Territories by the time she created her image for the embassy roof in Berlin. As the landscape architect for the Legislative Assembly Building and Capital Site in Yellowknife, Northwest Territories (1994), Hahn Oberlander had developed a design methodology for this site "based on the approach of the least intervention" (Hahn Oberlander, 2010). The experience of working in Canada's north inspired Hahn Oberlander to choose it as her subject for Berlin. She selected dark glass tiles, laid out in a form intended to evoke the path of the Mackenzie River and provide an image of Canada while simultaneously reflecting the Berlin sky (Hahn Oberlander, 2010) (Figure 14.3).

But this image of water had an ecological purpose. Rainwater, collected between the tiles, was to be used as gray water. On the "shores" of the river, Hahn Oberlander incorporated Mugo Pine, a tree that is known to Canadians, but indigenous to Berlin. This miniature yet ecologically accurate forest landscape projects an image of Canada as a place of majestic yet delicate nature, but it is not only an image. Just as an actual river would regulate the cycle of precipitation, Hahn Oberlander's glass river collects water that would otherwise evaporate, while the strategic planting of dense, low-growth shrubs helps to cool the roof in summer, and retain heat in winter, while attracting much-needed insects, birds and, eventually, a diversification of the roof flora.

Hahn Oberlander had already established these sustainable design principles via award-winning collaborations such as the C.K. Choi building in Vancouver (1996), the rooftop garden for Vancouver Public Library (1995), and the monumental, public rooftop landscape at Vancouver's Robson Square (1983). But as KPMB associate architect Luigi LaRocca explains, the Berlin project was a roof design without precedent. The security requirements of the embassy, the regulations concerning green roofs, and an ongoing struggle with local agents over materials meant that the innovative roof design required special planning permission. Several hundred thousand euros were spent on the roof itself, but "far more" were spent on concept and preparation (LaRocca, 2010).

This green roof is not a public park. Physical access is limited to the Canadian ambassador to Germany, whose chambers look northward over the landscape. Tour groups may view the green roof, from those same offices and from an iron walkway, by appointment. But at present the main audience for the green roof is the office workers on the upper floors of the towers on Potsdamer Platz, immediately to the west, or visitors to the nearby Marriott Hotel. The roof's lush vegetation and shimmering blue-black "river" may have also been selling features for the residents of the infill project located beside the embassy to the south-east. Yet despite the private—and privatized—aspects of the roof, the lack of physical access is a positive factor in terms of design. In designs where human safety and weight need not be taken into account, there are fewer restrictions on the form and the kinds of plants that may be chosen (Hahn Oberlander et al., 2002: 7). Acting as an environmental artist, Hahn Oberlander designed this roof as a substantial, symbolic, and sustainable part of the overall architecture of the embassy (Hahn Oberlander, 2010).

Figure 14.3 The green roof of the Canadian embassy in Berlin, 2010. The photograph shows the roofs of adjacent buildings and Tiergarten Park.

Source: Image courtesy of Shauna Janssen.

Conclusion: Fourth Nature or the Thin End of the Green Wedge

> Instead of being defined in terms of the rational city, fourth nature would reflect the intertwinement of cultural and natural environments, and would spontaneously grow into a truly urban bio- and geo-diverse area. Thus in the middle of a thoroughly built environment vitally important space for animals and plants would be set aside . . . it would then be possible for people to enter into an interactive, dialectical relationship with nature in urban culture.
>
> (Häkli, 1996: 132)

> Art is where life most readily transforms itself.
>
> (Grosz, 2008: 78)

The symbolism of Hahn Oberlander's roof is not unidirectional; just as forests and rivers are strong cultural icons for Germany, so too are images of forested landscapes significant within Canadian tropes of national identity; both may be recognized in the form of the tiny forest on the roof of the Canadian embassy in Berlin. But both tropes become more complicated here if the environmental aspects of the roof are taken into account, not so much within Germany's recent tradition of green-roof construction, but rather, within the longer history of ruderal landscapes in the city. The embassy roof is not only wedge-like in form; it is a green wedge that unites key aspects of Ebenezer Howard—and Brinkmann's—concept, as well as the green wedges of Gleisdreieck and Südgelände. On the one hand, it is a designed landscape intended to evoke significant symbolism, connecting viewers to idealized landscapes far removed from the gritty, on-the-ground realities of First Nations' land rights, or of Leipziger Platz's history of anti-Semitism. On the other hand, its remove from direct human contact makes it precisely the kind of site that could support an urban biotope, in time.

In the former sense the roof is not unusual—human cultural history is replete with landscapes designed to have an emotional impact. And in the latter sense, the roof is equally unremarkable—any satellite photograph of Leipziger Platz will quickly demonstrate that all new buildings in the area have green roofs. But what is unique is the way in which this green roof attempts to blur the boundaries between spaces for human actors, and spaces for non-human actors. While there is no denying the ways in which the roof acts as an image, as art, or the ways the roof intentionally summons the landscape tropes of both Canada and Germany, its human-oriented purposes are counterbalanced by the roof's inaccessibility to human users, and the design's irreducibly ecological purpose. In this apparent contradiction, the tiny urban forest on top of the Canadian embassy at Berlin points the way out of the conundrum that Häkli and Cronon identified in the mid-1990s, by which "nature" and "city" are seen to be mutually exclusive. As "fourth nature," the green roof proposes that people may enter into an interactive, dialectical relationship with nature in urban culture not through actual, physical access, but through the equally powerful thresholds of art and design.

In this, the green roof of the Canadian embassy suggests the thin end of a green wedge of a new sort, one that uses human intervention not to create resource-hungry monocultures, nor to leave detritus in which enterprising hemerochores may take root. As designed space, this new green wedge must work with the site-specific, biological knowledge that the Institute of Ecology produced in the 1970s and 1980s; with the cultural symbols and systems of signification around "nature," and with the capacity of art to affect and be affected "by life in its other modalities" (Grosz, 2008: 103). A shift in understanding the human role in urban ecology is necessary in order to meaningfully integrate the social and political histories that have shaped the particular form of any given site—any given wedge—with the biotropic specificity of that site. It is a tall order. And while we might not always be able to sit on the result and have a picnic, humans, if they are willing to see how nature has "enabled rather than inhibited cultural and political production," (Grosz, 2008: 2) might be able to play a more useful role in helping the biological world, to which we belong, do what it does best: proliferate.

Notes

1 Hahn Oberlander (b. 1924) collaborated with Canadian architects Kuwabara Payne McKenna Blumberg (KPMB), Gagnon Letellier Cyr, Smith Carter, and the German firm HOK/Urbana.

2 Compare with apophytes, which are native plants that grow in an area whose existing vegetation has been disturbed by human activity. It is significant that this terminology originated in West Berlin's botanical science community. Lachmund writes, "From the early 1960s, Berlin ecologists were already drawing upon the concept of 'hemerobia'—the extent to which an ecosystem is shaped by human influence—to make sense of the peculiar features of nature in the city" (2011: 214).

3 This complex partnership is due in part to the fact that the embassy is not the only occupant of the site; the building envelope includes offices, shops, and residences (Kanada Haus, 2008).

References

Bentley Mays, J. (2006) 'Diplomatically Speaking: The New Canadian Embassy in Berlin,' Canadian Architect, www.canadianarchitect.com/news/diplomatically-speaking/1000201972/?&er=NA (accessed 30 August 2011).

Brinkmann, F. (1910) 'Modern City Planning,' *Municipal Engineering*, 31: 257–260.

Cantor, Steven L. (2008) *Green Roofs in Sustainable Landscape Design*, W. W. Norton & Company, New York City.

Cronon, William. (1996) 'The Trouble with Wilderness: or, Getting Back to the Wrong Nature,' in William Cronon (ed) *Uncommon Ground: Rethinking the Human Place in Nature*, W. W. Norton & Company, New York City, pp. 69–90.

DFAIT. (1998) Request for Proposals/Proposal Information: New Canadian Embassy in Berlin, Government of Canada, Ottawa, ON.

Dörries, Cornelia. (2005) Thilo Lenz (ed) *Embassy of Canada, Berlin*, Robert Bryce (trans), Stadtwandel Verlag, Berlin.

Embassy of Canada in Berlin. 'Architectural Highlights of Canada's New Embassy in Berlin,' Canada's New Embassy Building in Berlin, Government of Canada, www.canadainternational.gc.ca/germany-allemagne/offices-bureaux/new_building-nouvelle-edifice.aspx?lang=eng&view=d (accessed 28 August 2011).

Freeman, A. (1997) 'Past Haunts Canada's Embassy Site in Berlin,' *The Globe and Mail*, 26 December, p. A1.

Freund, A. (2012) *Beyond the Nation? Immigrants' Local Lives in Transnational Cultures*, University of Toronto Press, Toronto, ON.

Grosz, E. (2008) *Chaos, Territory, Art: Deleuze and the Framing of the Earth*, Columbia University Press, New York City.

Hahn Oberlander, C. (2012) Interview with Yvonne Gall, 'The Grand Dame of Green Design,' *Ideas, Encore*, 8 November, 53 minutes, 59 seconds.

Hahn Oberlander, C. (2010) Personal interview. 9 October, Vancouver, BC.

Hahn Oberlander, C., Whitelaw, E. and Matsuzaki, E. (2002) *Green Roofs—A Design Guide and Review of the Relevant Technologies*, Public Works and Government Services Canada, Ottawa, ON.

Häkli, J. (1996) 'Culture and Politics of Nature in the City: The Case of Berlin's "Green Wedge,"' *Capitalism Nature Socialism*, 7(2): 125–138.

Holm, A. and Kuhn, A. (2011) 'Squatting and Urban Renewal: The Interaction of Squatter Movements and Strategies of Urban Restructuring in Berlin,' *International Journal of Urban and Regional Research*, 35(3): 644–658.

Huyssen, A. (1997) 'The Voids of Berlin,' *Critical Inquiry*, 24(1): 57–81.

Howard, E. (1902) *Garden Cities of To-Morrow*, Swan Sonnenschein & Co., London.

Kanada Haus. (2008) 'Guten Tag Kanada: Hello Berlin,' Kanada Haus: Haus Mit Botschaft, www.kanada-haus.de (accessed 20 July 2012).

Lachmund, J. (2011) 'The Making of an Urban Ecology,' in D. Brantz and S. Dümpelmann (eds) *Greening the City: Urban Landscapes in the Twentieth Century*, University of Virginia Press, Charlottesville, VA.

LaRocca, L. (2010) Telephone interview, 2 November.

Laurie, I. C. (1979) 'Introduction' in Ian C. Laurie (ed) *Nature in Cities*, John Wiley & Sons, Chichester, New York, Brisbane, Toronto, pp. xv–xix.

Lehrer, U. (2003) 'The Spectacularization of the Building Process: Berlin, Potsdamer Platz,' *Genre*, 36: 383–404.

Lekan, T. (2005) 'German Landscape: Local Promotion of the Heimat Abroad,' in K. M. O'Donnell, R. Bridenthal and N. Reagin (eds) *The Heimat Abroad: The Boundaries of Germanness*, University of Michigan Press, Ann Arbor, MI, pp.141–166.

Lekan, T. (2009) *Imagining the Nation in Nature: Landscape Preservation and Germany Identity*, Harvard University Press, Boston, MA.

Manus, M. and Rochon, L. (eds) (2005) *Picturing Landscape Architecture: Projects of Cornelia Hahn Oberlander as seen by Etta Gerdes*, Goethe Institute and Callwey Verlag for Edition Topos, Montréal and Munich.

Notzke, C. (1995) 'A New Perspective in Aboriginal Natural Resource Management: Co-Management,' *Geoforum*, 26(2): 187–209.

Rosol, M. (2010) 'Public Participation in Post-Fordist Urban Green Space Governance: The Case of Community Gardens in Berlin', *International Journal of Urban and Regional Research*, 34(3): 548–563.

Sabloff, A. (2001) *Reordering the Natural World: Humans and Animals in the City*, University of Toronto Press, London, Toronto, Buffalo.

Saint-Laurent, D. (2000) 'Approches biogéographiques de la nature en ville: parcs, espaces verts et friches,' *Cahiers de géographie d'University Québec*, 44(122): 147–166.

Schama, S. (1995) *Landscape and Memory*, Random House, Toronto, ON.

Schmaling, S. (2005–06) 'Masked Nostalgia, Chic Regression: The 'Critical' Reconstruction of Berlin,' *Harvard Design Review*, 23: 24–30.

Stinson, K. (2008) *Love Every Leaf: The Life of Landscape Architect Cornelia Hahn Oberlander*, Tundra Books, Toronto, ON.

Sukopp, H., Blume, H. P. and Kunick W. (1979) 'The Soil, Flora, and Vegetation of Berlin's Wastelands,' in Ian C. Laurie (ed) *Nature in Cities*, John Wiley & Sons, Chichester, New York City, Brisbane, Toronto, pp. 115–132.

Therrien, M.-J. (2005) *Au-delà des frontières: l'architecture des ambassades canadiennes, 1930–2005*, Presses Université Laval, Laval, QC.

Till, K. E. (2012) 'Wounded Cities: Memory-work and a Place-based Ethics of Care,' *Political Geography*, 31: 3–14.

VisitBerlin. (2013) 'Park at Gleisdreieck: Track beds into rose gardens,' VisitBerlin, Berlin Tourismus & Kongress GmbH, www.visitberlin.de/en/spot/park-at-gleisdreieck (accessed 30 October 2013).

Wilson, J. (2012) *The German Forest: Nature Identity, and the Contestation of a National Symbol, 1871–1914*, University of Toronto Press, Toronto, ON.

Part 3

Actions and Interventions in the Urban Forest

15 "A Few Trees" in Gezi Park

Resisting the Spatial Politics of Neoliberalism in Turkey

*Bengi Akbulut**

> This struggle is against all the practices that expropriate our coasts, forests and public spaces. It is against the hydropower, nuclear and coal plants that destroy nature. It is against the profit-oriented projects that target not only Taksim Square and Gezi Park, but also Goztepe Park, Kusdili Meadow, Haydarpasa Train Station, Ataturk Forest Farm and a variety of other commons. It is against the 3rd bridge project for which the groundbreaking ceremony was shamelessly held today, which will destroy all the forests and dry up all the water resources in Istanbul. It is against the 3rd airport project that will serve the customers for the 3rd bridge and will pave the way for rent-generating construction in northern Istanbul. It is against the New Istanbul Project that will ruin all our natural resources . . . The struggle here is for all our people whose houses and living spaces were confiscated . . . Reclaiming Taksim, the memory of struggle and solidarity, means reclaiming not only the square and the park, but also all these values and rights. For all these reasons, we call upon everyone who wants to stand up for their rights and liberties, city, living spaces and future to come to Taksim's Gezi Park.
>
> (Taksim Solidarity Press Release, May 29, 2013; author's translation)

Introduction

Late on the night of May 27, 2013, a message was circulated on numerous electronic list servers and social media networks among activists in Istanbul. It was a call for urgent action, more precisely the immediate formation of guard watching at the now-famous Gezi Park, since a bulldozer was spotted demolishing the park's outer walls and it was rapidly advancing toward uprooting the trees. After being confronted on the spot by a few activists and journalists, the officer in charge claimed that the demolition was part of a municipal zoning plan, a plan that neither the urban planners/architects nor the public at large were aware of. A tent was soon set up and some 60 people were guarding the park when the bulldozer returned the next morning to resume its operations, this time accompanied by the police. At dawn, 5 days later, thousands of people from the Asian side of the city were marching to cross the Bosphorous Bridge on foot in support of the resistance. Ten days later, the people reclaimed the park and organized a communal living space in the de facto absence of the state. At the same time, millions of people had been out in the streets in protest all around the country and countless were injured and some were killed.

That small patch of urban greenspace at Gezi Park came to serve as a common ground where seemingly isolated grievances were articulated by a society that was long deemed apolitical. Efforts to adequately analyze the causes and consequences of the events that transpired have been countless—and there are likely to be more. One widespread reading among these analyses has been to cast the Gezi revolts predominantly as a reaction to the ruling Justice and Development Party's (JDP) Islamist/conservative policies and authoritarian rule.[1] In this chapter, however, I argue that such an interpretation skirts the larger processes implied by the neoliberal developmentalist agenda/growth strategy intensified especially within the last decade. It also carries immense political danger as it limits the possibilities and potentialities opened up by the revolts by rendering invisible the broader political–economic setting and the reservoir of social opposition that predated and facilitated the uprising at Gezi Park.

What I will do here, in contrast, is utilize a political ecology perspective to locate Gezi Park within the spatialization of neoliberal capitalism under JDP rule in Turkey and the different threads of social opposition that have surfaced in response. This perspective helps to illuminate the making of socio-environments (including urban space) via interconnected processes and their contested, power-laden nature (Swyngedouw and Heynen, 2003). It urges us to see the socio-environmental changes as embedded within the economic, political, social relations that surround and produce them, and the inequalities that both create and are created by them. Through the lens of political ecology, it becomes possible to account for the historical and geographical political-economic processes that Gezi Park and its proposed transformation reflect—as well as the inequalities and injustices that they created.

In what follows, I first sketch out the contours of the neoliberal growth strategy that marks Turkey's economic miracle, emphasizing especially its restructuring of space. I then discuss the dynamics of urban transformation, particularly of Taksim—the neighborhood where Gezi Park is located—and situate the proposed demolition of the park within them. After providing a brief account of the emergence and evolution of the resistance, I (as a situated researcher-activist who had been engaged with the movement) describe the processes through which the park was made a space of collective living and commoning where, on the one hand, an alternative socio-economic order was enacted and, on the other hand, a commoning ground for seemingly separated struggles of reclaiming space, was created. The last section concludes by touching upon the ways in which the revolts have reshaped and motivated the collective political imagination and potential in Turkey.

The Backdrop: Turkey's Spatialized Growth Miracle

Since coming to power in 2003, JDP has not only retained the historically strong commitment to modernization via economic growth, a defining characteristic of the state–society relationships in Turkey, but also adopted an aggressive neoliberal agenda in implementing it. Dubbed a miracle and praised for its successful realization of neoliberalism with a Muslim face, Turkey achieved high-growth rates in this period. The miracle, however, had a visibly spatial twist: the capitalization of the natural environment, privatization of realms previously under public ownership, and the expropriation and redistribution of property through "legal" means have accelerated in the last decade. Monumental projects, such as a third bridge on the Bosphorous, a third airport to be built in the middle of Istanbul's northern forests, and a canal to connect Marmara to the Black Sea to name a few, have served as symbols of progress and welfare, and evidence that the state is working hard for its people, while their socio-economic consequences have been brushed aside.

Given that the state-facilitated (if not led) construction bubble and destructive energy investments have been the main pillars of Turkey's "miracle," such radical reconstruction of space is hardly surprising. The contribution of construction to GDP, for instance, rose from 3.8 percent to 4.8 percent in 2006, 5.6 percent in 2010 and 6 percent in 2012. The sector's expansion is best attested by the increase in the number of new buildings per year, which went from 40,430 in 2002 to 597,000 in 2006 and to 817,000 in 2010 (TCA, 2012). Massive urban gentrification and renewal projects account for most of this striking rise in construction, for which the government paved the way by undertaking a series of legal changes. In a nutshell, the legal structure was revised to allow (state) expropriation of urban land and its redevelopment and marketing. Most notable among them is the Law on the Transformation of Spaces under the Risk of Natural Disasters passed in 2012, which effectively consolidated the prior piecemeal legislation made for specific neighborhoods under the guise of policymaking for disaster risk. Prepared with no participation or consultation with the general public, the Law provides the Ministry of Environment and Urban Planning with the sole authority to assign risk status, and develop projects, purchase and sell assets, and initiate construction on land assigned under risk (UCTEA, 2012). In practice, it removed the few remaining impediments to accelerated urban transformation in the form of evictions, gentrification and expropriation of public and private property. In addition, by way of numerous government decrees from 2001 onward, the Housing Development Administration, a directorate tied directly to the Prime Ministry, was endowed with extraordinary powers ranging from urban land confiscation to profit-oriented project development, urban renewal and transformation in collaboration with local governments and private companies in establishing housing sector enterprises directly or jointly with the private sector.

Through the legal and administrative changes mentioned above, the state emerged as not only the regulator but also an active constructor of, and participant in, real-estate and urban land markets; an epitome of roll-out neoliberalism (Peck and Tickell, 2002). This process has opened up new areas of capital accumulation and thus rendered the construction sector the steam engine of economic growth, while displacing large working-class sections of the urban population. It is worth noting that neighborhoods populated by minorities, such as *Sulukule* (Roma) and

Tarlabasi (Kurds), or known to have a radical left history, such as *1 Mayis* and *Gulsuyu-Gulensu*, were the first targets. Opposition to the attacks on these marginalized neighborhoods was quickly delegitimized by their portrayal as crime-laden ghettos and threats to social order, and thus failed to garner wide public support. As gentrification processes sprawled to the broader urban landscape, however, reaction and dissent acquired a larger social base.

Even though its ramifications were predominantly reflected in rural areas, the state's role was equally pivotal in the case of the energy sector. The restructuring and liberalization of the energy markets started in early 2000s and opened fields of energy investments previously beyond reach, most notably coal and hydropower, to the private sector. This was buttressed by the provision of market assurance, as it was guaranteed that the energy produced could be sold to the energy pool constructed by the state, and the revision of environmental legislation that could potentially halt the development of the sector. As a result, socio-environmentally destructive energy investments increased radically—and so did the conflicts and resistances that emerged around them.[2]

Perhaps most notable among these resistances are those related to small-scale, river-type, hydro-electric power plants (HEPPs). Based on the rerouting of streams to a suitable height by covered channels, from where they are dropped into turbines, these plants have mushroomed especially since the late 2000s.[3] The HEPPs take up whole or parts of water bodies, radically alter ecosystems and landscapes, and bring about climatic change. In doing so, they not only damage rural livelihoods but they also impair living spaces of rural populations. Concerns with such detrimental effects of the HEPPs, coupled with the substantial symbolic value attached to water and the overall decline in the viability of the countryside, produced an unprecedented—albeit fragmented—environmental movement that joined together rural populations and urban environmental activists.

Despite grievances voiced against the spatial restructuring that went hand-in-hand with its neoliberal growth strategy, JDP has managed to draw up broad-based support for its rule. The idea of modernization via economic growth has been worked and reworked to form an indispensable basis of the spatial politics JDP has successfully mobilized: highways, power plants, skyscrapers and the mega projects mentioned above served to materialize the ideal of modernization in the most effective way and received admiration from a variety of different groups in society.[4] The construction-led growth model of the JDP reproduced the consent of large sections of the urban population, not only through the distribution of rents accruing from construction investments, but also by the effective persuasion of middle-lower classes that the growth model opened up opportunities to own housing property and higher consumption levels. The parallel silencing and de-legitimization of social struggles against ecological destruction and urban transformation, and the close association of construction with modernization in the social imagination, has buttressed this strategy (Akbulut and Adaman, 2013).

To sum up, Turkey stood in the midst of processes of radical urban transformation, massive urban displacement and aggravated ecological destruction in the summer of 2013. It was also the aftermath of a few months of escalated violence and intolerance against any hint of social grievance, such as protests against the demolition of the historical cinema theater *Emek*, and May Day celebrations.[5] In that sense, the revolts that occurred around the demolition of Gezi Park were hardly an isolated case from the larger dynamics of the restructuring of urban and rural space. Thus, the solidity and the rage that the opposition voiced through the revolts are partly due to an accumulated and/or anticipated discontent with numerous similar processes of urban and rural enclosures.[6] The park, however, held further significance as it was implicated in the transformation of Taksim, the most central and symbolic neighborhood of Istanbul.

The New Istanbul, the New Taksim

During JDP's general election campaign in 2012, the prime minister announced the party's plans for the redevelopment of Taksim Square, perhaps the single-most important square in Turkey. Some listened to him in admiration while others dreaded the news. The Taksim Project—as it has come to be known by the public—involved not only the pedestrianization of the Square and the re-channeling of traffic to underground tunnels, but also the demolition of Gezi Park and the construction of a replica of the historical military barracks, which the Park replaced in 1940, to serve as a shopping mall complex and a high-end residence.

Perhaps more importantly, the Project formally put into place the government's grander vision of a new, transformed Taksim, parts of which were already under way. Within the last few years, a massive urban gentrification project was initiated in the predominantly Kurdish-populated Tarlabasi, a residential neighborhood extending from Taksim Square towards the Golden Horn, running side by side with the famous Istiklal Street. Evictions and demolitions started in late 2011 to clear the way for the construction of luxurious offices, stores and residential units geared to the service of a visibly different class of the urban population. This was paralleled by the gentrification of Istiklal Street, both through explicit state intervention and the "natural" dynamics of the market economy. With the heightened attraction of the area as a center of urban middle-class consumption and the capitalization of its centrality for culture, (local) small businesses that could not afford rising rents were gradually replaced with larger ones, often linked to global capital.[7] This process of market-led gentrification was augmented by rapid (and shady) privatizations of public estates. One among these in particular, the sale of and plans to convert the Cercle d'Orient building and the historical *Emek* movie theater into a shopping mall, spurred a large campaign and a series of protests—in fact, it would not be unrealistic to call it a milestone in the process building up to the Gezi protests.[8] The Project was not only a step extending and accelerating these dynamics already underway, but added a unique ingredient: the reconstruction and the pedestrianization de facto stripped the Square of its capacity to host protests and large gatherings.[9]

In short, the Taksim Project implied a large-scale transformation of the whole Taksim area and a fundamental change in its character (let alone its users), from a public space with symbolic value (not least due to its historical identity as a place of protest and resistance) to a space of capital accumulation and consumption; from a neighborhood of workers, urban dwellers, and immigrants (among others marginalized by the political-economic system) to a sterilized, homogenized space emptied of collective memory and identity, marketed in different ways to tourists, shoppers, and real estate developers.

The Project triggered significant public outcry as it, among others, entailed the enclosure of a public space and the destruction of an important urban forest, Gezi Park. Once the Armenian cemetery (1560–1865), then the host of Ottoman artillery barracks (1806–1938), this space had been seized by the Turkish state in 1935 (together with the Armenian gravestones which were later used in various construction projects in the city) and was opened to recreational use. In 1940, with the implementation of plans commissioned to Henri Prost, the barracks—which were in ruins by then—were demolished and the first urban park of the republican history, Gezi Park, was built (Polat, 2013). The park does not only hold critical ecological value as a host of fauna and flora and provide invaluable environmental service as an urban source of oxygen, but it had also been one of the few remaining public spaces in Taksim—until its blockade by the police after the protests in the summer of 2013.

In response, various non-governmental organizations (NGOs), environmentalists, neighborhood associations, labor unions and alike, most of which had already been active in urban and/or ecological politics, came together to form the umbrella organization of Taksim Solidarity in December 2011.[10] In addition to the legal battle initiated against the Project in general, and Gezi Park's demolition in particular, the Solidarity undertook a wide campaign to raise awareness and support. When construction in the Square started in the fall of 2012, Taksim Solidarity organized Taksim Watch Duty, a 24-hour mass guard overseeing the construction site, to draw attention to the ramifications of the Project. The watch, lasting for a total of 4 months, was eventually reduced to symbolic weekly gatherings and later dropped altogether as winter arrived, but an information station replaced it. Most recently, a festival organized by one of Taksim Solidarity's constituents was held in the Park in April 2013 with the aim of revitalizing the struggle and to acquire visibility and support. A process of discussing and evaluating ways to move the struggle forward was ongoing when the Gezi revolt flared up.

"[T]he Fear Threshold is Long Passed. This is Going to String Out"[11]

After news of the Park's attempted demolition hit urban-activist circles on the night of May 27, around 50 people who had gathered—some of whom were affiliated with Taksim Solidarity while others had just heard the news and went to the park—decided to spend the night at the park. They quickly organized a watch duty and pitched their tents, while food and blankets were arranged via social media to be brought to the park. It is worth noting that the core of the group did not only share a history of involvement with urban-environmental politics, but they were also well-versed in crowd organizing, holding occupations/demonstrations and tactical strategizing in the face of possible confrontation with the police. Their prior experience with direct action was effectively transported to the movement in the park.

A few dozen more joined the original 50 the following day, reaching around 80 people when the bulldozers arrived once more, this time accompanied by the police. The scuffle between the protestors and the police ended when tear gas was used to disperse the crowd. The bulldozers started operating and uprooted more trees until they were stopped by the arrival of the MP Sirri Sureyya Onder, a former cinema director and a well-known figure, at the scene. Meanwhile, the evident insistence to demolish a public space despite wide public reaction and the use of tear gas on a peaceful group of protestors attracted more supporters. An open forum was held later that day to discuss ways to carry on the struggle and to devise strategies to deal with police violence and the organization of daily life in the park. Solidarity visits were paid by labor unions and NGOs later in the day, perhaps providing the first signs that the movement to defend the Park was to become a uniting ground for a variety of struggles.

Following a relatively quiet night and day, the protestors woke up to an intense cloud of tear gas on the morning of May 30 and they were forced to evacuate the park while the security forces burned their tents. The park was effectively under siege by the police while the demolition was started, once again, and stopped, once again, by Onder's intervention. Thousands poured into the park that night, marking the point when the resistance started acquiring a mass-movement character (Figure 15.1). An even harsher police intervention took place the next morning, only this time the park's exits were blockaded by the police. Many were injured trying to escape from the flying tear gas canisters, some people were actually hit by them. Efforts to reassemble were brutally repressed with the use of tear gas and pressurized water, as were later attempts at staging protests against the

Figure 15.1 Protestors in Gezi Park, Istanbul, Turkey, on the third day of protests.

Source: Lara Fresko, 2013.

persistence in the (unlawful) demolition of the park and escalated police violence. After a main union confederate's (Confederation of Revolutionary Worker Unions-*DISK*) press statement and a packed sit-in at Taksim Square was also dispelled by the use of vicious force, the crowd was chased along Istiklal Street and its back alleys. While the hunt of protestors was ongoing under a heightened dose of violence, thousands were flowing into Taksim to join the revolt. People were out on the streets en masse, literally marching where public transit was cut off. Protests were being organized in other cities in solidarity with the Gezi resistance—the revolt was spreading.

Clashes with the police, under clouds of tear gas, continued all night and into the next day in Taksim streets, which by then resembled a battlefield. Thousands of people were chanting "this is only the beginning, the struggle will continue" when they finally reclaimed Taksim Square and Gezi Park in the afternoon on June 1. Within the few days that followed, Taksim Solidarity communicated four demands to the government: that Gezi Park should stay as a park; that the police violence against the protestors should stop immediately and those officials responsible for the violence should be dismissed; that the detained protestors should be released; and that de jure and de facto barriers to freedom of speech and protest in all public spaces should be removed.

The government, however, tried to disparage and slander the resistance from the very beginning.[12] The protestors were portrayed as alcoholics, degenerates, and vandals; the accusations ranged from a conspiracy by foreign finance capital to destabilize the economy to an orchestrated move to take down the government. It was in this context that JDP's deputy chair claimed that the protestors were not concerned about Gezi Park or "a few trees within it," and their real aim was to debilitate the government. The unprecedented scope and support that the protests gained, however, forced the government to step back. Days after the park was reclaimed, a so-called dialogue was initiated and a meeting was called with representatives from Taksim Solidarity on the night of June 14 but it turned out to be futile as the prime minister abandoned the gathering early in the meeting. Meanwhile, the occupants of the park had self-organized open forums to discuss future demands and further steps that the movement should take. The following day (a Saturday), when the park was swamped with an unprepared, unsuspecting crowd including children and the elderly, police forces moved in once again and evacuated the park with the usual—but more brutal—use of tear gas.[13] The Park was held under siege by police forces for a few weeks, arbitrarily opened to public at times, and remains closed for large public gatherings as of mid-2014.

Until its final evacuation, the park was made a true common space and a concretization of collective living by its occupiers. It also became a commoning ground for seemingly separate struggles that had remained isolated, namely local ecological resistances, urban-neighborhood movements, and different agents of the labor movement. The park had not only brought together groups whose union would otherwise be unthinkable, such as the Kurdish movement, secularist-nationalist Kemalists and anti-capitalist Muslims, but also promoted a type of social relationship that enabled them to co-exist. In this double sense, the organization of collective life in the park was a commoning practice.

Enacting Park-as-Commons

The days when the park was maintained by the protestors demonstrated an extraordinary politics of space that, for many, fell nothing short of the Paris Commune: the main arteries stemming out of Taksim Square, normally congested with vehicles trying to make their way into or out of the neighborhood, were blocked by barricades named after those killed by police violence during the revolt (Figure 15.2). Emptied out of their usual occupants, these streets were walked by protestor-occupiers carrying necessities to the park as well as white collar professionals going to work in plazas nearby. Entering the park from the east entrance, one would first come across a communal café on the left and the infirmary next to it, whose surroundings had been cleared to ensure easy access in case of an emergency. Further on the left lay the patch of land turned into a vegetable garden by a guerilla garden collective, a group that has been active in a number of similar ventures around the city. One would then pass by the Gezi Kids workshop next to the garden and the communal cafeterias located in front of it. Scattered between the tents, announcements for various workshops would be spotted, which had become one of the regular activities in the park together with the open forums.

During this time, the park animated an organization of state-less, commodity-less collective life, with its communal library (Figure 15.3) and cafeterias, volunteer infirmaries, workshop corners and so on. The variety of goods and services produced in the Park, such as food, media, library services, social education, security, hygiene, firefighting, and more, were produced collectively by the unremunerated

Figure 15.2 Barricades on a main artery leading to Taksim Square, Istanbul. It is named after Abdullah Comert, who died after being shot by a teargas canister on the head during the protests in Antakya.

Source: Lara Fresko, 2013.

Figure 15.3
Gezi Park Library, Istanbul.

Source: Lara Fresko, 2013.

labor of the protestor-occupiers. Their distribution took place via non-market channels based on solidarity and sharing, rather than exchange. The organization of these activities, as well as the general organization of the space, did not rely on any centralized structure but was largely autonomous, and at times, spontaneous.[14] Attempts in realizing direct democracy in open forums emerged later on, when protestor-occupiers discussed experiences and needs as well as demands and possible ways of furthering the struggle.

It would of course be far-fetched to suggest that the park embodied a complete detachment from capitalist relations of production and distribution or the broader political system. After all, the sustenance of the life in the park depended heavily on gifts flowing from the capitalist "outside world" and the legitimacy provided by the support of the larger public. But it indeed represented a space in which an alternative socio-economic order was manifested and a different sense of community was constituted. This was a space where the market logic did not have purchase, use-value superseded exchange value, and an ethic of solidarity was operationalized instead of self-interest. The park was constituted as a non-capitalist urban commons by its collective use (rather than the specific property rights defined over it) and its re-production by the practices of its users (Akbulut and Soylu, 2013; Turan, 2013; see also Gambetti, 2013).

But the park has been constituted as a commons in a second sense: it had become a fertile commoning ground for social movements who had remained largely isolated and disconnected from each other despite the close resonance between their struggles. Numerous actors that populate the landscape of social opposition, covering a variety of fields—lesbian, gay, bisexual, transsexual and transvestite activists, feminists, socialist left, anarchists, Kemalists, ecologists, urban activists, Kurds, anti-capitalist Muslims are the first to come to mind—and of organizational structures—parties, NGOs, grassroots movements, collectives—have transported their accumulated experience as well as political/politicizing practices to the park.

Solidarity Kitchen, a communal neighborhood collective from Tarlabasi, started producing meals on the first day; grassroots environmental movements from outside of Istanbul organized discussions to publicize their struggles and convey their experiences; urban ecologists held a number of workshops on mega projects that threaten the socio-ecological sustainability of Istanbul (which sow the seeds of the Defense of Northern Forests, an Istanbul-based activist group fighting against the destruction of urban space on the Black Sea coast of the city); feminists took direct actions such as spraying over sexist street writings and posting signs against sexist slogans. In fact, this potluck of political repertoire is what accounts in part for the movement's ability to practice a more proactive politics of a solidaristic social and economic life. It has also laid the political groundwork, in a more fundamental sense, that enabled the largely rural and localized environmental resistances and the urban movements centered on "the right to the city" (Lefebvre, 1995).

Broadly speaking, the environmental movement in contemporary Turkey has been marked by its fragmented, localized character. Barring a few exceptional efforts, local environmental resistances have remained mostly independent from each other despite the close resemblance between the discourses, demands and strategies mobilized by them. Although this is somewhat understandable, as each struggle is inevitably conditioned by the specifics of micro-contexts and particularities, their inability to form durable alliances and produce a broader political agenda has been a major weakness. However, by transporting their histories and struggles to the park, local environmental resistances formed a milieu that revealed the political line connecting them as well as the Gezi resistance: the intricate relationship between the state, capital, and the environment (Erensu, 2013). The words of an anti-HEPP activist is telling in this respect: "If we win [the fight for] Gezi in Taksim, we will have won [the fight in our town]; when we win in [our town], we will have won in Taksim as well, we will have won all over the country."[15] On the flip side, the accumulated experience of these localized movements in fact played an invaluable role in producing the foundation on which Gezi resistance arose. Not only did the strategies and types of direct action employed during Gezi resistance (such as hugging or climbing up the trees and standing or lying in front of the bulldozers) bear the marks of this collective history, but also the key actors of the resistance, who have long been familiar with it. It is, therefore, hardly surprising that Taksim Solidarity has consistently made references to various environmental resistances when contextualizing its stance, as the quote at the beginning of this chapter attests (Sahin, 2013; Erensu, 2013).

Conversely, the park provided the opportunity for the (mostly rural) environmental struggles to come together with the urban struggles that focus mostly on defense of public spaces and the right to housing. Although it is almost intuitive for these two lines of movements to stand together, as both problematize and politicize the restructuring of space—whether urban or rural—by the capital–state nexus, now intensified in the contemporary era of neoliberalism, they had rarely joined forces in action in the past. Gezi resistance equipped them with a common ground and a shared language that revolves around the enclosure of collective living spaces that translates into urban parks, neighborhoods, squares in the city and forests, rivers, valleys in the country. It has invigorated the political potential of both by enabling them to connect their particular struggles to the more structural processes of capital accumulation and the facilitating role state action plays in opening up spaces for it.

"This is Only the Beginning! The Struggle Continues!"[16]

The days of the Taksim commune are over, but the struggle is certainly not. The political reservoir that this urban greenspace has produced continues to energize social movements and activism. The seeds of direct democracy sown in Gezi Park continue to grow in neighborhood forums, which sprang up in numerous locations all around Istanbul immediately after the park's final evacuation. These forums do not only tackle specific neighborhood issues, but they aim to keep the Gezi experience alive by fostering non-market ways of organizing economic life and problematizing urban transformation and ecological destruction in the city by holding workshops, forming working groups and commissions, and staging protests. They also engage with larger issues of the national political agenda such as the peace process initiated by the government with the Kurdish insurgency. One of the very first squatting experiments in the country took place in November 2013: Don Quixote Social Center, itself stemming from a neighborhood forum, owed much of its inspiration and legitimacy to the politics of reclaiming common space animated in Gezi Park. The Defense of Northern Forests, a network of activists, academics and urban planners, was formed shortly after the park days to launch a campaign against the destruction of greenspaces lying in the northern part of the city by the third bridge, the third airport, Canal Istanbul and urban redevelopment projects.

It is true that the Gezi resistance became somewhat of a multiplier that brought together different groups with different sources of discontent (from secular Kemalists to anti-capitalist Muslims) in an anti-government front. I have tried to demonstrate, however, that the resistance should not be seen as a mere vehicle through which discontent with the authoritarian politics of JDP has been mobilized, but rather be situated within a nuanced and historical reading of the broader political-economic context. I have offered here one such reading from a political ecology perspective. Building on the premise that socio-ecological movements cannot be understood outside their socio-spatial contexts (Kaika and Swyngedouw, 2011), I have argued that the Gezi revolts were never an isolated case from the larger dynamics of restructuring urban and rural space. Gezi Park and the politics built around it reflected the dynamics of capital accumulation, neoliberal enclosures of space and power relations; processes that are hardly unique to the case of Gezi Park, and are deeply imprinted by inequalities.

The fact that the collective grievances and rage of the losers of spatial dynamics of neoliberal capitalism, ranging from villagers fighting for their water in northeast Anatolia to residents of gentrified neighborhoods in Istanbul, found a common voice in the Park was arguably the most distinctive feature of the Gezi revolts. The languages of opposition found within the movement were hardly confined to the

Park but alluded to the larger, more structural processes and injustices discussed above. The movement also went beyond forming an oppositional front and animated the organization of a non-capitalist, communitarian, socio-economic sphere. In these senses, the Gezi Park protests presented a contrast to most socio-ecological movements trapped within a narrow not-in-my-backyard discourse which remain reactive rather than transformative (Kaika and Swyngedouw, 2011). In a related vein, they have been an invaluable experience for those of us—political ecologists—who have a political stake and an analytical interest in the emergence, maintenance and evolution of social movements.

How the long-term impacts of the Gezi revolts will play out in the political setting of the country remains to be seen. But it would not be far-fetched to suggest that they have left an irreversible mark on our collective political imagination and potential. Above all, they have enabled us to rediscover our capacity to self-organize and to reclaim our right to self-determination. The movement's motto, "this is only the beginning, the struggle will continue" holds more meaning for us today than ever.

Notes

* To the memory of Abdullah Comert, Ali Ismail Korkmaz, Ethem Sarisuluk, Medeni Yildirim and Mehmet Ayvalitas who were killed by state violence during the Gezi revolts.

1 See a piece by renowned writer Elif Safak for a glorified example of this position: www.theguardian.com/world/2013/jun/03/taksim-square-istanbul-turkey-protest (accessed March 26, 2014).

2 Certainly, this is not to suggest that the only destructive interventions undertaken under JDP rule are energy investments nor that the JDP government has been an exception in its disregard for the socio-environmental implications of developmentalism. See Akbulut (2011) and Akbulut and Adaman (2013) for a treatment of growth fetishism in Turkey as an integral part of state-making. See also www.ejolt.org/2013/06/turkeys-tree-revolution-part-2-everyday-im-chapulling/ (accessed 16 June, 2014).

3 In mid-2013, the number of SHPs at various operational and planning phases was over 1700.

4 The idea of modernization and catching-up with the West dominates the social imagination like no other in Turkey. More specifically, it has been the building block of the state–society relationships and as the Turkish state has historically achieved its power and legitimacy, first and foremost, from the promise of fulfilling the ideal of modernization (Akbulut, 2011; Akbulut and Adaman, 2013).

5 As will be discussed below, the government then declared a ban on holding meetings and protests in Taksim Square and those who tried to practice this very right were regularly subjected to police brutality.

6 That children from Tarlabasi have set construction scaffolds and huge advertisement boards of the Tarlabasi Renovation Project on fire while marching to Taksim Square is a telling example of this (Ozgur, 2013).

7 The municipality's "tables-and-chairs" operation was the finishing touch for the food services. In the fall of 2010, outside seating was banned in Taksim, which forced many small restaurants and cafés to down-size or shut down altogether. There has been a wide consensus among the urban activists and local business owners that this was indeed a move to open up space for large capital.

8 See www.theguardian.com/world/2013/apr/15/turkey-historic-emek-theatre-final-curtain (accessed March 26, 2014).

9 This particular dimension was later operationalized formally as well, first with the designation of Taksim Square off-limits for May Day celebrations and later with the government's more general ban on the holding of protests in the Square and its surroundings in May 2013.

10 Although political parties of the left as well as non-party leftist organizations were also included in Taksim Solidarity, they have generally been less than motivated. This has usually been interpreted as a sign that the left in Turkey essentializes the labor struggle and belittles social movements, and that they have not attached much political importance to the spatial politics of opposing the enclosure of a public space (Ozgur, 2013).

11 One of the many tweets that circulated on May 31, 2013.

12 The prime minister himself had called the protestors *çapulcular* (looters) during the first few days of the resistance; a name quickly reverted, embraced and used by the protestors when referring to themselves.

13 The protests and street clashes continued after the Park was dismantled. Istanbul, as well as other city centers, was once again turned into a battlefield during the night and day following June 15. Weekly demonstrations to condemn police brutality and the Park's completely unlawful blockade were held for a while with Taksim Solidarity's initiative, but they were heavily intervened by security forces as usual. The demonstrations were gradually stopped with the general realization that fighting against police violence was becoming futile and taking away from the movement's energy.

14 While Taksim Solidarity was still the main body holding a legitimate claim to represent the resistance, it had been considerably slower in responding to the needs and demands that arose out of the organization of the daily life in the Park—unsurprisingly so, as it was an umbrella organization that brought together very different actors, from organized left to grassroots movements, and required lengthy deliberation and discussion before taking action. The absence of a central organizing body, however, turned out to be a positive force as it enabled the protestors to discover their self-organizing potential.

15 Quoted in Erensu, 2013.

16 This phrase was the most commonly chanted slogan during the revolts, and became something of a trademark. It is also telling that the slogan was first chanted during the Emek movie theater protests mentioned earlier.

References

Akbulut, B. (2011) 'State Hegemony and Sustainable Development: A Political Economy Analysis of Two Local Experiences in Turkey'. PhD dissertation, University of Massachusetts, Amherst, MA.

Akbulut, B. and F. Adaman (2013) 'The Unbearable Charm of Modernization: Growth Fetishism and the Making of State in Turkey'. *Perspectives: Political Analysis and Commentary from Turkey*, 5.13.

Akbulut, B. and C.I. Soylu (2013) '¿Tragedia o cercamiento?: Imaginar y promulgar Parque Gezi como bien común' ('Tragedy or Enclosure?: Imagining and Enacting Gezi Park as Commons'). *Ecologia Politica: Cuadernos de Debate Internacional*, 45 (December 2013): 43–47.

Erensu, S. (2013) 'Ekoloji Mucadelesi: Gezi'den Once, Gezi'den Sonra' ('Ecological Struggles: Before Gezi, After Gezi'). *Uc Ekoloji: Doga, Dusunce, Siyaset*. Summer 2013.

Gambetti, Z. (2013) 'Gezi Resistance as Surplus Value'. www.jadaliyya.com/pages/index/12672/the-gezi-resistance-as-surplus-value (accessed March 26, 2014).

Kaika, M. and E. Swyngedouw (2011) 'The Urbanization of Nature: Great Promises, Impasse, and New Beginnings' in *The New Blackwell Companion to the City* (G. Bridge and S. Watson eds), Oxford: Wiley-Blackwell.

Lefebvre, H. (1995) *Writings on Cities*. Oxford: Wiley-Blackwell.

Ozgur, D. (2013) 'How the Gezi Revolt gave Birth to Park Democracy'. *Perspectives: Political Analysis and Commentary from Turkey*, 5.13.

Peck, J. and A. Tickell (2002) 'Neoliberalizing Space', *Antipode*, 34(3): 380–404.

Polat, I. (2013) 'Gezi: Merkezi Yonetimin Vesayetine ve Kentsel Donusume Isyan' ('Gezi: Revolt against the Tutelage of Central Administration and Urban Transformation'). *Uc Ekoloji: Doga, Dusunce, Siyaset*. Summer 2013.

Sahin, U. (2013) 'Ilk Soz: Gezi Direnisi: Bir Demokrasi ve Ekoloji Mucadelesi' ('Prelude: Gezi Resistance: A Democracy and Ecology Struggle). *Uc Ekoloji: Doga, Dusunce, Siyaset*. Summer 2013.

Swyngedouw, E. and N.C. Heynen (2003) 'Urban Political Ecology, Justice and the Politics of Scale'. *Antipode*, 35(5): 898–918.

Taksim Solidarity Press Release, May 29, 2013.

Turan, O. (2013) 'Gezi Parki Direnisi ve Armagan Dunyasi' ('Gezi Park Resistance and the World of Gift'). *Toplumsal Tarih*, 238, October 2013.

Turkish Contractors' Association (TCA) (2012) 'Analysis of the Construction Sector', April 2012.

Union of Chambers of Turkish Engineers and Architects (UCTEA) (2012) Press Release. www.mimarlarodasiankara.org/index.php?Did=4426 (accessed March 26, 2014).

16 Constructing New York City's Urban Forest

The Politics and Governance of the MillionTreesNYC Campaign

Lindsay K. Campbell

Introduction

In 2005–2006, bureaucrats at the New York City Department of Parks and Recreation (DPR) began to marshal quantitative evidence to argue for investment in tree planting as part of Mayor Bloomberg's long-term sustainability plan, PlaNYC 2030, launched in 2007. Concurrently, Bette Midler—the celebrity founder of the non-profit New York Restoration Project (NYRP)—announced her dream of planting one million trees in New York City. These two efforts were brought together as the MillionTreesNYC campaign, a formal public–private partnership to plant and care for one million trees citywide by 2017.[1] Realizing that this effort could not be sustained in isolation, leaders of the campaign created an Advisory Committee that engaged more than 100 environmental organizations. Various programs were then implemented to build public awareness about trees—focusing on the multi-functional benefits of the urban forest—and create a constituency of engaged citizens involved with the campaign. Through engagement in volunteer stewardship programs, residents' labor, support, and enthusiasm was cultivated and harnessed in the planting and maintenance of trees, without devolving strategic decision-making authority to them. At one level, the case demonstrates the role of public, civic, and private actors in the networked governance of a successful, large-scale, urban green-infrastructure campaign.

This chapter probes deeper to explore how a political ecology and discursive approach promotes a critical understanding of the politics surrounding the governance of the urban forest in New York City from 2007 to 2011. It asks: what agendas did actors involved in the governance of urban forestry set? What discourses of the environment and society did they deploy? It explores who participated in decision-making and who did not, what rationales were used, and with what consequences. After briefly situating the study in the literature on political ecology and environmental governance and reviewing my methods, I present an analysis of the MillionTreesNYC campaign to show the complex forces at play in the contemporary construction of the urban forest. The case reveals the presence of pragmatic responses to the neoliberal era simultaneously overlapping with commitments to quality of life, sustainability, and environmental justice. To critique wholesale the neoliberal context is to miss the real commitments to advancing equity and environmental quality within that space; while to universally laud the achievements of the campaign is to miss the opportunity for further advances toward those ends.

Urban Environmental Governance: Actors and Discourses

Political ecology brings attention to the dynamics among actors involved in environmental governance—including the state, civil society, and the public. Municipal urban forestry campaigns can be understood as strategies used by competitive, global cities investing in environmental quality as part of city image-making, within a political-economic context of rescaled, post-industrial, neoliberalism (While et al., 2004; Jonas and While, 2007). Keil and Boudreau (2006) describe these efforts as "roll out environmentalism," whereby the state creates new institutions and governance arrangements that address the results of previous, harsher forms of neoliberalism. Perkins (2011) applies this notion to the "roll out" of urban forestry, wherein the state has recently increased its support for urban forestry initiatives via grants and partnership with non-state actors—a practice which he critiques from a Gramscian perspective. Depending upon one's standpoint, the role of civil society organizations in environmental governance can be celebrated for their progressive potential (Robbins, 2004; Escobar, 2008), as well as critiqued on grounds of accountability, representation, and transparency (Swyngedouw, 2005). In an urban forestry case, Heynen and Perkins (2007) criticize the non-profit Greening Milwaukee for its selective serving of certain residents (e.g. planting trees with homeowners, but not with renters). Despite the debates over accountability, it is clear that civil society is working in networked governance arrangements with the state—including in the management of the urban environment (Bulkeley, 2005; Gustavsson et al., 2009; Connolly et al., 2013).

The public also play a key role in urban natural resource management. Forestry—in both rural and urban contexts—has a long history as a highly scientific, technical, and professionalized field that is managed by bureaucrats and licensed private sector actors, including arborists, foresters, and nursery growers (see, for example Scott, 1998; Ricard, 2005; Konijnendijk et al., 2006). Yet, traditions in community forestry and community-based, natural resource management that came originally from rural contexts and the global South have been adapted more recently for urban and global North settings (Burch and Grove, 1993; Weber, 2000; McCarthy, 2002; Schroeder et al., 2006; Murphy-Dunning, 2009). As well, there is a long tradition of public, voluntary engagement in municipal shade tree commissions and other local environmental stewardship groups (Ricard, 2005). Following a community forestry ethos involves a normative commitment to engaging the public in governance of the urban forest (Konijnendijk, 2013). In some cases, however, the public is engaged as a labor force rather than as decision-makers. Some critical scholars have argued that this practice is a direct result of neoliberalism at work. Because municipalities operating in austere fiscal times lack the resources to ensure the survival of investments in new green infrastructure, volunteers have been enrolled to assist with watering, pruning, and maintenance of trees as well as the maintenance of neighborhood parks (Perkins, 2009, 2013). Brand (2007) argues that neoliberal discourses work to create 'green subjects' whereby people adopt an individualized way of thinking about environmental problems. For example, he critiques this self-regulating environmentalism of personal responsibility, such as encouraging consumers to switch to CFL light bulbs—shifting our focus away from the need for collective action and structural transformation. These individualized appeals can similarly be seen in efforts to get the public to plant or care for trees. Other scholars explore the question of who participates, with what motivations, in the volunteer stewardship of the urban environment, finding stewardship to be a form of individual expression and collective action (Svendsen, 2009). Some have analyzed volunteer tree planting as a form of civic engagement and have found a reciprocal relationship whereby higher levels of volunteer stewardship are linked with higher degrees of participation in other forms of civic activity (Fisher et al., 2011).

Finally, discursive acts also construct the urban forest. Political ecologists examine the discursive construction of nature—urban and otherwise—and bring critical scrutiny to forms of 'official' knowledge codified in state practices (Demeritt, 2002; Robbins, 2004; Castree, 2005). Scott (1998) demonstrates the state's tendency toward calculation, quantification, and simplification in natural resource management, relying upon an instrumental and rationalized view of the environment as a bundle of goods and services. There have been numerous academic and applied efforts to quantify and commodify the value of the urban forest, including the United States Department of Agriculture (USDA) Forest Service models STRATUM and UFORE, both of which are now incorporated into the i-Tree software suite for analyzing urban forest characteristics and benefits. These models, as well as other approaches to urban tree canopy analysis and prioritization are used widely by public land managers and decision-makers to make the case for investing in trees and to advance urban forestry agendas (Grove et al., 2005; Peper et al., 2007; Nowak et al., 2010; Pincetl, 2010). At the same time, qualitative researchers have pointed to the limits of quantification in capturing the full range of values associated with the urban forest, nature, or any phenomena (Livingstone, 1992; Enticott, 2001; Heynen et al., 2007; Robertson, 2007). Perhaps some of the most prominent socio-cultural values associated with trees—such as beauty or aesthetic value, sacredness, personal and place attachment, and cultural heritage—are precisely those that are most difficult to represent in quantitative models (Westphal, 2003; Svendsen, 2009; Svendsen and Campbell, 2010). Further, critical scholars argue that current neoliberal discourses of quantification cast the urban forest as an amenity that creates value for landowners, which varies depending on one's status as a homeowner or renter, thereby creating inequalities (Heynen and Perkins, 2007). Finally, recent work in the urban political ecology tradition seeks to connect the discursive and the material: "The material production of environments is necessarily impregnated with the mobilization of particular discourses and understandings (if not ideologies) of and about nature and the environment" (Heynen et al., 2006: 7). Clearly, the way we understand 'nature' fundamentally shapes our policies and practices of how we build and manage our cities.

Methods and Data Analysis

The selection of New York City as a qualitative case study was grounded in my situated understanding of MillionTreesNYC, as well as my access to gatekeepers who could inform this research (Rose, 1997; Dowling, 2005; Dunn, 2005).[2] My primary method was to conduct semi-structured interviews with municipal and non-profit representatives in planning, urban forestry, and parks. I interviewed a total of 35 subjects: 20 from the public sector, 11 from civil society,[3] and 4 from the private sector. This composition came from using snowball sampling until reaching saturation in interview content. All participants gave informed consent to participate as confidential subjects and to be audio recorded (IRB # 11-714M). Interviews averaged 1–2 hours and covered involvement in urban forestry and sustainability planning; organizational values; program activities; and partnership networks. Recordings were transcribed in full and supplemented with field notes. As secondary methods, I conducted discourse analysis and participant observation. I reviewed documents, plans, and websites released between 2007 and 2011, including: PlaNYC (2007), PlaNYC 2.0 (the April 2011 update to the plan), and MillionTreesNYC documents. This analysis focuses on the broad ideological contours and attendant power relations of these documents (Waitt, 2005). I also utilized participant observation. As a member of the MillionTreesNYC Advisory Committee since 2007, I participated in meetings related to the implementation of the campaign. In addition, I shadowed urban foresters as they selected sites for new trees and received tree deliveries; participated in volunteer planting days; and attended a tree giveaway and a tree composting event. The qualitative analysis software, NVivo, was used to store, code, and analyze the transcripts and documents. My approach is one of grounded theory—where the thematic categories emerged directly from my data (Dey, 1999).

Case Study: Constructing New York City's Urban Forest

The Public Sector: The Mayor, Bureaucrats, and Quantification

New York City has long-standing capacity in creating and maintaining its urban forest and has more recently engaged in municipally led sustainability planning that helped trigger the MillionTreesNYC campaign. DPR is one of the largest urban natural resource management organizations in the world, with approximately 7,000 employees (including seasonal staff) and expenditures of $382 million in fiscal year 2010 (NYC MMR, 2011: 112). The agency manages 29,000 acres of open space and cares for approximately five million trees citywide (NYC DPR, 2012). Created in 2007, PlaNYC 2030 is New York City's long-term sustainability plan. It was the product of top-down leadership from the Bloomberg mayoral administration working across city agencies to identify measurable goals toward 'sustainability' in the areas of land, water, transportation, energy, air, and climate change (City of New York, 2007). DPR bureaucrats marshaled evidence to convince City Hall of the merit of investments in tree planting as a core component of PlaNYC. The agency used information it had spent years collecting and analyzing through data management systems, geographic information systems and remote sensing, and tree censuses. In particular, they used the STRATUM model, which offered a quantitative—and monetized—view of the urban forest. Then-DPR Commissioner Adrian Benepe was quoted in *The New York Times* as saying: "Trees are great for a variety of reasons, but how do you explain that to the Office of Management and Budget?" (Randall, 2007).

The STRATUM report to DPR is infused with the language of business sense and inter-city competition:

> New York City's street trees are a valuable asset, providing approximately $100.2 million or $172 per tree ($15 per capita) in net annual benefits to the community. Over the years, the city has invested millions in its urban forest. Citizens are now receiving a return on that investment—trees are providing $5.60 in benefits for every $1 spent on tree planting and care. New York City's benefit–cost ratio of 5.60 exceeds all other cities studied to date, including Fort Collins, Colorado (2.18), Glendale, Arizona (2.41), and Charlotte, North Carolina (3.25).
>
> (Peper et al. 2007: 2–3)

Much of the economic benefit identified is associated with increases in real estate value and commercial activity on tree-lined streets (Anderson and Cordell, 1988; McPherson and Simpson, 2002; Wolf, 2005; Donovan and Butry, 2010). This appealed to the mayor, who viewed investments in green infrastructure as part of a strategy to attract global talent to live and work in New York City.

Once so convinced, City Hall committed major financial and political support to tree planting. Approximately $400 million in capital funds went to DPR for urban forestry as part of PlaNYC in 2007 (Table 1). A DPR memo said "PlaNYC represents the most significant change in municipal urban greening since the Parks Department first funded citywide curbside tree planting under Robert Moses in 1934" (Watt, 2007). This campaign also became a mayoral priority, which drove an aggressive timeline, reporting, and tracking. This mayoral enthusiasm was catalyzed by the argumentation of long-time DPR bureaucrats; interviewees note that Bloomberg differed from his predecessors in empowering his agency staff to exercise leadership. Finding this space for entrepreneurialism and creativity in hierarchic institutions was no small feat.

Civil Society: An Elite Non-profit Sets its Sights

Concurrently and completely outside of the PlaNYC process, a professionalized non-profit organization with a celebrity founder became interested in developing a citywide tree-planting effort. NYRP is a greening group founded in 1995 by entertainer Bette Midler; at their 2006 spring picnic fundraiser, Midler announced that she wanted to "plant one million trees in New York City." The announcement surprised many of the staff at both NYRP and DPR, who did not view NYRP as

Table 16.1 DPR's PlaNYC funding, with urban forestry related capital funding in bold

8 regional parks	$386 million
290 open schoolyards	$96 million
36 field lighting sites	$42 million
25 synthetic turf fields	$22 million
800 greenstreets	**$15 million**
natural area reforestation	**$150 million**
220,000 street trees	**$226 million**
Total capital budget:	$906 million
7-year pruning cycle	$2.7 million
Stump removal	$2.0 million
Maintenance staff (227)	$10.4 million
Total annual expense budget increase	$15.1 million

Source: NYC DPR. 2011. "MillionTreesNYC-PlaNYC." Presentation. May 16, 2011.

having a forestry agenda or expertise. One interviewee elaborated on the way in which the founder's changing interests shaped the programmatic focus of NYRP:

> I think NYRP really didn't know what it wanted to be . . . This comes from the very leader of the organization, Bette, who—every day there's a priority and a new idea. And so this chaos that surrounds her is what the organization *is*. 'Cause when Bette picks up the phone and says, "Oh my gosh, I was just driving down 157th Street and there's plastic bags in the trees. Stop everything and get everybody up there." Now we're focused on plastic bags and then the next day it's something else . . . NYRP is doing too many things instead of being really good at two things.
>
> (Respondent 35)

But, Midler's celebrity offered her a platform for courting donors, attracting media attention, and gaining audience with public officials. One respondent said, "Bette can definitely pick up the phone and talk to the mayor" (respondent 35). Similarly, this billionaire mayor was comfortable working with high-powered elites, like business magnate and philanthropist George Soros, to support city initiatives with private dollars. The sequence of what exactly transpired between the announcement at the picnic and the formal announcement of NYRP as a partner in MillionTreesNYC remains murky, but it centers on high-level contact between Midler and City Hall—including Mayor Bloomberg and Deputy Mayor Patti Harris, known as one of the key gatekeepers to Bloomberg. Leadership at the Mayor's Office, DPR, and NYRP worked together—initially behind closed doors—to craft a joint tree-planting campaign.

Hybrid Governance: A Formal Public–Private Partnership

PlaNYC's process has been critiqued by some as lacking transparency, with public input seen as token or after the fact (Angotti, 2010). Non-governmental input was formally incorporated into PlaNYC through the Sustainability Advisory Board (SAB). Yet, because of the clear leadership and expertise of DPR in the PlaNYC goal-setting process, respondents noted that the SAB played little role in crafting the forestry-related goals. After the internal agency work was complete, a six-month, public-outreach process was held (ICLEI, 2010). Public engagement largely served to get the word out about PlaNYC to those who were most inclined to care, but did not substantively shift its goals. Some argue that the MillionTreesNYC campaign was used to help cement stronger public buy-in for the overall plan. In fact, the million trees goal was released to reporters one day in advance of the public release of PlaNYC on April 22, 2007 (Rivera, 2007). The tree-planting goal was viewed as a "feel good" issue, in sharp contrast to the political divisiveness of goals like the failed attempt at congestion pricing (ICLEI, 2010).

The MillionTreesNYC campaign fits Konijnendijk's (2013) description of "new hybrid models of urban forestry" (4). The rationale behind this partnership was threefold. First and foremost it was a funding strategy to leverage municipal funding with private dollars. MillionTreesNYC attracted $10 million in donations from Bloomberg Philanthropies and David Rockefeller—demonstrating how the mayor drew on his private wealth and networks to support his "signature initiatives." Then, NYRP secured lead corporate sponsors: Toyota, BNP Paribas, and Home Depot. Over the course of 2007–2010, NYRP grew from a $6 million to a $13 million organization—an astronomical leap for the local environmental non-profit sector. Second, the joint effort sought to employ the strengths of each partner. In an ideal view, the bureaucratic expertise in tree planting and economies of scale of the large municipal agency would be balanced by the nimble innovation of the non-profit organization—particularly given NYRP's savvy in outreach, marketing, and special events. Third, the campaign needed both partners to plant citywide, across land jurisdictions. DPR, working with private contractors, would plant street trees, in parks, and reforest thousands of acres of "natural areas." NYRP would plant on public housing grounds, schoolyards, "publicly accessible private lands," and give away trees to residents.

Both formal and informal "rules of the game" were articulated in this new hybrid partnership (Konijnendijk, 2013). The partnership was institutionalized through a formal Memorandum of Understanding (MOU) that articulated goals, roles, and responsibilities. Although not written into the MOU, partners agreed on a ratio of planting targets: 60 percent DPR and 40 percent NYRP. The MOU also established a $35 million fundraising goal for the campaign for NYRP (MTNYC, 2008: 7). The parties routinized their contact through a number of means: constant email contact, monthly tree-operations meetings about planting decisions, biweekly meetings between the directors of MillionTreesNYC at DPR and NYRP, and biweekly joint meetings called the MillionTreesNYC Taskforce. Some respondents saw the Taskforce as an important means of communication, but others argued that these meetings consisted of "reporting back" tree-planting metrics rather than making joint decisions. Despite the regular contact, some respondents leveled the critique that MillionTreesNYC was a divided partnership.

Figure 16.1 Launching the MillionTreesNYC campaign as a public–private partnership. From left to right: Drew Becher, former Exectuive Director, NYRP; Adrian Benepe, former Commissioner, DPR; Michael Bloomberg, former Mayor of the City of New York; Bette Midler, founder of NYRP; Patricia Harris, former First Deputy Mayor of the City of New York.

Source: Malcolm Pinckney, City of New York Department of Parks and Recreation, 2007.

The two sides differed in mandate and capacity. A sense of accountability or public trust was seen as being crucial to DPR's ability to successfully meet the targets of the campaign and appropriately spend public funds. NYRP, in contrast, was motivated by a "moral obligation"—the passions, commitments, and interests of its founder (respondent 28). One manager elaborated on these differences:

> I think basically [NYRP's] involvement is optional . . . They get the credit no matter what they do . . . Who's going to sit down and analyze who did what and who claimed what credit? And have they really met their goal? . . . I think [DPR is] held much more to account within the government structure partly 'cause its public money. We're entrusted with this money and how we spend it is deeply important to how . . . we're perceived . . . that's part of the public trust. (Respondent 27)

Finally, several respondents indicated a stark contrast in the capacity of these groups as tree-planting entities. These differences necessitated mutual learning across DPR and NYRP and required a re-working of tree-planting targets to 70 percent DPR and 30 percent NYRP.

Leaders on both sides of the partnership, however, brought a sense of distributional justice to the implementation of the campaign. DPR bureaucrats were cognizant of the way in which the urban forest had developed unevenly over the course of the past several decades, because street trees were previously planted via a request-based system. The managers saw this large-scale campaign as a chance to correct these inequalities, planting first in the neighborhoods that were most lacking in trees. This built on an existing DPR program that preceded PlaNYC called 'Trees for Public Health.' One respondent reflected on the city's process of setting PlaNYC's tree-planting goals:

> We started doing the math with [DPR] about where there were and were not street trees, [and] it became clear that this was an initiative that was actually as much or more about environmental justice as it was about creating elite property values. And so once we set the goal that says, "Look, every place that it is feasible to put a sidewalk tree, we would like to put a sidewalk tree," you immediately have a policy that fills in the valleys. And, frankly, there aren't that many places on the Upper East Side that you can put more street trees, but there are lots of places in the South Bronx. And so it was one of these things that turned a transition from a hard infrastructure plan into a sustainability plan, and an elitist, global competitive story into a quality of life-for-all story, [this] is to my mind a lot of the magic of what we did.
>
> (Respondent 49)

So, while certain rationales around the global, competitive city might have motivated Bloomberg and his City Hall staff, other sets of values and practices—including commitments to quality of life and environmental justice—were introduced by agency staff through the implementation of the tree-planting initiative. In parallel, the belief that everyone deserves access to clean, green neighborhoods was also one of the core driving forces behind Midler's creation of NYRP. Thus, the organization is committed to greening in "high need neighborhoods" (respondent 10). Even when managers didn't use the language of 'justice,' they sought to have a defensible rationale that they could present to the public for why they planted in what places and in what sequence.

Networked Governance: Advisory Committee

The politics of resource management in this case do not stop with the relations between DPR and NYRP; a broader set of stakeholders are involved with MillionTreesNYC. Realizing that the campaign would be prominent in the organizational landscape of New York City, its leaders deliberately crafted roles for their allies via the MillionTreesNYC Advisory Committee—a group of approximately 400 individual members from 109 organizations. The campaign sought to cast a wide net and include the broad network of experts, leaders, and line staff engaged in urban forestry and natural resource stewardship citywide in this effort. Research reveals the existing diversity and number of civic, public, and

Figure 16.2 Tree stewardship in East New York, Brooklyn, NY, USA—one of the Trees for Public Health neighborhoods prioritized for block planting due to low street tree stocking levels.

Source: Susan Kornacki, MillionTreesNYC, City of New York Department of Parks and Recreation, 2013.

private organizations involved in environmental work in New York City, with nearly 2,800 civic stewardship groups identified citywide as of 2007 (Fisher et al., 2012; Connolly et al., 2013). One respondent described the city's existing "baggage of people, personalities, organizations, and events" as something that had to be addressed in the creation of this campaign (respondent 28).

The Advisory Committee offered a threefold benefit: (1) it provides a means for outside experts to contribute ideas, resources, skills, and programs to the campaign. Leaders noted the procedural importance of the committee in light of critique of PlaNYC as top-down. One respondent noted a desire not to "reinvent the wheel" with this campaign, and to build on the existing expertise and investments of dozens of groups citywide. (2) The committee helped head off critique before it emerged. By being asked to participate, potential critics or competitors—especially in the crowded context of the non-profit world—would feel invested in the campaign. (3) The broad committee membership aimed to support the longevity of the campaign, particularly beyond the 2014 change in mayoral administration. One campaign leader used the language of creating a "movement" around urban forestry (respondent 15).

Although the intention of the Advisory Committee was one of shared governance, it did not always live up to this ideal. The body was organized into seven thematic subcommittees (Tree Planting; Stewardship; Education and Programs; Community Outreach; Marketing and Public Relations; Public Policy Initiatives; and Research and Evaluation). The structure of the subcommittee leadership was intended to include representatives from DPR, NYRP, and an outside organization. As entirely volunteer positions, the level of engagement of outside entities waxed and waned over time and across issue areas, depending on organizational circumstance and challenges each subcommittee faced. For example, the stewardship subcommittee ended up giving rise to a formalized, funded stewardship program called StewCorps. For the research and evaluation subcommittee, academic and governmental researchers were interested in accessing data and field sites, conducting research, and publishing articles about MillionTreesNYC; this productive self-interest helped keep the subcommittee active. Of the overall Advisory Committee, one respondent acknowledged the varying levels of engagement of different committee members: "There's always going to be great lists of organizations and only 5 percent are really going to do the work" (respondent 28). Moreover, while some participants valued the opportunity to give input, others argued that it felt 'token' and their expertise underutilized. One leader within the campaign asked: "Are they advising us or are we advising them? . . . The relationship between the Advisory Committee, Parks, and NYRP is very unclear" (respondent 11).

Engaging or Harnessing the Public? Outreach and Stewardship Programs

MillionTreesNYC also sought to educate the public about the urban forest through outreach and public relations. Building from the language of PlaNYC around greening the growing city, MillionTreesNYC advanced a storyline that trees help make communities more liveable—and competitive—in the face of that growth. An excerpt from the campaign's website demonstrates this framing:

> **Why Plant a Million Trees?**
>
> New York City is **growing**! You can see it—and **feel** it—in every neighborhood in every borough. It's exciting, and it's what makes New York the greatest city in the world. But, like in any thriving metropolis, it's important to make sure the Big Apple and its residents—meaning **you**!—are healthy and happy while adjusting to the growth and the many changes it will bring with it.
>
> Planting trees is one of the most beneficial [hyperlink to a page about urban forest benefits] and cost-effective ways to help ease these growing pains. Trees help clean our air, and reduce the pollutants that trigger asthma attacks and exacerbate other respiratory diseases. They cool our streets, sidewalks, and homes on hot summer days. Trees increase property value, and encourage

neighborhood revitalization. And trees make our City an even more beautiful and comfortable place to live, work, and visit.

(MillionTreesNYC website, 2010: milliontreesnyc.org)

The multiple benefits of trees were celebrated in public remarks and outreach campaigns. Citing research that trees help mitigate urban heat island effect, improve air quality, create liveable streets, and provide broad psycho-social benefits (Nowak et al., 2010), the campaign could leverage different arguments with different constituencies. For example, a large-scale print advertisment featured images of trees and people in the city, touting trees as 'Zen masters' ("Trees do more than you think. They promote relaxation and fitness, enhance our emotional and mental health, and even encourage us to drive a little slower") and 'exercise partners' ("While protecting us from the sun, they encourage outdoor play and exercise—helping in our fight against obesity"). Interviewees felt that outreach was successful in making MillionTreesNYC one of the most visible efforts of PlaNYC, but converting public awareness to public engagement required developing hands-on stewardship programs, as illustrated by three examples.

First, street trees in New York City are planted by private firms that are contractually obligated to guarantee the trees for two years after planting. For all of PlaNYC's capital commitments to tree planting, managers, critics, and advocates felt that it needed a greater commitment of maintenance funds. This became more acute following cuts to the maintenance budget in 2008 after the global financial crisis. One municipal employee said, "Nobody wants to be the mayor of a city with a million dead trees" (respondent 1). Cognizant of this danger, MillionTreesNYC developed stewardship programs including an online Adopt-a-Tree website and free tools giveaways for gardening in tree pits. A StewCorps program offered formal tree-care training and certification. But, as of November 2012, just 5,506 trees were adopted citywide and approximately 1,500 stewardship actions were reported online. A campaign leader noted that many more trees were adopted than were reported online, but there is no clear mechanism for tracking that activity. Moreover, developing sustained grassroots stewardship of street trees remains a challenge, not just in New York City, but in many cities across the country.

Second, approximately two-thirds of the first 500,000 trees planted were on so-called 'natural areas' through afforestation and reforestation practices.[4] The 2008 recession led to 30 percent budget cuts in all city agencies. This led DPR to shift its reforestation from working with city employees and contractors to using volunteers. Now, each fall and spring, approximately 20,000 trees are planted in single-day, volunteer-planting events. Despite the significant professional preparatory work that is required, using volunteers for their physical labor still remains a significant cost-cutting measure. Although cost-savings was one driving rationale, DPR also hoped that these volunteers would come to feel more invested in park sites that are often overlooked or less visible than traditional recreational sites. From the perspective of leaders developing MillionTreesNYC as a volunteer program, active engagement with tree-planting events is one of the key points of contact

Figure 16.3 Volunteer stewards at a MillionTreesNYC reforestation planting.
Source: Malcolm Pinckney, City of New York Department of Parks and Recreation, 2013.

between the public and the campaign, and, indeed, many volunteers have become sustained participants in MillionTreesNYC events each season.[5]

Third, NYRP organizes free, public tree giveaways in order to build awareness about trees, facilitate planting on private land, and develop support for the campaign amongst the public. Finally, 'influence plantings' are perhaps the grayest area in terms of tracking the effect of the campaign. Included in these counts is an estimate that 25 percent of trees sold at area tree retailers, such as Home Depot, are planted by New York City residents.

Discussion

This study follows Swyngedouw and Heynen (2003) and others in the project of bringing political ecology to the global North and urban realms, by tracing the politics of actors involved in the governance of the urban forest in New York City. While some critical scholars have considered current state–civil society arrangements around urban forestry as neoliberal hegemony at work (Perkins, 2011), I reveal a more complicated picture. First, the municipal government commitment of leadership, funding, and staff in support of PlaNYC's tree-planting goals reflects the strong, on-going role of local government in urban forestry. We can critique that these capital funds were a one-time infusion in the face of on-going cuts to the maintenance budget, particularly after the 2008 financial crisis. But the net effect of the strategic goal setting and implementation of the campaign was increased investment in both the extent and equity of the urban forest over a multi-year campaign. The spending of those public funds and the targeting of neighborhoods with low tree canopy was first guided by the bureaucrats within city government pursuing a distributional justice agenda. Indeed, as Keil and Boudreau (2006) have noted, municipal workers can be agents of progressive change.

MillionTreesNYC also fits Konijnendijk's (2013) description of "hybrid models of urban forestry" with the thorough entanglement of public and private, planting across spatial turf and over time. We see the pragmatic leveraging of public resources with private resources via the partnership. There is space in this political arena for certain, well-positioned, non-profit actors to engage. We must attend to the variation in the degree of professionalization of civil society groups (Carmin, 1999; Andrews and Edwards, 2005; Fisher et al., 2012), as not all civic groups are grassroots or represent 'the public.' We can certainly raise questions about the accountability of non-profits and call attention to the need for transparency and responsiveness in governance (Swyngedouw, 2005). In addition, I note differences in: mandate, motivation, capacity, and expertise of the two sides of the public–private partnership. DPR, motivated by a sense of the public trust and accountability to spend public monies, brought technical expertise and economies of scale in urban forestry operations. NYRP, motivated by the commitments of its founder, brought the nimble ability to raise private dollars, do marketing and outreach, and hold innovative events. This case shows, however, that leaders of the campaign attempted to create a structure for consultation and advising through the Advisory Committee. I therefore complicate the notion of hybridity to one of *networked* governance—noting the dozens of civic, public, and private actors who were brought into this campaign as advisors. The role of those advisors varied over the course of the campaign and across the issues their subcommittee was addressing.

More work remains to be done to fully engage the public as citizens of the urban forest. Despite normative calls for the engagement of the public as stakeholders in urban forestry, we are not seeing the devolution of decision-making authority to residents (Perkins, 2009). For the most part, the public is viewed as recipients or *consumers*: of messages, of educational activities, of stewardship programs, of trees, and of ecosystem services. Indeed, MillionTreesNYC successfully built a constituency of allies for the urban forest, while at the same time harnessing their labor in its planting and care. This model can be contrasted with other forms of community forestry (such as New Haven's Urban Resources Initiative) where decision-making is devolved; where tree planting is the means to community empowerment, not the end goal (Murphy-Dunning, 2009). But we must continue to explore the public's role as active agents or *producers* even when a community forestry model is not explicitly in use—including their positive experience as volunteers and as stewards. For example, in a study of MillionTreesNYC volunteer stewards, Fisher et al. (2011) found that participation in volunteer tree stewardship correlates with other forms of civic engagement. We need to more fully explore what 'participation' and 'meaningful involvement' in a tree-planting campaign entails, as well as who benefits in that process and how.

The case also reveals how a large-scale, urban tree-planting campaign discursively frames the environment and the city, both through internal agenda-setting and external public relations. The urban forest is quantified, monetized, and promoted for its multiple benefits. And trees and other green infrastructure are seen as part of the strategy for building a 21st-century sustainable city. This discourse fits, in part, with a neoliberal understanding of the urban forest as an amenity that gets harnessed into image-making of the city as 'green' (Heynen and Perkins, 2007). It is important to note, however, that while leaders at City Hall were convinced by economic arguments, part of the success of the campaign comes from its flexible discourse that was broad enough to incorporate diverse sets of actors with very distinct motivations to engage with the urban forest. The messaging of the campaign focused on the multi-dimensional benefits of the urban forest, including sociocultural benefits that are much harder to monetize. Moreover, throughout the implementation of the campaign, we also see discursive and material practices that show a commitment to environmental justice and quality of life.

We can raise questions of what is missing or lost in the promulgation of these discourses across countless plans and campaigns. In terms of governance: how do we square the notion of 'infrastructure' as something that the state provides with the neoliberal (and pragmatic) notion that residents must be involved in the care and maintenance of green infrastructure in order to ensure that it survives and functions? What complexity is obscured in the quantification of benefits and services? What are the reasons for planting trees that are simply not quantifiable? Finally, we can pose the question of: sustainability for whom? Future research should more fully examine the question of who benefits from these investments in green infrastructure and who is excluded. And we must explore how those benefits and beneficiaries evolve over time as the multi-year campaign unfolds and as the city changes politically, ecologically, and demographically. It is clear that leaders of the MillionTreesNYC campaign are attuned to the need for distributional justice at the local scale in the implementation of the campaign. But also, particularly if the campaigns are using rhetoric around climate change and global environmental phenomena, we need to examine justice in a multi-scalar way to ask if there any trade-offs within the region or globally (Heynen, 2003). What are the benefits, costs, and opportunity costs of such a campaign? Normatively, why should we plant one million more trees in New York City? What other programs must go hand-in-hand with our tree-planting efforts? Many critical scholars note the way in which local environmental programs can enhance quality of life, but caution against pursuing such policies in the absence of or as a substitute for explicit social justice policies.

Notes

1 The original completion date was scheduled for 2017, but because the campaign is ahead of schedule as of October 2013, MillionTreesNYC leaders expect to finish by the end of 2015.

2 I am uniquely situated to reflect on the politics of natural resource management in New York City. I have been working as a researcher with the USDA Forest Service at the New York City Urban Field Station since 2002 (See: www.nrs.fs.fed.us/nyc/).

My professional role puts me in direct involvement with leaders in the environmental field in New York City. I am situated as both a participant in and a researcher of the MillionTreesNYC campaign, as I have served on the Advisory Committee and the Research and Evaluation Subcommittee in addition to conducting this study. As such, a critical stance that simply dismisses the effort as neoliberalism at work is not sufficient to me; I am working to explore and unravel the full complexity of the case from critical and embedded standpoints.

3 Interviewees included employees of formal non-profits and members of informal community groups involved in the campaign, but did not include members of the general public.

4 These 51 sites include woodlands, meadows, marshes, and wetlands; they are larger in acreage than typical recreational parks and comprise a total of more than 8,700 acres citywide (City of New York DPR, 2012).

5 Having participated in MillionTreesNYC volunteer plantings and thoroughly enjoyed it, I recognize the dissonance between the abstract notion of my labor being 'harnessed' and the felt, affective experience of planting trees as a voluntary, leisure, or civic engagement practice. This study does not explore the motivations and experiences of volunteer stewards engaging with the campaign, including the multi-dimensional benefits of participation (but see Fisher et al., 2011).

References

Anderson, L.M. and Cordell, H.K. (1988) 'Influence of Trees on Residential Property Values in Athens, Georgia (USA): A Survey Based on Actual Sales Prices', *Landscape and Urban Planning*, 15: 153–164.

Andrews, K. and Edwards, B. (2005) 'The Structure of Local Environmentalism', *Mobilization*, 10(2): 213–234.

Angotti, T. (2010) 'PlaNYC at Three: Time to Include the Neighborhoods', *Gotham Gazette*, April 2010, www.gothamgazette.com/article/Land%20Use/20100412/12/3239 (accessed 7 May 2010).

Brand, P. (2007) 'Green Subjection: The Politics of Neoliberal Urban Environmental Management', *International Journal of Urban and Regional Research*, 31(3): 616–632.

Bulkeley, H. (2005) 'Reconfiguring Environmental Governance: Towards a Politics of Scales and Networks', *Political Geography*, 24(8): 875–902.

Burch, W.R., Jr. and Grove, J.M. (1993) 'People, Trees and Participation on the Urban Frontier', *Unasylva*, 44: 19–27.

Carmin, J. (1999) 'Voluntary Associations, Professional Organizations, and the Environmental Movement in the United States', *Environmental Politics*, 8: 101–121.

Castree, N. (2005) *Nature*, Routledge, New York City.

City of New York. (2007) 'PlaNYC: A Greater, Greener New York', www.nytelecom.vo.llnwd.net/o15/agencies/planyc2030/pdf/full_report_2007.pdf (accessed 25 March 2014).

City of New York. (2011) 'Preliminary Mayor's Management Report', www.nyc.gov/html/ops/downloads/pdf/mmr/0211_mmr.pdf (accessed 9 April 2013).

City of New York Department of Parks and Recreation. (2012) 'About Parks', www.nycgovparks.org/about/faq (accessed 9 April 2013).

Connolly, J.J., Svendsen, E.S., Fisher, D.R. and Campbell, L.K. (2013) 'Organizing Urban Ecosystem Services through Environmental Stewardship Governance in New York City', *Landscape and Urban Planning*, 109: 76–84.

Demeritt, D. (2002) 'What is the "Social Construction of Nature"? A Typology and Sympathetic Critique', *Progress in Human Geography*, 26(6): 767–790.

Dey, I. (1999) *Grounding Grounded Theory: Guidelines for Qualitative Inquiry*, Academic Press, San Diego, CA.

Donovan, G.H. and Butry, D.T. (2010) 'Trees in the City: Valuing Street Trees in Portland, Oregon', *Landscape and Urban Planning*, 94(2): 77–83.

Dowling, R. (2005) 'Power, Subjectivity and Ethics in Qualitative Research', in I. Hay (ed.) *Qualitative Methods in Human Geography, 2nd Edition*, Oxford University Press, Oxford, pp. 19–29.

Dunn, K. (2005) 'Interviewing', in I. Hay (ed.) *Qualitative Methods in Human Geography, 2nd Edition*, Oxford University Press, Oxford, pp. 79–105.

Enticott, G. (2001) 'Calculating Nature: The Case of Badgers, Bovine Tuberculosis and Cattle', *Journal of Rural Studies*, 17; 149–164.

Escobar, A. (2008) *Territories of Difference*, Duke University Press, Durham, NC.

Fisher, D.R., Campbell, L.K., and Svendsen, E.S. (2012) 'The Organizational Structure of Urban Environmental Stewardship', *Environmental Politics*, 21(1): 26–48.

Fisher, D.R., Connolly, J.J., Svendsen, E.S., and Campbell, L.K. (2011) 'Digging Together: Why People Volunteer to Help Plant One Million Trees in New York City.' Environmental Stewardship Project at the Center for Society and the Environment of the University of Maryland White Paper no. 2. 36 pp. www.nrs.fs.fed.us/nyc/local-resources/downloads/Digging_Together_White_Paper.pdf (accessed 20 June 2013).

Grove, J.M., O'Neil-Dunne, J, Pelletier, K., Nowak, D., and Walton, J. (2005) 'A Report on New York City's Present and Possible Urban Tree Canopy' USDA Forest Service Northeastern Research Station. Report Prepared for: Fiona Watt, Chief Forestry & Horticulture, Department of Parks & Recreation, City of New York. 25pp. www.nrs.fs.fed.us/nyc/local-resources/downloads/Grove_UTC_NYC_FINAL.pdf (accessed 22 October 2012).

Gustavsson, E., Elander, I., and Lundmark, M. (2009) 'Multilevel Governance, Networking Cities, and the Geography of Climate-change Mitigation: Two Swedish Examples', *Environment and Planning C: Government and Policy*, 27(1): 59–74.

Heynen, N. (2003) 'The Scalar Production of Injustice within the Urban Forest', *Antipode*, 35(5): 980–998.

Heynen, N. and Perkins, H. (2007) 'Scalar Dialectics in Green: Urban Private Property and the Contradictions of the Neoliberalization of Nature', in N. Heynen, J. McCarthy, S. Prudham and P. Robbins (eds) *Neoliberal Environments: False Promises and Unnatural Consequences*, Routledge, London, pp. 190–201.

Henyen, N., Kaika, M., and Swyngedouw, E. (2006) 'Urban Political Ecology: Politicizing the Production of Urban Natures', in N. Heynen, M. Kaika, and E. Swyngedouw (eds) *In the Nature of Cities: Urban Political Ecology and the Politics of Urban Metabolism*, Routledge, London, pp. 1–20.

Heynen, N., McCarthy, J., Prudham, S., and Robbins, P. (2007) 'Introduction', in N. Heynen, J. McCarthy, S. Prudham and P. Robbins (eds) *Neoliberal Environments: False Promises and Unnatural Consequences*, Routledge, London, pp. 1–22.

ICLEI. (2010) 'The Process Behind PlaNYC. How the City of New York Developed its Comprehensive Long-Term Sustainability Plan', April 2010. www.nytelecom.vo.llnwd.net/o15/agencies/planyc2030/pdf/iclei_planyc_case_study_201004.pdf (accessed 26 March 2014).

Jonas, A.E.G. and While, A. (2007) 'Greening the Entrepreneurial City?: Looking for Spaces of Sustainability Politics in the Competitive City', in R. Krueger and D. Gibbs (eds) *The Sustainable Development Paradox: Urban Political Economy in the United States and Europe*, Guilford Press, New York City, pp. 123–159.

Keil, Roger and Boudreau, Julie-Anne. (2006) 'Metropolics and Metabolics: Rolling Out Environmentalism in Toronto', in N. Heynen, M. Kaika, and E. Swyngedouw (eds) *In the Nature of Cities: Urban Political Ecology and the Politics of Urban Metabolism*, Routledge, London: 41–62.

Konijnendijk, C. (2013) 'From Government to Governance: Urban Forests as New Commons', Paper presented at *Urban Forests and Political Ecologies* conference. 19 April 2013. Toronto, Canada.

Konijnendijk, C.C., Ricard, R.M., Kenney, A., and Randrup, T.B. (2006) 'Defining Urban Forestry: A Comparative Perspective of North America and Europe', *Urban Forestry & Urban Greening*, 4: 93–103.

Livingstone, D. (1992) 'Statistics Don't Bleed: Quantification and its Detractors' in *The Geographical Tradition*, Blackwell, Malden, MA, pp. 304–328.

McCarthy, J. (2002) 'First World Political Ecology: Lessons from the Wise Use Movement', *Environment and Planning A*, 34: 1281–1302.

McPherson, E.G. and Simpson, J.R. (2002) 'A Comparison of Municipal Forest Benefits and Costs in Modesto and Santa Monica, California, USA', *Urban Forestry and Urban Greening*, 1: 61–74.

MillionTreesNYC. (2008) 'Memorandum of Understanding between New York City Department of Parks and Recreation and New York Restoration Project', September 19, 2008, 13pp.

Murphy-Dunning, C. (2009) 'From Front Yards to Street Corners: Revitalizing Neighborhoods through Community-based Land Stewardship', in L. Campbell and A. Wiesen (eds) *Restorative Commons: Creating Health and Well-Being through Urban Landscapes*, General Technical Report, US Department of Agriculture, Forest Service, Northern Research Station, Newtown Square, PA, pp. 154–163.

Nowak, D.J., Stein, S.M., Randler, P.B., Greenfield, E.J., Comas, S.J., Carr, M.A. and Alig, R.J. (2010) *Sustaining America's Urban Trees and Forests: A Forests on the Edge Report.* Gen. Tech. Rep. NRS-62, US Department of Agriculture, Forest Service, Northern Research Station, Newtown Square, PA.

Peper, P.J., McPherson, G.E., Simpson, J.R., Gardner, S.L., Vargas, K.E. and Xiao, Q. (2007) 'New York City, New York Municipal Forest Resource Analysis', Center for Urban Forest Research, USDA Forest Service, Pacific Southwest Research Station.

Perkins, H. (2009) 'Out from the (Green) Shadow? Neoliberal Hegemony through the Market Logic of Shared Urban Environmental Governance', *Political Geography*, 28(7): 395–405.

Perkins, H. (2011) 'Gramsci in Green: Neoliberal Hegemony through Urban Forestry and the Potential for a Political Ecology of Praxis', *Geoforum*, 42(5): 558–566.

Perkins, H. (2013) 'Consent to Neoliberal Hegemony through Coercive Urban Environmental Governance', *International Journal of Urban and Regional Research*, 37(1): 311–327.

Pincetl, S. (2010) 'Implementing Municipal Tree Planting: Los Angeles Million-Tree Initiative', *Environmental Management*, 45(2): 227–238.

Randall, D.K. (2007) 'Maybe only God Can Make a Tree, but only People can Put a Price on it', *New York Times*, 18 April 2007. www.nytimes.com/2007/04/18/nyregion/18trees.html?_r=0 (accessed January 2, 2013).

Ricard, R.M. (2005) 'Shade Trees and Tree Wardens: Revising the History of Urban Forestry', *Journal of Forestry*, 103(5): 230–233.

Rivera, R. (2007) 'A Shady Plan by the Mayor that's Likely to be Popular', *New York Times*, April 22, 2007.

Robbins, P. (2004) *Political Ecology: A Critical Introduction*, Blackwell, London.

Robertson, M. (2007) 'The Neoliberalization of Ecosystem Services: Wetland Mitigation Banking and the Problem of Measurement', in N. Heynen, J. McCarthy, S. Prudham and P. Robbins (eds) *Neoliberal Environments: False Promises and Unnatural Consequences*, Routledge, London, pp. 114–125.

Rose, G. (1997) 'Situating Knowledges: Positionality, Reflexivities and other Tactics', *Progress in Human Geography*, 21(3): 305–320.

Schroeder, R., St Martin, K. and Albert, K. (2006) 'Political Ecology in North America: Discovering the Third World Within?' *Geoforum*, 37(2): 163–168.

Scott, J. (1998) *Seeing Like a State: How Certain Schemes to Improve the Human Condition Have Failed*, Yale University Press, New Haven, CT.

Svendsen, E. (2009) 'Cultivating Resilience: Urban Stewardship as a Means to Improving Health and Well-Being', in L. Campbell and A. Wiesen (eds) *Restorative Commons: Creating Health and Well-Being through Urban Landscapes*, General Technical Report. US Department of Agriculture, Forest Service, Northern Research Station, Newtown Square, PA, pp. 58–87.

Svendsen, E. and Campbell, L. (2010) 'Living Memorials: Understanding the Social Meanings of Community-Based Memorials to September 11, 2001', *Environment and Behavior*, 42: 318–334.

Swyngedouw, E. (2005) 'Governance Innovation and the Citizen: The Janus Face of Governance-beyond-the-state', *Urban Studies*, 42(11): 1991–2006.

Swyngedouw, E. and Heynen, N. (2003) 'Urban Political Ecology, Justice and the Politics of Scale', *Antipode*, 35: 898–919.

Waitt, G. (2005) 'Doing Discourse Analysis', in I. Hay (ed.) *Qualitative Research Methods in Human Geography, 2nd Edition*, Oxford University Press, Oxford, pp. 163–191.

Watt, F. (2007) 'PlaNYTree', NYC Department of Parks and Recreation internal memorandum, June 26, 2007.

Weber, E.P. (2000) 'A New Vanguard for the Environment: Grass-Roots Ecosystem Management as a New Environmental Movement', *Society and Natural Resources*, 13: 237–259.

Westphal, L.M. (2003) 'Social Aspects of Urban Forestry: Urban Greening and Social Benefits: A Study of Empowerment Outcomes', *Journal of Arboriculture*, 29(3): 137–147.

While, A., Jonas, A. and Gibbs, D. (2004) 'The Environment and the Entrepreneurial City: Searching for the Urban "Sustainability Fix" in Manchester and Leeds', *International Journal of Urban and Regional Research*, 28(3): 549–569.

Wolf, K.L. (2005) 'Business District Streetscapes, Trees and Consumer Response', *Journal of Forestry*, 103(8): 396–400.

17 Reimagining Ecology in the City of Cape Town

Contemporary Urban Ecological Research and the Role of the African Centre for Cities

Pippin Anderson

An Ecological Journey: Arriving at Contemporary Cape Town

Crosby (1986) notes the slow conquest of Africa by European colonialists when compared to other regions of the world and attributes this to the challenging nature of African ecosystems, being as they were 'too fecund, too untamed and untamable' (Crosby, 1986: 137). Cape Town, South Africa, the focus of this chapter, and the most south-western city on the African continent, presented just such an untamed wilderness to its early settlers. A consideration of the contemporary ecology of the City of Cape Town must be viewed in the context of a long history of indigenous occupation, recognition of the region as a centre of exceptional biodiversity, and a trying passage to European colonization marked by social and environmental resistance, and more recently a half century of apartheid planning. In the context of post-apartheid Cape Town, the African Centre for Cities (ACC) urban ecology CityLab at the University of Cape Town aims to reframe urban ecology research as more than just a biological endeavour but as a function of its unique social and political circumstance. This chapter reflects on the role of this research entity in drawing together research and researchers and in striving towards greater interdisciplinary engagement and more appropriately framed urban ecology. Contemporary understandings of the ecology of the City are presented and particular attention is given to three cases, two forest-related and one around community indigenous gardening, all of which highlight the complexities and challenges of meeting conservation and anthropogenic needs in a developing city.

The region of present-day Cape Town has been inhabited for millennia, first by San hunter gatherers and then more recently in the last 2000 plus years by Khoi herders (Deacon, 1992). The Dutch were the first European settlers when they set up a victualing station at the Cape to support the fleets of the Dutch East India Company as they plied the spice route. The route past the Cape was arduous but the profits of the spice trade made this journey worth the high associated risks. Indeed an early trip past the Cape described at the time as a profitable journey saw the loss of two of the original four ships and the lives of some 80 to 100 men, being half of the original crew that set out (Crosby, 1986). The Cape was readily noted as a useful mid-way point between Europe and the East that offered a most-desired bundle of ecosystem services, in particular in the form of its perennial water and abundant wildlife (Anderson and O'Farrell, 2012). Settlement long eluded the Europeans, however, as a rocky shoreline and hostile seas severely limited access, and once land was attained, the aboriginal population was resistant to incursion. Crosby (1986) holds that Africa was untamable, and the African people made a significant contribution in doing what taming they could and that this in turn aided colonization. African people had a superior understanding of the benefits, challenges and potential usefulness of their landscapes frequently typified by considerable dangerous wildlife and harsh conditions (Crosby, 1986). Certainly early writings from the Cape note the particular environmental knowledge of the local people (Anderson and O'Farrell, 2012). Slow agricultural establishment, continued threat from dangerous animals, long wet and stormy winters and dry summers characterized by drought and gale-force winds and natural veld fires all retarded a thriving European settlement at first (Anderson and O'Farrell, 2012). Through a process of 'taming' and reordering of nature, with the systematic eradication of predators, the canalization of rivers, and the uptake of indigenous environmental knowledge, settlement took off and the City of Cape Town slowly came into being (Anderson and O'Farrell, 2012). This reconfiguration to a useable, benign landscape, informed by different agents in time, saw the early and subsequent consistent, degradation of the natural environment.

The ecology of the City of Cape Town and its spatial arrangement in and around the City has been variably used in the process of settlement, for example for the systematic removal of waste, for provisioning both the site itself and a considerable body of passing people and trade, and in denoting rank and then race. In his model of the apartheid city, Davies (1981) notes a white residential core with suburban extension in sectors of desirable environment sharply differentiated by socio-economic status. While reflecting on the apartheid city of Durban, he describes white settlement as clustering with respect to 'desirable elevated land with seaward aspects, good views and ventilation and water front localities' (Davies, 1964: 27). The City of Cape Town has battled to shake off its apartheid form and still echoes Davies' model. Historically white and wealthier suburbs cluster around the lower slopes of Table Mountain and the seaboard areas, where they undoubtedly enjoy some environmental shelter and the most environmentally aesthetic areas of the City, while historically black and coloured[1] areas are found on the lowlands, characterized by mobile sands and seasonal inundation in winter due to an exceptionally high water table. Here there is certainly a landscape of the wealthy characterized by an environmental aesthetic, in close proximity to Table Mountain National Park and the sea, and could be construed as a 'landscapes of awe' (Duncan and Duncan, 2004). Cast in striking relief are those environmentally marginal or sacrificial 'blight' landscapes relegated to the poor (Duncan and Duncan, 2004). In Cape Town, who accesses and uses what landscapes remains largely racially informed. Within this crude differentiation, at a finer-scale, environmental determinants are seen to play out, for example, in the wealthier areas where housing prices are 'trimmed' by the strong south-easterly winds ubiquitous to the Cape

Town summer, with the most expensive dwellings in the most sheltered areas (Small, 2012). Today about 40 per cent of Cape Town's households are classified as poor without the means to access basic necessities of food and shelter (CoCT, 2007). The City is characterized by an inefficient and fragmented spatial framework that preserves racial inequalities inherited from the apartheid era. Post-apartheid spatial changes have tended to exacerbate this fragmentation and segregation (Lemanski, 2007; Turok 2001).

Conservation Agendas

Just as history has informed current spatial form and perceptions of nature and ecology in the City of Cape Town, so too has a singular evolutionary history given rise to the exceptional biodiversity that marks the Cape Town region. Cape Town is situated in the Cape Floristic Region, the smallest and most diverse of the earth's six floral kingdoms (Myers et al., 2000; Holmes et al., 2008). The Fynbos vegetation which typifies the Cape Floristic Region is low in stature, shrubby, and fire adapted. Characterized by tussock-like members of the *Restionaceae*, small heath-like *Ericaceae*, and shrubby *Proteaceae*, this vegetation type, which has numerous sub-types, seldom exceeds 2 metres in height with the rare exception of some of the larger *Proteaceae*. True forest, indigenous Afromontane forest, is rare in the Cape. While the area is bioclimatically suitable for trees, slow maturity rates in response to nutrient-poor soils combined with a fire frequency in the region of 7–15 years means trees are largely excluded and only naturally present in fire refugia, such as deep kloofs (Rebelo et al., 2006). The region is home to some 9000 plant species on just 90 000 km^2, accounting for 44 per cent of the flora of the subcontinent on a mere 4 per cent of the land area (Rebelo et al., 2006). Shifts in global ocean currents and a resulting period of aridification in the Miocene (26 million years ago) were likely responsible for the historical changes in vegetation cover of the region from a more luxuriant cover of sub-tropical thicket to the shrubbier Fynbos vegetation of today. The exceptional diversity of the Fynbos vegetation is attributed to a long period of evolution, high edaphic diversity, and a complex physiography presenting unique micro-climates and intermittent isolation due to fire-patch driven fragmentation resulting in species isolation (Linder, 2005). The metropolitan area of the City of Cape Town is currently home to an estimated 3,350 plant species of which 190 are endemic to the City. Despite these apparently high numbers, urbanization has taken its toll on this diversity and the metropolitan region of Cape Town has 11 of 21 critically endangered vegetation types listed nationally (Rebelo et al., 2011). The combination of high natural biodiversity and the considerable threat posed by transformation to urban form gives the region its status as a biodiversity hotspot (Myers et al., 2000). Challenges to conservation include ongoing land conversion to meet housing and economic-development demands at odds with a biodiversity-conservation agenda, and in a form that reinforces the existing sprawling nature (Goodness and Anderson, 2013). The suppression of native vegetation by non-native invasive species, and altered fire regimes with too-regular accidental fires or entirely suppressed fire in response to human settlement see altered ecological drivers in this urban setting that obstruct biodiversity conservation in the City (Holmes et al., 2012). Conservation areas, and open greenspace, while enjoyed can also pose challenges to city dwellers. Fuel for the spread of fires poses a particular threat to those in informal settlements and those bordering parks and conservation areas. Canalized and poorly managed riverine areas result in flooding posing a significant health and safety risk to those living in close proximity. Cape Town has a high crime rate (City of Cape Town, 2007) and open greenspace and wilderness areas are often sites of misdeed and as a result perceived as dangerous and undesired (Holmes et al., 2008; Goodness and Anderson, 2013).

Greenspace in Cape Town today takes various forms and includes formal parks and recreation sites, brown field sites of abandoned land or degraded remnant patches yet to be formally transformed, and conservation areas. Ownership and responsibility for conservation in the City is devolved to various tiers of government and as a result a single vision for conservation in the City is hard to secure (Goodness and Anderson, 2013). National Government, through their conservation agency South African National Parks (SANPARKS), manages the Table Mountain National Park which is the largest conservation entity in the City comprising of 25,000 hectares of predominantly high-lying land. The lowland areas of the City, where the bulk of the diversity of vegetation types lie, do not share the same degree of protection and are under considerable threat from development (Holmes et al., 2012). Vegetation types confined to these lowland areas are poorly conserved and currently fall below nationally derived conservation targets (Rebelo et al., 2006). In almost all cases insufficient remnants remain to conserve representative diversity. In these lowland areas there are a number of smaller nature reserves and most of these come under the management of the City of Cape Town local government.

While the case for nature, and associated ecosystem services, in cities is being better and more effectively demonstrated and articulated, there is still a perception of nature, and in particular conservation, as a rural endeavour. Biodiversity conservation is rarely considered central to the urban fabric, but perceived as a device for capturing and securing wild nature of a type generally excluded by urban form and function (Sandberg and Wekerle, 2009). As urban dwellers we often believe we are less dependent on immediate and locally delivered ecosystem services as many of our provisioning needs are outsourced beyond the city (O'Farrell et al., 2012; Ernstson et al., 2008). By this way of thinking, what we choose to do with our remnant greenspace in cities becomes a matter of taste or culturally constructed values (Ernstson, 2008). Growing empirical research demonstrates the multifaceted importance of urban nature. However, in the developing context the preservation of, and funding for, open greenspace, is still pitted against demands for housing and economic opportunities. There is still a need to make the linkages between greenspace and human well-being a universally recognized value, and not the preserve of the rich. Whose visions and needs are met in these instances becomes an important political question (Ernstson, 2008). Recently, there is growing recognition that nature conservation is not a pure-bred activity and nowhere is this truer than in cities (Haila, 2012). Just as intertwined economic, political, social and ecological processes together form our highly uneven and unjust urban landscapes

Figure 17.1 The flora of Cape Town, South Africa, is predominantly Fynbos, a low and scrubby vegetation type. The footprint of the City has had a devastating impact with the resultant loss of large portions of now critically endangered vegetation types. The most significant losses have been in the low-lying areas of the City, in particular on what are termed the Cape Flats.

Source: Pippin Anderson, 2013.

(Swyngedouw and Heynen, 2003; Robbins, 2004), so must these same factors be engaged in forging new, scientifically robust, sustainable and socially just urban landscapes.

Contemporary Ecologies and the African Centre for Cities CityLab Project

The rate of urbanization in Africa is currently the highest in the world (UN-Habitat, 2010a). Here, urbanization is characterized by a disjuncture between urban population growth and growth in the formal economic sector, and urban poverty is typical (Freund, 2007; UN-Habitat, 2010b). The spatial form is sprawling and predictions of the associated increase in land cover with urban growth in Africa put this at an astounding 700 per cent over the 2000–2030 period. Untethered to traditional economic factors, and with no concomitant increase in agriculture or industrial development, urbanization in Africa tends to be anchored to natural resource exploitation and export. As much as 62 per cent of the urban form in sub-Saharan Africa is characterized by informal settlement. Also singular to African urbanization are the rapid growth of smaller towns, persistent rural–urban linkages in recently urbanized communities (Anderson et al., 2013a), and a dependency on the natural environment among the urban poor (Schaffler and Swilling, 2013). This rapid urbanization tends to be poorly governed. The ACC, a University of Cape Town research entity, was established in 2006 to serve as a platform for research on complex urban issues in the global South from an African perspective (Anderson et al., 2013b). Deviating from a narrow scholarship focused on development challenges, the ACC endeavours to understand and seek solutions to reverse growing urban inequality and environmental degradation. The ACC aims to build knowledge networks between research institutions across Africa and at the same time grow appropriately trained urban professionals who can manage Africa's cities.

In 1997, Parnell suggested that South African cities provide a laboratory for urban studies towards understanding African cities with a view to identifying systemic responses. Based on this notion of city as laboratory, in 2008 the ACC set up the Cape Town CityLab programme. The intention of the programme is to broker interdisciplinary engagement, both across academic disciplines and between the academy and broader society, towards new knowledge generation, knowledge sharing, and the creation of partnerships to engage with and address the issues pertinent to the African urban situation (Anderson et al., 2013b). The programme hosts a number of CityLabs and these are thematic, or geographically defined, and take various forms and paths. CityLabs have been formed around, for instance, the issue of densification, climate change, violence and urban health. While each CityLab has the broad mandate of producing a publication that reflects the state of knowledge for the area in question, each undertakes to expose, understand, problematize, contribute to, and be generative in agenda setting and directing active research by varying degrees and in different manners. The growing international evidence of the linkages between urban nature and social well-being and sustainability, the particular vagaries of the ecologies of African cities as characterized above, and the ostensible links between biodiversity infrastructure and poverty alleviation (Schaffler and Swilling, 2013) all make an understanding of, and proactive approach to, urban ecology a critical part of any attempt to engage in the urban crisis presented in Africa. These broad African ecological considerations—and the particular challenges of a developing continent, and indeed often conflicting views and agendas, outlined above with respect to matters relating to urban ecology in the City of Cape Town—make it an obvious thematic area for deliberation and an urban ecology CityLab was added to the stable of CityLabs emerging in the ACC in 2010.

What is presented below is an overview of urban ecology work in the City of Cape Town broadly between 2010 and 2012, with reflections on the role of the ACC urban ecology CityLab in this research and practice terrain. The urban ecology CityLab took the route of a series of academic seminars and some more informal knowledge-sharing sessions and field trips, and used these to inform a special issue publication on the social and ecological research in the City of Cape Town. While in its first two years the urban ecology CityLab did not carry out any research per se, it served to expose certain issues and pieces of research currently underway (reframing these in an urban ecology framework) and served as a platform for knowledge sharing and critique. The urban ecology CityLab, guided by the focus of the ACC on the particular nature of urban Africa, sought to situate the urban ecology of the City of Cape Town in its true social and political context. The CityLab was catalytic in other areas, giving rise to the conceptualization and creation of a postgraduate urban ecology course at the University of Cape Town, hosting the UN-Habitat Cities and Biodiversity Outlook Africa workshop towards their recent book publication which features a chapter characterizing the urban ecology of the African continent (Anderson et al., 2013a), and more broadly in promoting partnerships and knowledge sharing in taking the urban ecology research agenda forward. It was also undoubtedly informative in shaping the professional expertise and career of the convener.

Little urban ecological research, in the strict sense, has been done in South Africa (Cilliers and Siebert, 2012). Two notable exceptions are an active research group at the University of the North West, and work emanating from the City of Durban (Cilliers and Siebert, 2012). As a biodiversity hotspot and the only city which includes a national park, urban ecology studies in Cape Town have been largely driven by an urban-nature conservation agenda (Cilliers and Siebert, 2012). With the initiation of the ACC, and the establishment of a dedicated urban ecology CityLab, there has been an increase in the amount of urban ecology-focused research in the City. Work predating the start of this urban ecology group in 2009 was predominantly, but not entirely, dominated by biological and conservation considerations. Much of this is in keeping with Corbyn's critique of ecological research carried out in cities where frequently the particularly urban character of the setting is overlooked and not engaged, on the basis of being too complex, and therefore historically shunned by traditional ecologists (Corbyn, 2010). More recently, researchers have started to engage in the exciting challenges presented by the urban context and also, though perhaps lagging a little, the social dimension of ecology in the City of Cape Town. The ACC's urban ecology CityLab certainly endeavoured to make the political dimensions of urban ecology more evident.

A reflection on recent and current work, and that presented in the special issue culminating from the urban ecology CityLab and its seminar series in its first two years since its inception, shows a substantial understanding of the ecology in the City of Cape Town (Anderson and Elmquvist, 2012). There is evidence of the significant role of the 'spatial' with respect to ecological and social processes where distances to resources, connectivity and boundaries are all important drivers of social engagement and ecological function (O'Farrell et al., 2012; Hoffman and O'Riain, 2012; Pauw and Louw, 2012). Growing recognition of the relevance of historical patterns of land use and urban planning emerging through various eras of governance in presenting a particular spatial form which dictates contemporary landscape-level function and engagement is in keeping with current views in political urban ecology (Swyngedouw and Heynen, 2003; O'Farrell et al., 2012). Spatial form dictates how species respond to the City where proximity to people has given rise to, for example, people–animal conflict situations with baboon (*Papio*) troops entering the City (Hoffman and O'Riain, 2012), and clear fragmentation and isolation effects among certain bird species (Pauw and Louw, 2012). Ecological research in Cape Town has demonstrated a specificity around species response, or at least functional type, in the urban that points to a need for considerable additional research echoing similar calls in the international literature (Pauw and Louw, 2012; Hoffman and O'Riain, 2012). While we know enough to demonstrate the importance of the role of spatial form, empirical case studies demonstrating how this plays out are limited to a few species and serve to caution against generalizing.

A study in 2012 by the University of Cape Town's urban ecology class—a class established on the impetus of the urban ecology CityLab—explored the relationship between urban greenspace and distanced travelled by users. An anticipated trend was evident, where users tend to be those living closest to the greenspace. An interesting exception, however, relates to the Green Point Urban Park. This park, a soccer world cup legacy project, designed and built in 2010, is a significant recreational space untrammelled by apartheid segregationist history. The park offers playgrounds for different ages designed with input from relevant health and children's development professionals, and fitness equipment for adults. The centre of the park hosts a biodiversity garden, cast very much in a pragmatic and user-focused manner, demonstrating, for example, which indigenous plant species might grow best in different parts of the City. When looking at users of this park and the distance people travel to access it, what is evident is that for an amenity such as this, people are prepared to travel much further. This demonstrates both the value of, and desire for, creatively designed urban greenspace. The evident interest and apparent appeal of open greenspace demonstrated in this study flags the environmental injustice of findings from a recent study by O'Farrell et al. (2012) exploring distances between open greenspace and schools in Cape Town which shows poor access to urban greenspaces on the Cape Flats, the historical apartheid suburbs of

Figure 17.2 The Green Point Urban Park, constructed as a soccer world cup legacy project, presents a significant urban greenspace in post-apartheid Cape Town. The park presents a multitude of recreational options and includes a biodiversity showcase garden. Research shows the park is accessed by people from further afield than in the case of other recreational open greenspace around the City, demonstrating the desire for well-designed public open space in Cape Town. Here school children explore the biodiversity garden with the Cape Town Stadium in the background.

Source: Jane Battersby, 2011.

Cape Town. We know greenspace is much desired, in particular by the youth (Ashwell, 2010), and that access remains severely limited in particular in historically racially marginalized and current poor communities. This is not an area of redress picked up in post-apartheid South Africa, and clearly one worthy of attention (Turok, 2001).

We know that ecosystem services in the City of Cape Town generate significant income every year in both the formal and informal economies (de Wit et al., 2009; Petersen et al., 2012), and knowledge of where these ecosystem services 'play out' in the landscape, and who might benefit from them, is becoming more apparent and has the potential to better inform spatial planning and management (O'Farrell et al., 2012). Societal expectations and perceptions inform ecological process, for example in the management of invasive plant species, and understanding and managing these attitudes is one of the greatest challenges to promoting a functional ecology in the city (van Wilgen et al., 2012). Some interesting cases around ecosystem services, forests, and the social use of indigenous flora in Cape Town in recent years demonstrate some of the contested nature of ecology in the City. While trees are planted in abundance in private space, they are rare in large quantities in public space and, as mentioned previously, virtually absent in the indigenous vegetation. An interesting contestation that has arisen in recent years is in the management of old plantation forestry areas in Cape Town. Land under large tracts of pine plantation forestry in the vicinity of Table Mountain National Park, planted in late 1800s, was recently handed over to SANPARKS. This shift in ownership from private concession to National Park saw a move to clear these plantations and restore indigenous vegetation of a type that is critically endangered (Rattle, 2012). A fire exposed a surprisingly intact and fertile indigenous seed bank in clear-felled areas despite the 100-odd years of forestry (Rattle, 2012). Following extensive public consultation and a plan that allowed for a diversity of activities and land use within the park, the process of clear-felling of these areas began with the simultaneous burning and some seed augmentation for restoration back to indigenous Fynbos. This brought on an outcry from some of the middle-income members in the surrounding suburban areas who had used the site for dog-walking, horse-riding, mushroom collecting, and hiking for a considerable period of time. The argument put forward by those opposed to the felling of the trees was for the need for shade for certain recreation activities, something that the indigenous Fynbos flora of the Cape, being short and scrubby, generally does not provide, putting forward a strong socially informed ecosystem services argument for the preservation of this cultural ecosystem service (Rattle, 2012). The cases put forward for the felling were based on arguments for the restoration and preservation of indigenous biodiversity. Following a long consultative public process led by SANPARKS, at present a process of clear felling, burning for restoration, and an allowed re-establishment period to ensure a good seed bank is secured in select places, is being followed. The suggestion is to in fact replant pine trees in some places to meet these demands. So often in developing regions there is an assumption that conservation is the preserve of the wealthy and opposition will come from the poor. Here an interesting case of activism in favour of the retention of a non-indigenous, plant-dominated landscape versus one informed by the unique local biodiversity played out among a largely middle-income community. The case demonstrates interesting perceptions around forests in an otherwise largely unforested landscape, and serves to expose the role of history and experience in informing people's views and values, and awakens questions about which ecosystem services are sought in an urban landscape (Ernstson et al., 2008).

Another interesting case of trees or forested areas relates to the introduction of non-indigenous acacia trees. Originally these trees were introduced in the early 1800s as a means of stabilizing extensive mobile sand dunes across the Cape Flats. The trees of choice were informed by the fact that, at the time, the Cape came under the rule of the British and was exposed to numerous biological introductions from its sister colony Australia. Since their original introduction, many of these acacia species have spread extensively. How these large wooded stands are viewed

Figure 17.3 Cape Town is renowned for its exceptional plant diversity, and views around biodiversity vary among the populace. Recently, the clearing of extensive tracts of exotic pine forestry and the systematic restoration of these areas back to indigenous Fynbos brought outcries from local users of these areas who have enjoyed using them for recreation, noting in particular their shade in an otherwise largely un-treed, indigenous flora. The 'Shout for Shade' campaign sought to halt clearing, citing cultural ecosystem service arguments.

Source: Jonathan Schrire, 2012.

now varies across interest groups and serves to highlight the often complex and 'messy' nature of urban ecology. In some camps these acacia stands are viewed as highly problematic. Termed 'alien invasives' these trees are held accountable for significant loss in local indigenous biodiversity (Rebelo et al., 2006), water-thirsty in an arid region with demonstrated 30–70 per cent reduction in run-off in these dense stands of trees compared to uninvaded Fynbos (Richardson, 1998), contributors to altered fire regimes with higher biomass loads contributing to larger and hotter fires (van Wilgen, 2012), and are typically noted as sites of anti-social behaviour where almost impenetrable thickets allow for undesirable activities out of the public gaze. While the biodiversity and conservation concerns might be middle class, though not necessarily so, the associated social negatives are felt by the urban poor, in particular on the Cape Flats where these stands occur. Many of these Australian acacias are among those species targeted by the government public works programme, Working for Water, developed to improve run-off and develop biodiversity-related jobs (Turpie et al., 2008). As a result these acacia species have now also become synonymous with employment opportunities and social upliftment (Turpie et al., 2008). Wood is traditionally collected in these wooded areas by the urban poor both for personal fuel use and for sale to the public. While current clearing efforts allow this ongoing use, the full eradication (an unlikely outcome, but one that is sought) would pose a livelihood threat to the urban poor who use these stands as woodlots. A further layer of complexity, and not one that is much considered, is added by the fact that these trees still serve to secure otherwise mobile sand dunes which when remobilized become highly problematic in the urban setting. These cases related to what might loosely constitute urban forests show the significance and relevance of trees in an otherwise largely tree-free, indigenous flora and the diverse views and perceptions, and indeed uses, of trees through time. Cohen (1999) suggests that as a species we are predisposed to trees and the potential for fetishization, with trees pegged as an environmental solve-all, is enormous. The case of Cape Town in this respect warrants closer examination, where there is a significant tension between social agendas promoting urban forest and those in favour of indigenous vegetation.

A further interesting case of urban ecological activism is the planting of indigenous flora in the Bottom Road neighbourhood (Ernstson, 2012). Here, community members in a neighborhood on the Cape Flats have followed a route of civic-led restoration, cleaning up a wasteland area covered in acacia trees and dumped rubbish. Tapping in to one of the public works programmes, Working for Wetlands, and working with managers from an adjacent conservation area, people from this Cape Flats neighbourhood were able to access plant material. Very much framed in the rhetoric of reclaiming space and an indigenous heritage, and making beautiful and perhaps in some respects healing past injustices in post-apartheid Cape Town, this small group of community members developed a large tract of indigenous garden along the shores of one of the Cape Flats wetland areas. Work exploring the ecological value has shown this small, socially engaged indigenous planting scheme to have a small, but significant, impact on broader measures of ecological functioning such as soil health and functional diversity (Avlonitis, 2011). In this case ecology is used to include and heal and stands in contrast to traditional exclusionary conservation efforts. These case studies presented show how perceptions and engagements around urban ecologies are swayed by political, economic, social, and ecological arguments and influences that vary through time and space and present contemporary conundrums without clear solutions. We are reminded of the messiness of conservation, especially in the urban context (Haila, 2012).

While the work presented here is not the sole preserve of the ACC's urban ecology CityLab, the Lab has served to pool these merging understandings, bringing together practitioners and scientists towards common understandings framed in contemporary theories. Striving for interdisciplinary work did see successful knowledge sharing and writing partnerships forged across hierarchies of practice. That said, the somewhat divided discipline of ecology, and the multifaceted aspects of urban ecology, did not lend itself to cross-discipline collaborations. So we had government and city officials working with academic researchers, indeed these types of partnerships proved particularly productive with respect to writing where those in government had less time and access to the necessary resources to write, but brought experience and novel insights to the process. We failed to secure good relationships between, for example, water ecologists and those working more broadly on ecosystem services or historical ecologists and environmental economists. This terrain of true interdisciplinary work, critical to understanding the social, political, economic, and ecological aspects of the city, remains a challenge to the CityLab into the future. The CityLab also saw the significant personal growth of the convenor of the Lab who was exposed to all aspects of the urban ecology research arena, and went on to design and now runs an urban ecology course at the University of Cape Town. In these first two years, the urban ecology CityLab has presented a special feature publication outlining the current state of knowledge of urban ecology in the City of Cape Town. In the process, the CityLab has served to facilitate engagements between the academy and broader society, hopefully contributing to durable relationships between different entities working in this space. The CityLab has certainly served to problematize urban ecology in the City and has been an agent in re-casting old framings, and providing a space for conversations around new imaginings of ecology in the City—better informed by contemporary theories and thinking. The CityLab aims to revisit the urban ecology research agenda in the City and to guide and facilitate active research in this arena into the future.

Conclusions

The City of Cape Town has a remarkable natural environment that is interwoven in the everyday lives of the people of the City and plays out at multiple scales. In many respects, this is a complex love–hate relationship, but one that must persist for the resilience of the City (Cartwright et al., 2012). The ecology of the City of Cape Town today has been forged by global and local political processes (Swyngedouw and Heynen, 2003; Robbins, 2004), played out on a particular canvas painted by a unique set of evolutionary, and more recently, historical forces. What emerges is an array of conservation and utilitarian issues bounded together in the functioning ecology of the City of Cape Town. Social perceptions around nature in the City are hugely variable, and this is compounded by inequitable access to greenspace, and nature more generally. Administering to humanitarian issues frequently takes precedence over biodiversity conservation and conservation agendas are hard to meet. Political ecology suggests that environments and the associated populace are mutually produced, leading to complex strengths and vulnerabilities of these ecological systems (Crosby, 1986; Swyngedouw and Heynen, 2003). What emerges is a management challenge with significant cases of scale where local populations desire certain, and not always agreed upon, landscapes and have particular development and social needs and demands, and global conservation concerns, sometimes but not always met with sympathy locally, which must be met at the local level. The management challenges of meeting these multiple, and frequently conflicting, anthropogenic and conservation goals are readily apparent. Amidst a growing understanding of the functional ecology of the City of Cape Town emerges complex and varied visions of what urban nature should be and whose vision that should be remains unclear. The challenge moving forward is to establish how we make an ecology that is fair, robust, has an appropriately South flavour, and yet also contributes to global conservation initiatives.

Note

1 While the label 'coloured' was used as part of a system of oppression during apartheid, it is also a lived identity in South Africa and Cape Town today and we use it here in that manner, but in full recognition of its complex and frequently problematic history (Ernstson, 2012).

References

Anderson, P.M.L. and Elmqvist, T. (2012) Urban Ecological and Social-ecological Research in the City of Cape Town. *Ecology and Society*, 17(4): 23. DOI: http://dx.doi.org/10.5751/ES-05076-170423.

Anderson, P. and O'Farrell, P. (2012) An Ecological View of the History of the Establishment of the City of Cape Town. *Ecology and Society*, 17(3): 28. www.ecologyandsociety.org/vol17/iss3/art28/ (accessed 29 November 2013).

Anderson, P., Okereke, C., Rudd, A. and Parnell, S. (2013a) Regional Assessment of Africa. In T. Elmqvist, M. Fragkias, J. Goodness, B. Guneralp, P.J. Marcotullio, R.I. McDonald, S. Parnell, M. Sendstad, M. Schewenius, K.C. Seto, and C. Wilkinson (eds) *Urbanization, Biodiversity and Ecosystem Services: Challenges and Opportunities. A Global Assessment.* Springer, New York City.

Anderson, P., Brown-Luthango, M., Cartwright, A., Farouk, I. and Smit, W. (2013b) Brokering Communities of Knowledge and Practice: Reflections on the African Centre for Cities' CityLab Programme. *Cities*, 32: 1–10.

Ashwell, A. (2010) Identity and Belonging: Urban Nature and Adolescent Development in the City of Cape Town. Unpublished PhD thesis, University of Cape Town.

Avlonitis, G. (2011) Understanding Urban Ecology: Exploring the Ecological Integrity of Small Scale Greening Interventions in the City of Cape Town. Unpublished thesis, University of Cape Town.

Cartwright, A., Oelofse, G., Parnell, S. and Ward, W. (2012) Climate at the City Scale. In A. Cartwright, S. Parnell, G. Oelofse, and S. Ward (eds) *Climate Change at the City Scale: Impacts, Mitigation and Adaptation in Cape Town.* Routledge, London, UK.

Cilliers, S.S. and Siebert, S.J. (2012) Urban Ecology in Cape Town: South African Comparisons and Reflections. *Ecology and Society*, 17(3): 33. http://dx.doi.org/10.5751/ES-05146-170333 (accessed 9 September 2013).

City of Cape Town. (2007) *City Statistics and Population Census: City Statistics.* Strategic Information, Strategic Development Information and GIS Department, City of Cape Town, South Africa.

City of Cape Town. (2010) Demographics Scenario—City of Cape Town Discussion Paper. Strategic Development Information and GIS Department, Cape Town.

Cohen, S. (1999) Promoting Eden: Tree Planting as the Environmental Panacea. *Cultural Geographies*, 6: 4–24.

Corbyn, Z. (2010) Ecologists Shun the Urban Jungle. *Nature News*.16 July. www.nature.com/news/2010/100716/full/news.2010.359.html (accessed 29 March 2014).

Crosby, A.W. (1986) *Ecological Imperialism: The Biological Expansion of Europe, 900–1900.* Cambridge University Press, Cambridge, UK.

Davies, R.J. (1964) Social Distance and the Distribution of Occupational Categories in Johannesburg and Pretoria. *South African Geographical Journal*, 46: 24–39.

Davies, R.J. (1981) The Spatial Formation of the South African City. *GeoJournal*, 2: 59–72.

Deacon, H.J. (1992) Human Settlement. In R.M. Cowling (ed.) *The Ecology of Fynbos: Nutrients, Fire and Diversity.* Oxford University Press, Cape Town.

De Wit, M.P., Van Zyl, H., Crookes, D.J., Blignaut, J.N., Jayiya, T., Goiset, V. and. Mahumani, B. K. (2009) Investing in Natural Assets. A Business Case for the Environment in the City of Cape Town. Report prepared for the City of Cape Town, Cape Town, South Africa, 18 August.

Duncan, J. and Duncan, N. (2004) *Landscapes of Privilege: The Politics of the Aesthetic in an American Suburb*. Routledge, New York City.

Ernstson, H. (2012) Re-translating Nature in Post-apartheid Cape Town: The Material Semiotics of People and Plants at Bottom Road. In R. Heeks (ed.) *Actor-Network Theory for Development: Working Paper Series*, Institute for Development Policy and Management, SED, University of Manchester, Manchester, UK.

Ernstson, H., Sörlin, S. and Elmqvist, T. (2008) Social Movements and Ecosystem Services—The Role of Social Network Structure in Protecting and Managing Urban Green Areas in Stockholm. *Ecology and Society,* 13(2): 39 [online] www.ecologyand society.org/vol13/iss2/art39/.

Freund, B. (2007) *The African City: A History*. Cambridge University Press, Cambridge, UK.

Goodness, J. and Anderson, P. (2013) Local Assessment of Cape Town: A 'Rainbow Nation's' Social-ecological Landscape—Navigating Management Complexities of Urbanization, Biodiversity, and Ecosystem Services in the Cape Floristic Region. In T. Elmqvist, M. Fragkias, J. Goodness, B. Guneralp, P.J. Marcotullio, R.I. McDonald, S. Parnell, M. Sendstad, M. Schewenius, K.C. Seto, and C.Wilkinson (eds) *Urbanization, Biodiversity and Ecosystem Services: Challenges and Opportunities. A Global Assessment.* Springer, New York City.

Haila, Y. (2012) Genealogy of Nature Conservation: A Political Perspective. *Nature Conservation*, 1: 27–52.

Hoffman, T.S. and O'Riain, M.J. (2012) Monkey Management: Using Spatial Ecology to Understand the Extent and Severity of Human–baboon Conflict in the Cape Peninsula, South Africa. *Ecology and Society*, 17(3): 13. www.ecologyandsociety.org/vol17/iss3/art13/ (accessed 9 September 2013).

Holmes, P., Wood, J. and Dorse, C. (2008) LAB Biodiversity Report: City of Cape Town. City of Cape Town, Cape Town. http://archive.iclei.org/fileadmin/template/project_templates/localactionbiodiversity/user_upload/LAB_Files/Final_Biodiv_Reports/Biodiversity_Report_CCT-LAB_2008.pdf (accessed 4 September 2013).

Holmes, P., Rebelo, A.G., Dorse, C. and Wood, J. (2012) Can Cape Town's Unique Biodiversity be Saved? Balancing Conservation Imperatives and Development Needs. *Ecology and Society*, 17(1): 8. http://dx.doi.org/10.5751/ES-04552-170228 (accessed 4 September 2013).

Lemanski, C. (2007) Global Cities in the South: Deepening Social and Spatial Polarisation in Cape Town. *Cities*, 24(6): 448–461.

Linder, H.P. (2005) Evolution of Diversity: The Cape Flora. *Trends in Plant Science*, 10(11): 536–541.

Myers, N., Mittermeyer, R.A., Fonseca, G.A. and Kent, J. (2000) Biodiversity Hotspots for Conservation Priorities. *Nature*, 403: 853–858.

O'Farrell, P.J., Anderson, P.M.L., Le Maitre, D. and Holmes, P.M. (2012) Insights and Opportunities Offered by a Rapid Ecosystem Service Assessment in Promoting a Conservation Agenda in an Urban Biodiversity Hotspot. *Ecology and Society*, 17(3): 27. www.ecologyandsociety.org/vol17/iss3/art27/ (accessed 29 November 2013).

Parnell, S. (1997) South African Cities, Perspectives from the Ivory Tower of Urban Studies. *Urban Studies*, 34(5–6): 891–906.

Pauw, A. and Louw, K. (2012) Urbanization Drives a Reduction in Functional Diversity in a Guild of Nectar-feeding Birds. *Ecology and Society*, 17(2): 27. www.ecologyandsociety.org/vol17/iss2/art27/, (accessed 29 November 2013).

Petersen, L., Moll, E., Collins, R. and Hockings, M. (2012) Devlopment of a Compendium of Local, Wild-harvested Species Used in the Informal Economy Trade, Cape Town, South Africa. *Ecology and Society,* 17(2): 26.http://dx.doi.org/10.5751/ES-04537-170226 (accessed 29 March 2014).

Rattle, J. (2012) Ways of Knowing Nature: Conflicting Notions of Land Use and Ecosystem Services in Cecilia Forest, Cape Town. Unpublished thesis, University of Cape Town.

Rebelo, A.G., Holmes, P.M., Dorse, C. and Wood, J. (2011) Impacts of Urbanization in a Biodiversity Hotspot: Conservation Challenges in Metropolitan Cape Town. *South African Journal of Botany*, 77: 20–35.

Rebelo, A.G., Boucher, C., Helme, N., Mucina, L. and Rutherford, M.R. (2006) Fynbos Biome. In L. Mucina and M.R. Rutherford (eds) *The Vegetation of South Africa, Lesotho and Swaziland*, South African National Biodiversity Institute, Pretoria, South Africa, pp. 158–159.

Richardson, D. (1998) Forest Trees as Invasive Aliens. *Conservation Biology*, 12(1): 18–26.

Robbins, P. (2004) *Critical Introductions to Geography: Political Ecology*. Blackwell Publishing, Oxford.

Sandberg, A. and Wekerle, G. (2009) Reaping Nature's Dividends: The Neoliberalization and Gentrification of Nature on the Oak Ridges Moraine. *Journal of Environmental Policy and Planning*, 12(1): 41–57.

Schaffler, A. and Swilling, M. (2013) Valuing Green Infrastructure in an Urban Environment Under Pressure—The Johannesburg Case. *Ecological Economics*, 86: 246–257.

Small, E. (2012) The Segregation City and Cape Town: A Biophysical Approach to Cape Town's Spatial Formation. Unpublished thesis, University of Cape Town.

Swyngedouw, E. and Heynen, N.C. (2003) Urban Political Ecology, Justice and the Politics of Scale. *Antipode*, 35(5): 898–918.

Turok, I. (2001) Persistent Polarisation Post-Apartheid? Progress Towards Urban Integration in Cape Town. *Urban Studies*, 38: 2349.

Turpie, J.K., Marias, C. and Blignaut, J.N. (2008) Payments for Environmental Services in Developing and Developed Countries. The Working for Water Programme: Evolution of a Payments for Ecosystem Services Mechanism that Addresses both Poverty and Ecosystem Service Delivery in South Africa. *Ecological Economics*, 65(4): 788–798.

UN-Habitat (United Nations Human Settlement Programme) (2010a) *The State of African Cities 2010: Governance, Inequality and Urban Land Markets*. UN-Habitat, Nairobi.

UN-Habitat (United Nations Human Settlement Programme) (2010b) *The State of the World's Cities, 2010/2011: Bridging the Urban Divide*. Earthscan, London.

van Wilgen, B.W., Forsyth, G.G. and Prins, P. (2012) The Management of Fire-adapted Ecosystems in an Urban Setting: The Case of Table Mountain National Park, South Africa. *Ecology and Society*, 17(1): 8. http://dx.doi.org/10.5751/ES-04526-170108 (accessed 29 March 2014).

18 Cultivating Citizen Stewards

Lessons from Formal and Non-Formal Educators

Gregory Smith

Introduction

How can we, as a society, educate children so they will value and wish to care for the natural features and systems of their local environments—like urban forests and airsheds—and do so in ways that benefit all people rather than only the privileged? This is a question I have been asking myself for the past two decades with regard to education and sustainability. Part of the answer lies in preparing young people who understand the nature of humanity's current circumstances with regard to our species' impact on natural systems and resource use, who feel connected enough to their place and community to want to get involved, who have the capacity to analyze and then solve problems and who possess the confidence to take action.

In an effort to enact that answer, for the past two decades a group of colleagues and I have been advocating for an approach to education that is grounded in children's bioregion (Smith, 1993, 1995; Woodhouse and Knapp, 2000; Smith, 2002; Gruenewald, 2003; Sobel, 2004; Gruenewald and Smith, 2008). We initially called this approach place-based education. More recently some of us have added community to place to make it absolutely clear that what we are talking about is the human–nature interface and the fact that neither can be separated from the other (Smith and Sobel, 2010).

My colleague, David Sobel at Antioch University New England, has defined place- and community-based education as "the process of using the local community and environment as a starting point to teach concepts in language arts, mathematics, social studies, science, and other subjects across the curriculum" (Sobel, 2004: 7). This kind of cross-curricular focus on the local is central to place- and community-based education. It is important to note as well that the local is the starting and not the ending point of this approach. Place- and community-based education does not require a retreat into parochialism. It instead argues that by beginning with the local, educators give students a chance to base their learning upon their lived experience of the world. The local also provides a venue within which it is possible to make small, and sometimes larger, changes in that world.

This is why the definition of place-based education developed by the Rural School and Community Trust provides an important addition to Sobel's vision.

For the Rural Trust, place-based education "is learning that is rooted in what is local—the unique history, environment, culture, economy, literature and art of a particular place. The community provides the context for learning, student work focuses on community needs and interests, and community members serve as resources and partners in every aspect of teaching learning" (Rural School and Community Trust, 2003). What place- and community-based educators recognize is that the effective transmission of a culture is the responsibility of an entire community—not only teachers. To be effective, this approach requires collaboration and partnerships with agencies, businesses, non-profit organizations and all manner of professionals—including architects and planners, and foresters and stormwater engineers.

The cross-disciplinary nature of place- and community-based education is especially well-suited to the concerns of political ecology. As political ecologists explore the linkages between the environment, power and decision-making, place- and community-based educators reach beyond the sciences as they present environmental issues to students, considering as well the ways these concerns are reflected in the social sciences and the humanities. David Gruenewald's seminal article, "The Best of Both Worlds: A Critical Pedagogy of Place" (2003), further posits that place-based education allows for, if not requires, a convergence of social justice and environmental perspectives. He argues that this approach ideally will involve a consideration of both decolonization and reinhabitation. Decolonization entails on the part of teachers an effort to introduce students to the ways in which a variety of social, economic and political factors contribute to the oppression of some people and privileging of others. Challenging such oppression must be central to efforts aimed at improving environmental conditions for everyone. Reinhabitation involves engaging students in activities aimed at restoring natural, social and cultural environments in ways that will support the health of humans, other-than-human species and the land, air and water that support living beings. Together this two-pronged approach, what Gruenewald calls the best of both worlds, could potentially lead to educational experiences that contribute to social and environmental betterment.

What I would like to share in this chapter are stories about ways in which formal and non-formal educators are providing opportunities for people in their communities to participate in activities that in all instances foster reinhabitation and in some, decolonization. All represent what can happen when educators direct young peoples' attention to the local. Because of this volume's exploration of the relationship between urban forests and political ecology, I will be focusing on natural resource issues—with an emphasis on projects that have involved forests, airsheds and watersheds—but it is important to realize that place- and community-based educators concern themselves with the full range of human issues—from the collection of oral histories to student participation in the development of long-term economic development plans, from the publishing of local newspapers to the development of student-run tax services for the poor and elderly. At issue is giving young people opportunities to become fully participating and contributing members of their communities, able to challenge oppressive practices and institutions and to create more restorative and regenerative alternatives in a multiplicity of human domains. The aim of place- and community-based education is to break down the barriers that often exist between classrooms and the broader world by channeling the energy of children and youth into needed community projects and giving them the chance to see that they have much to offer to others.

Urban Ecology Center, Milwaukee, Wisconsin

My first story is about a project that started in the early 1990s when a group of citizens decided to reclaim an urban park that had been abandoned by the city of Milwaukee, Wisconsin, and had become a gathering place for drug dealers and criminals. That program, the Urban Ecology Center, has grown into a $3.5 million operation with three sites throughout the city that in 2011–2012 provided school and community learning experiences and adventures for 90,000 children and adults (Figure 18.1). Education is central to its mission, and it has worked closely with neighboring schools, most of which serve primarily low-income Black and Latino students, to provide an array of formal and non-formal educational opportunities for Milwaukee's children and youth. The Urban Ecology Center is now also playing a major role in finding ways to restore some of the city's industrial brownfields and transform them into an arboretum and additional parklands in neighborhoods primarily populated by groups too often denied access to such amenities.

The story began at Riverside Park, designed in 1892 by Frederick Olmstead next to a lake behind a dam that had been constructed in the 1850s to provide ice and power for local breweries and industrial mills (Leinbach, 2008). For decades it had been a popular gathering place for city residents. By the middle of the 20th century, however, the lake had silted up and become a site of agricultural pollution. Given its declining use by the public, when the county was faced with financial problems in the 1970s, the park budget was cut entirely. By the mid-1980s, the park had become unsafe for both adults and children. Neighbors, who remembered what it had been like before the lake became polluted and its paths too dangerous to use, organized to correct the situation and return this urban forest once more to the public.

They formed the Friends of Riverside Park and began attracting volunteers from local religious, civic and scouting organizations to clean it up. Over a few years, volunteers made real progress in picking up trash, removing graffiti and taking apart homeless encampments. But people still weren't using the park. As organizers brainstormed strategies to address this issue, one idea rose to the surface and captured their attention. What might happen if neighborhood schools began using the park as a site for hands-on environmental education and service-learning opportunities? Would the regular presence of children make the park an undesirable location for drug dealing and other criminal activities? And as the criminal activity declined, would park use then increase? The organizers thought it was worth a try. In 1993, they purchased a used double-wide trailer and set up a portable classroom 30 yards from one of the park bathrooms.

Figure 18.1 Urban Ecology Center, Riverside Park, Milwaukee, WI, USA. Urban Ecology Center, 2013.

Source: Courtesy of John Suhar.

Volunteer naturalists presented programming to students from nearby schools. The effect on crime rates in the vicinity of the park was nearly immediate. In the three years prior to the creation of this program, there were on average 750–800 crimes per 10,000 people. By 1996, this had dropped to approximately 550 crimes per 10,000 people. In 2003, it was down to 50 crimes per 10,000 people. From early on, park activists knew they had uncovered an approach that had real promise (Leinbach, 2008).

By the late 1990s, Friends of Riverside Park were ready to hire their first executive director, a public school teacher with a background in environmental education, named Ken Leinbach. Committed to creating environmentally aware and active citizens through a combination of regular contact with nature and environmental mentors, he established the Neighborhood Environmental Education Project (NEEP). This project had seven key elements:

- It would only serve students within a two-mile radius of the Center. If the aim was to increase public use of the park outside of the school day, it was imperative that children and their families be able to get there on foot or by bicycle. This was especially important because 85 percent of the students who lived close to the park qualified for free or reduced-fee lunches.
- Each partnering school was required to pay $4,000 for at least 24 field trips over the course of the academic year. The remaining $5,000 needed to pay for the full expense would be covered by individual or corporate sponsors.
- Services to each partner school would include the minimum number of field trips but also access to all of the Center's resources including activity trunks, exhibits and its library. Professional-development workshops would also be provided for teachers.
- Recognizing that school bus fees can often place serious constraints on a school's ability to offer field trips, the Center purchased a small collection of 15-passenger mini-vans that are used to transport students to the park. All Center teachers must become qualified school bus drivers.
- Formal environmental education programs are geared to grade-specific curricular requirements and standards.
- A student–teacher ratio of 14:1 is guaranteed.
- Schools must send a liaison twice a year to the Center to learn about the program and help assess its impact on students.

(Leinbach, 2008)

During the 2011–2012 academic year, over 35,000 students participated in either NEEP offerings or other programs for Milwaukee Public Schools students. This is out of the approximately 78,000 students enrolled in the city's public schools. Another 2,000 students participated in after-school activities, nearly 3,000 in children and family programs, 2,600 in summer camps and 4,800 in non-formal outdoor adventures (Leinbach, 2013). The Urban Ecology Center is clearly reaching a significant number of young people in the Milwaukee area.

Central to the success of the NEEP program, as well as other Urban Ecology Center offerings, is its extensive use of volunteers to assist with student programming.

In 2011–2012, nearly 2,000 individuals put in 16,700 hours of volunteer work, equivalent to 11.6 percent of the 2011–2012 operating budget (Urban Ecology Center Board Report, 2013: 2). Access to these volunteers as well as individual and corporate financial support is linked to the Center's commitment to serving the entire community, not only children. Programs in citizen science, sustainable food, and outdoor activities such as rock climbing, snowshoeing and canoeing make the Urban Ecology Center a hub for a range of different nature-oriented activities in the city and a draw for broad-based public participation and support.

From the standpoint of place-based education, the Urban Ecology Center exemplifies what can happen when the boundaries between the classroom and region are broken down. Students have the opportunity to learn in ways that are grounded in their own communities. They become familiar with the place where they live and develop an interest in caring for it. Preliminary findings are indicating higher student achievement and levels of ecological literacy thanks to these place-based learning experiences (Leinbach, 2011). The Center is also modeling what it means to live in a more sustainable manner. The building constructed in 2004 to house the Riverside Park Center collects and uses its own rainwater, is powered by the sun and is built of many recycled materials. The newer Washington Park facility and the 2012 Menominee Park building reflect the same concerns. People throughout the city are being acquainted with how it is possible to meet human needs but do so in a manner that reduces our species' negative impacts on the natural world. People are also being shown how local citizens can play a major role in the restoration of degraded urban environments and the design of cities that work with, rather than against, nature.

With regard to political ecology, this work is significant in that it is making natural areas available to traditionally underserved populations. In their 2006 *Urban Affairs Review* article, Heynen et al. explore the degree to which the urban canopy in Milwaukee is inequitably distributed across different census tracts, with the residents of affluent census tracts having access to more trees than residents who possess fewer resources. They argue that Milwaukee officials need to find ways to address this inequity by developing policies that support the planting and care of more trees in low-income neighborhoods. The Urban Ecology Center is demonstrating another approach that has comparable merit by intentionally reclaiming and restoring abandoned parks and industrial areas close to less-affluent residents of the city. It is an approach that could be emulated in other cities and that further demonstrates the social and educational benefits of doing so.

Greater Egleston Community High School

The second example is located in Boston, Massachusetts, and grew from the work of a teacher at the Greater Egleston Community High School in Roxbury. This school was created in the early 1990s by a group of parents who hoped that an education focused on preparing adolescents to become community leaders might be more attractive than gangs and street life for their primarily Black and Latino sons and daughters. Founded initially as a charter school with Department of Labor funding, the Greater Egleston Community High School quickly became one of Boston's pilot schools, public schools given more autonomy and flexibility over curriculum, budgetary decisions and hiring than other district schools. Although administrative changes a few years ago have led to a shift away from its original mission, for more than a decade the Greater Egleston Community High School's students and teachers demonstrated what an urban orientation to place, community and activism can accomplish.

I am most familiar with the work of a science teacher named Elaine Senechal who joined the school's faculty in the mid-1990s. When she first interviewed at the school, she was struck by a vacant lot across the street that was littered with garbage and dominated by a billboard. She thought to herself, "How can kids want to learn when they've got to look at this everyday when they come to school?" She decided then that if she got the job, "That would change" (Senechal, 2008, personal communication). Much like Ken Leinbach and the activists in Milwaukee, she took what was in front of her and decided to do what she could to change it. At the outset, Senechal explored what would need to be done to turn the space into a park where people could find beauty and experience community. She began by developing and implementing a course on garden design to help students gain the skills needed to transform the vacant lot into a park. This class led to a preliminary plan that was then presented to the local neighborhood association. Students continued their inquiry by now focusing on plant growth and soil quality. Senechal was able to get a small grant to continue this work and secured the services of a landscape architect who worked with students to refine their design. During the first summer after beginning the project, she was also able to hire three students to interview neighborhood residents, tapping their ideas about what the park should look like. During the next year, students finalized the park design and presented it to different community groups. Throughout this time, the neighborhood association sought a local organization willing to lease the land from its owner, AK Media. This process took two years to complete, but once finalized, the garden and park were constructed, including a student-created mural that forms the backdrop to the park (Figure 18.2). When I visited Greater Egleston in the spring of 2007, the park was being cared for by an association of neighborhood businesses and had become the site of regular community events including musical concerts.

Senechal sought to address environmental problems that affected public health, as well. To find ways to create meaningful curriculum opportunities for her students, she started attending meetings of community groups that addressed environmental concerns. Two of these, Roxbury Environmental Empowerment Program (REEP) and Alternatives for Community and Environment (ACE) had recently become involved in efforts to reduce the rising rates of asthma in their part of Boston. Staff in both groups suspected that large numbers of diesel-burning vehicles were the source of the problem, and both organizations are also committed to youth development. The first step in their campaign to address this issue involved finding ways to measure air pollution on the ground. The Environmental Protection Agency operated air-monitoring equipment in central Boston but not in Roxbury. One of the first assignments Senechal gave her students was to simply

Figure 18.2 Student-built and designed Peace Park, Egleston Square, Boston, MA, USA.
Source: Courtesy of Gregory Smith.

count the number of trucks and buses that went past the school on Washington Street in an hour. They counted over one hundred. Students then worked with the environmental non-profit organizations to raise money to purchase air-monitoring equipment for local public health agencies. Their success led to the creation of an online source of information about air quality—AirBeat—and a student-created signaling system that alerted the public to pollution levels by raising different colored flags outside their classroom. The monitoring demonstrated that pollution levels in Roxbury were indeed higher than those in Harvard Square. Following up on this, students also surveyed the community to assess residents' knowledge about asthma and its causes.

At about the same time, ACE staff discovered that an unenforced Massachusetts statute regulated the amount of time vehicles could idle at a single location. The bus lot for the Massachusetts Bay Transit Authority (MBTA) was located just six blocks away from the school. Every morning drivers would start the engines of over 250 buses before beginning their routes, often warming up the vehicles for 30 minutes or more, creating a toxic mixture of carbon monoxide and particulates that threatened people's health. Working with the ACE organizers, students in Senechal's class began an anti-idling campaign that in time caught the attention of Boston's newspapers, mayor and city council. After six years, their efforts contributed to a court decision that fined the MBTA, required the transit authority to continue converting their buses to use natural gas rather than diesel and resulted in a city-wide enforcement of the anti-idling law.

These experiences had a profound impact on students who participated in the class and on the people who had an opportunity to watch them at work. As one young woman noted:

> I am proud of my accomplishments in environmental justice this trimester. Most importantly I have been able to gain confidence to speak in front of large groups of people. Before presenting to the City Council I was very nervous. But after watching them and my classmates somewhat debate I realized they are regular people just like my family, my teachers, and my friends, and I should not be nervous when it comes to speaking my mind.
>
> (Senechal, 2008: 100–101)

Over the years, students' skills as researchers and organizers led others to request their assistance when confronted with a need for information or challenges that threatened them or their neighborhood. Members of the Boston City Council, for example, developed so much respect for students' work that they asked them to investigate the impact of city policies on its low-income residents.

Teacher Elaine Senechal believes that this kind of educational process reduced the alienation and isolation students had encountered in previous schools by situating learning within a culturally familiar setting. It furthermore increased their engagement and motivation by taking on tasks that were immediately relevant to the life of their community. When students were given the opportunity to participate in work and develop skills valued by others, their sense of self and competence was enhanced. Finally, as these young people acquired knowledge about the kinds of strategies that are successful in bringing about beneficial change, they came to recognize their own power as social actors (Senechal, 2008). Students in the environmental justice course at the Greater Egleston Community High School discovered that they had the capacity to understand and challenge oppressive policies and practices and make significant contributions to both the aesthetic and environmental quality of their own community. This set of educational experiences demonstrates the potentiality of conjoining both decolonizaton and reinhabitation, and enacting the principles of political ecology in local communities.

West Linn/Wilsonville School District

The final examples of place- and community-based education I will describe are from the school district that serves my own primarily middle- to upper-middle-class community, south of Portland, Oregon. Although not a location where one might expect to find the forms of defensive activism associated with political ecology, young people in this community are subject to other powerful forces that contribute to alienation and disempowerment. Like many American children, their absorption in electronic media diminishes their connection with the natural world (Louv, 2008; Schei and Merrill, 2010). And their isolation from the social life of the broader community prevents them from developing the skills of civic participation required to become effective social actors. In 21st-century US society, it is not only non-EuroAmericans and the poor who are oppressed. Under the auspices of neoliberal policies, a growing proportion of the population across all but the most affluent groups is being cut out from decision-making and subjected to growing economic insecurity.

In 2009, I started teaching a semester-long course about sustainability and place-based education for district educators. Over a three-year period, approximately 75 teachers either participated in the course or a summer institute that focused on finding ways to integrate sustainability issues into their work with students by engaging them in local research or service-learning activities. Teachers at two primary schools and a middle school designed learning experiences that seem especially relevant to the topic of urban forestry and related concerns.

As part of their final project for the course, a third-grade teacher and the librarian at Boones Ferry Primary School in Wilsonville decided to create a unit that would invite their students to write a land-use history of property adjacent to the school that was being transformed by Portland's regional government, METRO, into an oak savannah. Prior to EuroAmerican settlement, much of the Willamette Valley had been shaped by the burning practices of the local Kalapuyan bands to provide habitat for deer, elk and a wide range of other species, as well as abundant acorns. Burning prevented Douglas fir trees from overshadowing and then eliminating Oregon white oak. METRO intended to restore a former farm with only a single remaining oak to its pre-EuroAmerican settlement state.

As they sought to gather material for the project, April Brendan-Locke and Margaret Wattman-Turner, the teacher and librarian, interviewed the daughter of one of the previous landowners. She was excited about their plans and gave them a banker's box filled with photographs and news clippings about the local activists whose commitment had led to the park rather than a housing development. This support and interest fired up the educators even more.

They introduced students to the idea by taking them on a walk to visit the oak. They explained that there was no current history of the park and that they could be the ones who researched and told that story. The children were excited, and when they got back to the classroom decided to narrate that story from the point of view of the oak tree.

The class was divided up into teams and given responsibility for investigating how the land was used by the Kalapuyan bands prior to 1830; the pioneers, homesteaders and river people in the 1850s and 1860s; the early town residents and farmers from the 1870s through the 1900s; the nature of the farming community between 1900 and the 1960s; and the modern suburbs that had been built from then until the present. Students examined history books, original source material and old photographs, and interviewed people in the community. Each team wrote a chapter for the book based upon what the children had learned. Then working with the artist son of the librarian, they developed ripped paper illustrations to depict the period that they had studied. All of this was then assembled into a book that was published locally (Figure 18.3).

Figure 18.3 Student-written and illustrated land use history of Graham Oaks Nature Park, Wilsonville, OR, USA.

Source: Courtesy of Gregory Smith.

When Graham Oaks Park was dedicated the following September, the work these students had accomplished was given a central role. In addition to formally presenting their book to the Wilsonville Public Library as the only published land-use history of the property, they were also given an opportunity to read sections from what they had written. As they did so, listeners were able to look off to the north where not far away stood the ancient oak tree they had chosen to be the narrator of their story.

During the dedication ceremony, the mayor of Wilsonville spoke of the important role that children at Boones Ferry Primary School could assume in being stewards of the park. They would play next to it for many years, since their middle school also adjoins the park. He said that they would be the eyes and ears of the land and help the rest of the community make sure it was treated well.

The restoration and preservation of both urban and suburban forests—to say nothing of all forests—will be contingent on people's experience of similar commitments and the belief that their own voices and actions matter.

A number of teachers at nearby Bolton Primary School in West Linn are creating learning experiences that are encouraging the development of similar forms of affiliation and responsibility. Fourth- and fifth-grade teachers, Lisa Terrall and Kelley Jones, are among the most active of these educators. They have developed a set of classroom and field-study experiences aimed at introducing their students to the local watershed, from the crest of the Cascades to the Pacific Ocean. Students learn about regional flora and fauna and take fieldtrips to the coast and to the Columbia Gorge where they have an opportunity to see multiple aspects of their home region. During the spring of 2012, the teachers asked their fifth-grade students what contribution they would like to make to the school and the watershed before they moved on to middle school. After exploring different options, the students decided they wanted to address an asphalt-covered alleyway between two wings of the school building that could be seen from the windows of a number of classrooms. The school is located on a bench-like formation next to a steep hillside leading down to the Willamette River. It is a natural place for stormwater to collect, and this alleyway was especially problematic. Water would pool there on rainy days faster than it could be released through the handful of drains placed in the asphalt.

The students determined that they wanted to remove the asphalt and replace it with a more permeable surface capable of allowing the water to percolate into the soil. Accomplishing this project would require the students to develop a plan, get the permission of the district and the city to implement it, raise money and organize a workday. The teachers gave them significant responsibility in completing all of these tasks. Experiences such as these—like the opportunity to investigate an important local topic and write a book—can do much to demonstrate to children how much power they have to make good things happen. These kinds of experiences can instill a taste for the sense of satisfaction and competence that comes with the accomplishment of meaningful tasks. Citizenship and stewardship take root in experiences that affirm anyone's worth, regardless of age.

The final story involves a project a sixth-grade teacher, Lisia Farley, has taught at Rosemont Ridge Middle School for the past two years. West Linn has been engaged in a number of controversies with regard to its aging water system, including the building of a significantly expanded water plant owned by two nearby cities on land located on West Linn's northern border. Farley's students spend a month toward the end of the school year investigating these issues and then present their findings during a day-long West Linn Water Summit to which the mayor, other city officials, parents and community members have been invited. In small teams, students describe what they have learned, their assessment of the problems and their suggestions for dealing with both the expense of updating water and sewage pipes, some of which are well over a century old. During the first year I served as a respondent to their ideas, and I was struck by the fact that this group of 12-year-olds probably had a better grasp of the nature of these problems than most of the city's citizens. They were furthermore being provided with the opportunity to share their thinking in a forum not that different from public hearings conducted by the city planning commission or city council. As with many of the previously described learning experiences, this project is preparing children for citizenship in ways that are uncommon in most schools. If young people are to grow into adults capable of protecting local ecosystems, it will be imperative that they possess the ability to research local issues, synthesize their ideas into brief presentations and possess the confidence and willingness to share their ideas with decision-makers.

Conclusion

These stories about strategies community members and educators have adopted to immerse young people in local natural and social environments will have hopefully sparked possibilities worth pursuing in readers' own towns and cities. At their core, these efforts are designed to engender a sense of connection and affiliation between children and their regions. If young people's affiliation is primarily tied to digitized experiences with few opportunities to participate in real-world problem solving, it seems unlikely that they will invest much of their energy in the preservation and maintenance of nearby environmental features like urban forests, waterways or airsheds.

Although the work of teachers and students in the schools described above is encouraging, it is also fragile and often dependent upon the commitments and energies of specific teachers who by their nature are risk-takers, willing to push the boundaries of schools as they are. In many respects, schools in general serve as agents of domestication (Dewey, 1938; Apple, 1982; Freire, 2000; Finn, 1999), preparing young people to accept and participate in mainstream economic, political and cultural institutions rather than challenging those institutions. The pressure to conform to the demands of the standards and accountability movement associated with neoliberal school-reform agendas has only intensified this process. To act in ways that counter these powerful trends requires courage and the willingness to sometimes sidestep or challenge school norms. The number of teachers willing to take steps like those described in the preceding pages will generally be small, and most will need to work tactfully and strategically in whatever institutional spaces they can find to pursue this more-liberatory educational agenda.

Such teachers also run the risk of engaging their students in learning activities that may be contradictory in nature. Schools that start school gardens and distribute produce to local food banks (another project begun by teachers at the Bolton Primary School) for example, may in the end support societal adaptation to neoliberal policies that call for cutting state and federal programs that previously provided food to economically disadvantaged individuals and families (McClintock,

2014) but now insist on individual or local self-reliance. The same thing is true of students who participate as volunteers in environmental mitigation projects aimed at diminishing the impact of environmentally destructive industrial practices in other parts of their community. As McClintock observes, however, even though such work may serve in some ways to advance neoliberal objectives, experiences such as these can also lay the groundwork for the adoption of new perspectives and possibilities that are more radical.

Erik Olin Wright's exploration of different approaches to social transformation in his volume *Envisioning Real Utopias* (2009) is especially helpful in demonstrating how in contemporary societies innovation on multiple fronts, even when some initiatives are contradictory in nature, may be essential. Wright suggests that social transformation occurs in three ways: (1) change agents can work to effect a rupture that leads to the replacement of one collection of institutions by another, something that happens during major revolutions; (2) they can work within the interstices of current institutions in ways that embody transformative possibilities under the radar of those charged with the protection of the status quo; (3) or they can engage in symbiotic activities that advance both transformative and mainstream goals and purposes. The projects described in this chapter seem primarily symbiotic, providing learning opportunities for young people that enhance their engagement in academic tasks and/or deepen their sense of affiliation with and responsibility toward natural and social features of their own communities. At the same time, these activities contribute to reduced crime rates, higher levels of student achievement, and the creation of potentially increased property values in neighborhoods with desirable parks and improved air quality, all changes that would meet with the approval of neoliberal policy advocates. It is in fact difficult to imagine schools assuming anything other than this symbiotic role given their role as state-funded institutions.

But while associated with activities that could be condemned as "reformist reforms," educational experiences like these also contribute to the cultivation of citizens who perceive themselves as problem solvers and social actors, exactly the kind of people needed to help participate in the creation of more sustainable and just cultural practices, and social, political and economic institutions. By giving young people the opportunity to identify and address problems of different scales within their own communities, assuring that they are able to experience some level of success in their activities and partnering with outside-school agencies or organizations to deepen and extend the impact of students' work beyond the classroom, educators can create a set of learning experiences capable of reversing the sense of isolation, alienation and powerlessness that constrain the civic participation of too many younger and older adults in developed societies. Young people prepared to act in these ways may decide that among the multitude of things that require their attention are the urban forests, watersheds, and airsheds that help sustain the health, beauty, and livability of the places where the majority of human beings are now making their homes.

References

Apple, M. (1982) *Education and Power*, Routledge & Kegan Paul, Boston, MA.

Dewey, J. (1938) *Experience and Education*, Macmillan, New York City.

Finn, P. (1999) *Literacy with an Attitude: Educating Working-class Children in their own Self-interest*, State University of New York Press, Albany, NY.

Freire, P. (2000) *Pedagogy of Freedom: Ethics, Democracy, and Civic Courage*, Rowman & Littlefield, Lanham, MA.

Gruenewald, D. (2003) 'The best of both worlds: A critical pedagogy and place,' *Educational Researcher*, 32(4): 3–12.

Gruenewald, D. and Smith, G. (2008) *Place-based Education in the Global Age: Local Diversity*, Taylor & Francis, New York City.

Heynen, N., Perkins, A., and Roy, P. (2006) 'The political ecology of uneven urban green space: The impact of political economy on race and ethnicity in producing environmental inequality,' *Urban Affairs Review*, 42(2): 3–25.

Leinbach, K. (2008) '"It's kind of fun to do the impossible": The story of Milwaukee's Urban Ecology Center,' *Children, Youth and Environments*, 18(2): 180–196, www.colorado.edu/journals/cye/18_2/18_2_08_UrbanEcology.pdf (accessed 27 March 2013).

Leinbach, K. (2011) 'Cities as agents of change,' panel presentation at Lewis & Clark College Environmental Affairs Symposium, October 12, Portland, OR.

Leinbach, K. (2013) 'Urban Ecology Center overview,' unpublished PowerPoint presentation.

Louv, R. (2008) *Last Child in the Woods: Saving Children from Nature-deficit Disorder*, Algonquin, New York City.

McClintock, N. (2014) 'Radical, reformist, and garden-variety neoliberal: Coming to terms with urban agriculture's contradictions,' *Local Environment: The International Journal of Justice and Sustainability*, 19(2): 147–171, http://dx.doi.org/10.1080/13549839.2012.752797 (accessed 1 April 2014).

Rural School and Community Trust (2003) *Rural Policy Matters*, September, p. 3.

Schei, T. (director) and Merrill, M. (producer) (2010) *Play Again* (motion picture), Ground Productions, United States.

Senechal, E. (2008) 'Environmental justice in Egleston Square,' in D. Gruenewald and G. Smith (eds) *Place-based Education in a Global Age: Local Diversity*, Taylor & Francis, New York City, pp. 85–112.

Smith, G. (1993) 'Shaping bioregional schools,' *Whole Earth Review*, 81 (Winter): 70–74.

Smith, G. (1995) 'The Petrolia School: Teaching and learning in place,' *Holistic Education Review*, 8(1): 44–53.

Smith, G. (2002) 'Place-based education: Learning to be where we are,' *Phi Delta Kappan*, 83(8): 584–594.

Smith, G. and Sobel, D. (2010) *Place- and Community-based Education in Schools*, Routledge, New York City.

Sobel, D. (2004) *Place-based Education: Connecting Classrooms and Communities*, Orion Books, Great Barrington, MA.

Urban Ecology Center (2013, January) 'Urban Ecology Center Board Report,' Urban Ecology Center, Milwaukee, WI.

Woodhouse, J. and Knapp, C. (2000) 'Place-based curriculum and instruction: Outdoor and environmental education approaches,' ERIC Clearinghouse on Education and Small Schools. Appalachia Educational Laboratory (ERICDocument Reproduction Series No. ED448012), Charleston, WV.

Wright, E. (2010) *Envisioning Real Utopias*, Verso, London, New York City.

19 Learning and Acting through Participatory Landscape Planning

The Case of the Bräkne River Valley, Sweden

Helena Mellqvist and Roland Gustavsson

Introduction

Landscape planners have for some decades struggled to handle tensions and trade-offs between policies for nature and landscape development, such as the EU Common Agricultural Policy, the Rio Convention and the European Landscape Convention. They have relied on centralization/standardization and facts provided by experts on the one hand, and perspectives from public participation/local knowledge on the other hand (Pinto-Correia et al., 2006). In parallel, landscape planners have struggled to combine a general position as 'outsiders' with the need to achieve positions on place-specificity as 'insiders' (Bauman, 1995; Gustavsson and Peterson, 2003; Blicharska et al., 2011).

Landscape planning and management face a major challenge resulting from increased urbanization, with over 50 per cent of the world's population and 72 per cent of Europe's population living in urban areas (WHO, 2014). The trend has led to cities getting bigger and denser and asserting a growing influence on the countryside, making it more uniform and less multi-functional. In the peri-urban landscape, there is a growing segment of residents and landowners who lack local knowledge on place-related landscape management. Fewer hands are engaged in shaping the landscape, taking care of it, and ensuring its multi-functionality. This affects ecological systems, attractiveness and visual impact.

Much has been written on how the urban has influenced a shift in rural land-use practices from traditional food and timber production to recreation and tourism activities. Selman (2012), for example, points to how small-scale family farming in Europe is supported more for touristic appeal than economic viability. Yet, many rural residents find it difficult to access and navigate support systems for such a transition and choose either to stop tending and using their land or abandon their properties altogether. In this chapter, we address this conundrum in our role as landscape planners and architects who are concerned about preserving the visual diversity and historical legacy of the landscape as well as presenting possibilities for building a different future. We are interested in facilitating the resurrection of the social cohesion and collaborative spirit of the old village society. We are also inspired by landscape urbanism's way of dealing with the peri-urban landscape where cities/villages both grow and shrink as a response to economic and social factors (Waldheim, 2006). We agree with Selman (2012) who calls for a new system of landscape governance that pays respect to both 'space and place' (Selman, 2012: 45), steering the direction of landscape change towards reconnection and regeneration of natural systems supporting local residents. This, he maintains, can be supported by green infrastructure planning, agro-tourism, and renewable energy production, new trends that bring new impacts to the landscape, but that are also often unpredictable to plan for given that the demand is generated by outside drivers and demands.

A complementary development plan is presented by Ostrom's (1990) concept of common pool resources (CPR). Ostrom emphasizes social institutions as determinants of what directions human–environment relations take in agricultural regions. For a sustainable use of CPR, Ostrom claims that one needs to work with local democracy, a high degree of autonomy and some kind of supervision. Ostrom suggests bringing in people living and working in the landscape, emphasizing the importance of collaboration between different layers where decisions on people's landscapes are formulated. The established norm on representative democracy becomes challenged in favour of participatory ideas (Arler, 2008).

In this chapter, we explore the peri-urban and rural landscapes of the Bräkne River Valley (BRV) in southeastern Sweden, which touch on elements of new governance and economic activities as articulated by both Selman and Ostrom. The BRV is part of the municipality of Ronneby, one of 289 municipalities in Sweden that are responsible for land use within their borders. Ronneby is in turn one of five municipalities in the county of Blekinge, the smallest of 21 counties in Sweden. The county is the national government's representative, implementing and supervising national policies but with specific responsibility for regional conditions. While the county has a broader interest to preserve and develop natural and cultural values, the municipality is responsible for planning and social issues, such as schools, elder care, housing, drainage and water supply. For the municipality it is at times easier to permit a depopulation of the rural parts as it is expensive to supply all services they are responsible for in remote areas. The county has above all been interested in biological values in the BRV. If the river valley is depopulated, it is likely that the county will become a manager for biodiversity, placing less focus on the care of the visual signs of human legacies and presences in the landscape.

Ronneby, the main town in the municipality, is famous for a spa and its tourist infrastructure. The spa was established in 1700 after a mineral-rich spring was discovered in the area. At the beginning of the 1800s, the Ronneby spa became a successful and popular resort for health tourists. A reason for the success was the thorough design of the spa park, including the forest with winding paths and a boat taking guests out into the exclusive archipelago to have a walk on the islands or visit a bath. Rural qualities made this urban spa-tourism popular. Despite this local history and the specific qualities in the area, Ronneby's mayor has stated that the countryside is destined to become empty.

In the following, we assess the prospects and problems of BRV in developing new governance structures and new sources of income, including the role of greenspace.

The European Landscape Convention (ELC) and other Possible Tools

Article 6 in the European Landscape Convention (ELC) describes measures of awareness-raising, learning objectives for professionals and students, and how to define local landscape qualities based on public consultation (Europe, 2000). The article provides universities with the opportunity (Swedish National Agency for Higher Education, 1992) to engage with what two Danish action researchers call 'democratizing our universities', that is, create room for public and community engagement (Nielsen and Nielsen, 2006).

However, meeting the content of Article 6 has hardly been done in countries like Sweden and Denmark (Sandström et al., 2008). Furthermore, the status of landscape in Swedish law is unclear and diffuse. But there are reasons for some optimism: a rural development programme in Sweden was introduced in 2007 and a revised version is due in 2014.[1] The purpose of this programme is to promote economic and ecological sustainability in the countryside, with a focus on social sustainability. The latter is connected to the EU's 'LEADER' programme for rural development, which comprises a partnership between the public, voluntary and private sectors. LEADER supports ideas and projects for rural development that benefit a larger group than just the individual. But the actual responsibility for landscape is causing a split between several sectors. The state is responsible for national interests such as environmental goals and national parks, while the 21 counties in Sweden are responsible for natural and cultural reserves. The county administrative board is responsible for implementing the government's decisions but plays a less important strategic part in Swedish planning. Municipalities are responsible for land use and operate with the master plan as the main tool. Master plans in their turn focus on densely populated areas leaving the rural areas quite untouched. Peri-urban areas are in a transition zone. However, new tools are prepared in the prolongation of 'deepened master plans', also including a focus on countryside interests.

Sweden ratified ELC on 1 May 2011, and possibly a change is occurring with ELC emphasizing the importance of the countryside for the greater landscape (Europe, 2000). The use of soft means of control, such as some kind of general education instead of legal controls, offers the best opportunities to achieve progress.

The Role of the Landscape Planner and Architect

As landscape planners and architects, we have been preoccupied with research projects, partnerships between municipalities and the university, and in teaching on a Swedish, as well as European, scale. A first step towards this was taken very early by Roland Gustavsson. Born in the BRV in the 1940s, Gustavsson published his master's thesis 'Foundation for a nature management plan in the Bräkne River Valley' in 1975. It was used by the county administration board to formulate the existing nature-management plan and, at the same time, became a bestseller among people working and living in the valley. In 2005, Helena Mellqvist published her master's thesis 'Making problematic cases successful in traditional landscape management—use of repetitive photographing in a communicative planning context', partly grounded in Gustavsson's work. The series of pictures from 1975 was repeated in 1995 and pairs of photographs were used in interviews on land-use changes and people's attitudes towards these.

We are inspired by Flyvbjerg's (2001) work on the importance of researchers' continuing presence in an area and the quest to understand patterns of power, traditions and local policies. This is in order to establish a relationship and a participatory engagement with local residents and stakeholders. An important impetus for the public engagement occurred when a medical doctor in the valley sent out a meeting invitation to all residents in the river valley under the heading 'Farewell to open landscapes?' The last dairy farmer had quit his business and people were worried that all open land in the area would be overgrown due to the absence of grazing animals. At the meeting, we were invited to present our work. The meeting room was filled to capacity and the local residents organized themselves into a number of theme groups, interacting in a way that became a new participatory-led process.

In 1998, *Bygd i samverkan* (Community in Cooperation) (BiS), was started in the valley's main village, Bräkne-Hoby, as a reaction to a series of closures of local service establishments, including a school, a bank and a library. The people who joined forces in BiS decided to do what they could to make administrators and politicians in the main town of Ronneby listen to their demands. In 2006, BiS was awarded the countryside award as the 'most successful countryside group of Sweden' by the national organization Hela Sverige ska leva (The Swedish Village Action Movement).

We started to work with BiS in connection with the major task of formulating an application to LEADER. Gustavsson was involved as a senior advisor but was also engaged within the project, creating a local interpretation centre, guiding excursions and giving lectures in the river valley. Mellqvist has followed the association's work and conducted interviews before and after major events and happenings. BiS is still an active and engaged local association in the BRV.

Our Political Ecological Approach

We adhere to three premises of political ecology (Robbins, 2011): (1) the positivist human–environmental social sciences have reached a limit in explaining issues related to environmental change; (2) there is an urgency to find and identify questions previously 'unasked and unanswered'; and (3) there is a growing concern over climate change and natural catastrophes that behoves us to consider the close relationship between politics and ecology (Robbins, 2011: 80). Political ecology, we believe, is something people do. We also agree with Bruno Latour's statement that where we previously modernized we now must ecologize (Blok and Jensen, 2011), meaning that local landscape planners must adopt new methods that care for ecosystems but also consider sources of economic income for local residents as well as their engagement in local planning and decision-making. Ecological threats, environmental vulnerabilities and the concept of the risk society are defining elements of society today that are relevant in the BRV (Beck, 1992; Blaikie et al., 1994).

Risks and vulnerabilities differ in time and place. In the BRV, processes of change are slow. The importance of local knowledge on land-use practices is losing ground in the peri-urban, small-scale landscape. The threat is outmigration, farm closures, loss of local know-how—all strongly affecting the livelihoods of those who remain. On a small scale, this puts much-cherished landscape features at risk, as well as local ecosystems connected to sources of income and people's identity, even though people's lives are rarely threatened in this part of the world (Swaffield and Primdahl, 2006; Cadieux, 2008; Widgren, 2012).

This brings us to the next important step in our contribution to political ecology: landscape democracy (Europe, 2000). We believe that a way of working with landscape democracy is to stay close to the people living and working in the area, show interest in future landscape transformation and stay close to the political as well as administrative systems with particular focus on place-based knowledge. We agree with Latour's priorities in political ecology, 'uncertainty, precaution and gradual experimentation' (Blok and Jensen, 2011: 90). We emphasize a focus on raising awareness of development perspectives and drawing on the local knowledge of the residents in the BRV without pre-judging the outcome (Mellqvist, Gustavsson and Gunnarsson, 2013).

The Bräkne River Valley (BRV)

The BRV illustrates a shift from a rural to a more urban society and landscape. This situation is not unique for the BRV. However, it is very distinct within the type of small-scale, multifunctional, mosaic landscape where one can see how economy, politics, subsidies, demography, traditions and to some degree local collaboration are intertwined. One small change, for example, can lead to important effects. An illustrative example from the BRV is when southern Sweden was severely hit by storm Gudrun in January 2005. The spruce forest in the northern part of the valley was felled by the storm, which raised concerns over water quality due to the loss of the forests' filtering effect. These concerns were dealt with collectively and resulted in a feeling of shared identity connected to the water.

The BRV is located in the transition zone of the middle-European and the Scandinavian landscapes. It has been recognized as being of national and European interest for nature conservation, cultural heritage, tourism, outdoor recreation and fishing. Greenspace forms an important part of the landscape. Currently, the 84 km-long valley hosts six nature reserves and the county administrative board has suggested creating even more. The local residents' attitudes towards nature reserves are not exclusively positive. When the regional authority takes over responsibility for managing green areas, the land is typically managed on the basis of time and money rather than when the flora and fauna indicate it is time to harvest crops.

There are, however, still attractive remnants of the old economy left in the landscape, such as flour- and sawmills and old mill ponds. Pollarding and extensive grazing are still present and the many small roads create a special atmosphere. Despite current rational approaches, even forestry has its special characteristics rooted in an older, small-scale farming system. In parallel, a new type of 'wilderness' and an increasing amount of 'city gardens' have generated a new 'urban' character within the rural setting. Exurban home and cottage owners in the peri-urban landscape, just like in the suburbs, keep well-managed gardens with horticultural plants surrounded by hedges. Some of the new wildernesses have emerged in abandoned places and contaminated sites which the experience industry and local tourist agency now advertise as biodiversity areas. Meanwhile, countryside dwellers generally consider such areas as abandoned or unworked land, echoing Cronon's (1996) point that wildernesses are often places where there is a prominent human presence.

In the 1970s, many of the active farmers retired and other local residents moved away. By the 1980s, 'new' inhabitants moved into the area. New patterns of housing started to merge within older, existing patterns. County administrators, considered 'outsiders', emerged more powerful as landowners became more dependent on subsidies to maintain their landscapes. The former multifunctional landscape turned into a more standardized one where local characteristics were absent (Figures 19.1, 19.2 and 19.3).

In this context, peri-urban landscapes are often described rather negatively, as something in-between two phases or 'as something that destroys a preferred land use' (Crankshaw, 2009: 219; see also Qviström, 2013). Outside smaller and larger cities, peri-urban landscapes show similar loss in details, such as hedges, tree rows,

Figure 19.1 Old stone bridge south of Bräkne-Hoby, Sweden, the main village along the valley. At this old stone bridge, the main road to Denmark passed for several hundred years. At the side of the bridge, there was a place where horses could drink water. Today, the bridge is closed.

Source: Stig Svensson, 1962; R. Gustavsson, 1975; Mattis Gustavsson, 1995.

Figure 19.2 Landscape change at Örseryd, Sweden. The photographs illustrate the landscape dynamics of a small-scale mosaic feature along the river valley where the land use is ongoing but less and less articulated and maintained. Even though the small road is sensitively integrated in the landscape, and the small fields are still grazed, the photographs show a strong simplification of the landscape between 1975 and 1995. The change has continued since 1995.

Source: R. Gustavsson, 1975; Mattis Gustavsson, 1995.

paths and small water bodies. Around bigger cities many inhabitants move outside of town because of a positive and attractive countryside atmosphere, while still working in the city. Outside smaller cities, many farmers who still live on the family farm feel that their way of life is being undermined (Crankshaw, 2009). In Sweden the growing amount of nature reserves in peri-urban landscapes complicates the picture. Through transfer of responsibility for these greenspaces, the municipality is more present and powerful than previously. Emphasizing recreation and aiming to attract visitors from near and far, municipalities risk running into conflicts with local inhabitants as care and management of these greenspaces is changing and is often neglected.

Figure 19.3 A combined flour and saw mill at Björstorp, Sweden, 2 km north of the coast line. The photographs from 1975 and 1995 show how much visual landscape qualities are dependent on local land use. Gardens and small-scale fields still contribute much to the scenic character of the place, but they are both disappearing. The visual accessibility of the old open landscape is in decline, especially in the summer, when the landscape looks "overgrown" with vegetation.

Source: R. Gustavsson, 1975, Mattis Gustavsson, 1995.

It was in the above scenario that we became involved as actors in the strategic and tactical layers of planning and management of the BRV. In the face of continued outmigration, we were faced with the prospects of the county administrative board taking over land and turning it into nature reserves with the preservation of biological diversity as the main focus. Our task was to work with the local residents to counter such a trend in order to build a viable rural economy that respects the legacies of human uses and presences in the landscape.

The Birth and the Development of the Periscape Group

In 2000, Gustavsson launched a new collaboration together with Teresa Pinto-Correia, a geographer from Portugal. The main idea was to deepen cooperation between European landscape researchers to improve bridging between disciplines and fields such as landscape planning and architecture, landscape management, landscape engineering, geography, nature conservation, agronomy and forestry, in order to support small-scale, multifunctional landscapes in the periphery of Europe.

Soon there were eight people in this European network, all focusing on landscape, land use, landscape management, and the relationship between people and landscape. The Periscapers share a wish to improve, affect or prevent undesired abandonment or depopulation. They identify threatened or overlooked values in landscapes and try to understand the driving forces behind landscape change. They look for new dimensions of existing landscape functions and then communicate their messages to local actors and other stakeholders. An interest in pedagogic challenges, a shared desire to experiment with classic methods and a need for creating shared activities characterizes the Periscapers' cooperation. Its Landscape Ambassador (LAMB) course is an interaction between theory and practice and a search for new ways of applying theories. The Periscapers feel that without the dimension of practical application, no new theories can be developed, and vice versa. Shared 'real life projects' need to be incorporated in their collaboration to keep evolving and to keep their 'unifying sparkle' alive.

The Landscape Ambassador Course

The LAMB course embodies a two-week-long seminar given in different places each time. Each teacher brings along five students from his or her home university. Originating from France, Hungary, Norway, Portugal, Slovenia and Sweden, the LAMB students represent richness in shared-landscape references. The Periscape group take advantage of their differences and 'overlaps' in competences. The involvement of different universities permits students to meet and learn together and from each other. Eight LAMB courses have been held to date. Teaching methods aim at raising students' understanding of the complexity of landscapes. Their different specialities make them 'translators' of different features and events in the landscape. That is why courses have always been organized around a case study (Michelin et al., 2008). It is also why teachers and students work closely together in an explorative setting. During the two weeks of a LAMB course, students are trained in perception and representation, detecting and understanding values of the site, drawing, investigating maps and spending time in the landscape. Investigating the deeper meaning with dialogue and dialogue-led planning methods, and communication with local inhabitants and stakeholders are given special attention.

Real-life Studios as the Pedagogic Approach

Academic teaching varies between disciplines, countries and time. Teachers have to follow general principles with few possibilities to experiment and look for new pedagogic ways. Focusing on 'real world studios' (Bruns et al., 2010) as the main pedagogic approach, it is important to note that most of this approach is not unique for the LAMB courses, having been quite widely used in landscape architecture and planning programmes throughout Europe and elsewhere. The training of planning and design in different contexts has been emphasized. What is special with the LAMB course is that teachers and professors leave the university and meet other students and professors from other disciplines in a place that is foreign to the majority of them. Moreover, the choice of course venue shifts and new cities or municipalities are chosen that are not well known in advance to the teachers. In the Periscape group, the teachers comprise a team, using meetings, references and literature to continue sharing and learning, year after year. Furthermore, the programme is always prepared 'in the field' during a special preparation week when the group sees the landscape and meets local key actors. To some extent LAMB meets many teachers' wish to connect to a new, current and/or 'burning' topic, and even the 'alien' but exciting can be considered. This adheres to common academic behaviour: a wish to act as an outsider, with an outsider's integrity, observing and analysing through general thinking. To a certain extent, the LAMB concept follows this approach. However, as expressed by researchers in landscape and participatory learning (Woodhill and Röling, 2000), it is also important to strengthen human dimensions and contextual aspects, as well as communicative and rhetoric methods.

Real-life studios, as we use them, rest on references like 'the understanding of knowledge', which can be linked to John Dewey as the father of the American pragmatism (Dewey, 1991; Schusterman, 2000), as well as early works of Husserl and Wittgenstein (Orlowski and Ruin, 1997). The latter emphasize understanding as something much more than just collecting facts, as the process by which an anonymous place is made into something personal. It also gives special attention to the search for the position of insiders, and context, referring to classic works such as those of Heidegger (1997) and Vygotsky (2012).

Results

Preparations

The LAMB course in the BRV brought an international group of students to the river landscape and let them discover it together with us and with representatives from the different layers of society. The students were to present ideas for future development of parts of the BRV. In the following account, we assess the success of the course in bringing awareness to the concept of landscape change in the BRV. The story of the ambassadors passing through the BRV is divided into three parts: a preparatory, an execution and a follow-up phase.

Six months before the LAMB course started, preparations in the river valley were intensified. Gustavsson spent many hours presenting the course for different audiences. The BRV residents were definitely the most important audience but it was also essential to 'anchor' the course with the regional and municipal authorities. The intention to build the LAMB course upon communication led to the enormous task of preparing the ground before the students' arrival. These preparations included: informing local stakeholders about the LAMB-students' arrival and encouraging stakeholders to share stories and participate in various events; teaching the students to actively bring in people's stories and areas of knowledge in their work; and stressing mutual learning.

A partnership between the Agricultural University and the municipality of Ronneby was signed containing an agreement that the students' material from the course should be compiled, discussed and elaborated in dialogue with the Swedish teachers and the municipality.

Workshops at the municipality were organized to see how administrators as well as local residents could use the students' material and turn it into their own projects. Many issues arose from these meetings: we met surprise, ignorance, doubt, excitement and curiosity provoking us to reflect about differences and similarities when planning in countryside or city contexts. The rural–urban dichotomy and the invasion of the urban into the rural has been a central topic in all LAMB courses: rural societies and landscapes fighting to be respected and gaining their full potential while dealing with processes which transform them into something urban or peri-urban (Michelin et al., 2008).

Course Realization

The course's communicative intentions were launched by inviting a range of actors to be responsible for some excursions or teaching situations. Morning lectures introduced each day, inspiring the students to think outside the box using their particular competences while exploring new fields of landscape research together with their fellow students, teachers and invited local stakeholders. After lunch, the entire course, or smaller groups of students, went on shared excursions to explore the landscape. Planned and unplanned interviews and meetings took place during the excursions. Students were troubled by the strong focus on dialogue and how to do all interviewees justice in their final proposal. The course organizers' original thought was to inspire the interviewees to get involved in the actual design process together with the student, but it turned out to be too complicated for students as well as for the local residents.

On the third day, students prepared a presentation of their first thoughts and analysis of the study area. This presentation was called the 'naive presentation', surprisingly with some of the most important concepts already identified. The mayor was invited and definitely contributed to the students' basic understanding of the valley. After the presentation, the students spent several days building up a deeper knowledge.

During some evenings, students and teachers met local associations in the BRV. The northern association in the forested part served moose casserole and other local specialties, while the students found themselves in a relaxed atmosphere for continuing their mission to understand not only the situation of people in this particular part of the world, but also to recognize differences within the valley.

In the middle of the course, the BiS invited everybody for a full day. The day started with attending a local church ceremony followed by a traditional 'church coffee'. The BiS did not only do this to show the importance of the church in many rural parts of Sweden, but also as a step for them to approach and establish a relation with the church. Later, students and teachers went out to work physically in the landscape and lunch was offered outdoors. Practical landscape management continued in the afternoon, and in the evening all were invited for dinner far away up along the river valley.

Final presentations were given twice in front of different audiences, with a slightly different twist. The first aimed to reach politicians and others working or living in the municipality. It became the dominant part of a political meeting for the board of the municipality in the city of Ronneby. This presentation was performed in English, followed by a discussion in English and Swedish. The final presentation brought everyone back to the warehouse of Pagelsborg where most of the work had taken place, and this time the focus was on those living and working in the river valley. Presentations were made by one Swedish student from each group followed by a discussion and commenting session. The non-Swedish students were silent but amazed how the Swedish people in the audience they had experienced as very calm and silent, suddenly had opinions and thoughts on their work.

Follow Up

Because of the partnership, municipal administrators were bound to participate in discussing the LAMB material in a series of workshops. We presented the material for interest organizations in the region. Gustavsson got even more involved in strategic roles acting as secretary and senior advisor for BiS. In September 2012, the Swedish rural parliament was arranged in Ronneby with about 1000 visitors. As part of this national event, the municipality of Ronneby arranged three bus tours, one of which went to the BRV. We took part in the excursions which were arranged to meet with local entrepreneurs and other actors. We also gave an introductory lecture indoors on the water course.

The Deliberative Process

Landscape planners and architects seek to cross geographical as well as administrative borders, formulating development plans customized for a particular place.

Tracking values and identifying affected actors is one of our areas of expertise and it is fascinating to see how the deliberative climate in the BRV resulted in shared actions for mutual learning. We believe the place-based learning process designed for the LAMB students is just as important for professionals as people living and working in the landscape. To meet and confront different stakeholders' concepts of the world is important in peri-urban landscapes. The Periscape group believes that 'landscape ambassadors' can help balance the expert-oriented planning of today with a stronger focus on places, people and potential values in specific landscape situations. This can provide a more fruitful approach to learning (and acting) in complex landscape contexts, as in the case of peri-urban areas and areas that are transformed from rural working landscapes to greenspaces serving the recreational needs of urban populations.

Evaluating the importance of the LAMB students' passage in BRV raises the question of who was interested in participating in the course (and learn) and who wanted to take part in the results (and act).

The local residents' interest in participating originated from their vulnerability. The BiS encouraged the community to see the availability of individual EU subsidies for grazing, keeping land well managed, and the establishment of nature reserves as a way to collectively maintain the landscape and raise the profile of the community. A distinct characteristic among BRV residents is how they learned to see their river valley as special from a European perspective. 'Six European professors and 30 students can't be wrong, we have something special here', is a frequently used quote from the chairman of the BiS when referring to the LAMB course. The different meetings and local arrangements around the LAMB students contributed to the raising of local awareness.

The BiS enjoyed media attention and took advantage of the sudden focus on their river landscape. But the Swedish students also worried about doing justice to all the varied and subtle perspectives of the interviewees in the river valley, especially the differences between the southern and northern residents of the valley. They also struggled to visualize the stories they chose not to bring into their respective projects These stories included the dreams of some residents to realize ecological farming that yields an economic return; a stronger presence of the municipality in support of rural villages; new forests that are not uprooted by wild boars; school buses; lights in the neighbour's house in the winter; new neighbours; and countryside living along a river with fish, swimming places and businesses. The local audience on the last day of the course focused on how the students' proposed design of new housing areas would attract more people to live in the area while maintaining its small-scale atmosphere. Even though student proposals were not well developed, it made the local inhabitants 'grow' in the role of thinking about and constructing new possibilities. The students' young age appeared to have been a positive influence on the deliberations. The residents felt that it was a powerful mutual learning experience to work with an age group (20–25 years) that is hard to reach except through the education system.

Local politicians became more aware of the importance of green resources outside of the main town, as well as becoming more interested in participating in discussions about policy actions through the LAMB course and follow-up activities. The politicians picked up practical and real ideas like 'the Bräkne basket—a basket full of landscape' (a student proposal to market small-scale, locally produced farm products in a shared basket). They have not, however, been able to find economic support for future development though they are now more aware of the national as well as European incentive programmes for the peri-urban landscape.

The county administrative board, by contrast, felt that the residents in the BRV were getting too much attention, both in terms of collaboration with local associations and financially supported projects. The board was more interested in identifying small areas of significant natural values, and felt local residents lacked a feeling of community connectedness to the landscape.

Yet, during the last five years, the peri-urban landscape has been upgraded little within urban society. Changes have occurred in the dialogue between urban and rural associations but this is a slow process. For the BRV residents, landscape and multifunctional perspectives are central—and have been used in a number of successful projects since the LAMB course. These have included several bicycle-tour events and a long-term aim to increase the attractiveness and accessibility of the whole valley to tourists. The municipality of Ronneby is funding a new economic local organization to raise issues related to peri-urban landscapes and green resources. This group comprises a mixture of representatives of local organizations and municipal administrators. At the same time, the regional government is still largely absent, as it continues to concentrate on specific projects and tasks given by the Swedish government in a top-down manner. Many residents in the BRV continue to disapprove of their childhood landscape being turned into nature reserves. The LAMB process worked within the tension between rural and urban(ization) interests in the BRV and managed to enhance local residents' pride in belonging to a network of new, emerging greenspaces and old, vanishing farming landscapes and then seeing and imagining the possibilities of these hybrids in building a better future.

Notes

1 www.jordbruksverket.se.

References

Arler, F. (2008) 'A true landscape democracy' in S. Arntzen and E. Brady (eds) *Humans in the Land*, Academic Press, Oslo.

Bauman, Z. (1995) *Postmodern Etik*, Daidalos, Göteborg.

Beck, U. (1992) *Risk Society: Towards a New Modernity*, Sage, London.

Blaikie, P., Cannon, T., Davis, I. and Wisner, B. (1994) *At Risk: Natural Hazards, People's Vulnerability, and Disasters*, Routledge, London.

Blicharska, M., Isaksson, K., Richardsson, T. and Wu, C.-J. (2011) 'Context dependency and stakeholder involvement in EIA: the decisive role of practitioners' *Journal of Environmental Planning and Management*, 54(3): 337–354.

Blok, A. and Jensen, T. E. (2011) *Bruno Latour: Hybrid Thoughts in a Hybrid World*, Routledge, New York City.

Bruns, D., Ortacesme, V., Stiles, R., de Vries, J., Holden, R. and Jorgensen, K. (2010) 'ECLAS Guidance on Landscape Architecture Education. The Tuning Project. ECLAS–LE:NOTRE', www.eclas.org/accreditation-advice.php (accessed 25 November 2013).

Cadieux, K. V. (2008) 'Political ecology of exurban "lifestyle" landscapes at Christchurch's contested urban fence', *Urban Forestry & Urban Greening*, 7(3): 183–194.

Crankshaw, N. (2009) 'Plowing or mowing? Rural sprawl in Nelson County, Kentucky', *Landscape Journal*, 28: 218–234.

Cronon, W. (ed.) (1996) *Uncommon Ground: Toward Reinventing Nature*. W. W Norton & Company, New York City.

Dewey, J. (1991 [1910]) *How We Think*, Promotheus Books, New York City.

Europe, Council of (2000) 'The European Landscape Convention', Florence.

Flyvbjerg, B. (2001) *Making Social Science Matter: Why Social Inquiry Fails and How it Can Succeed Again*. Cambridge University Press, Cambridge.

Gustavsson, R. and Peterson A. (2003) 'Authenticity in landscape conservation and management—the importance of the local context' in H. Palang and G. Fry (eds) *Landscape Interfaces*, Kluwer Academic Publishers, Dordrecht, pp. 319–356.

Heidegger, M. (1997) 'Poetry, language, thought', in J. Iverson Nassauer (ed.) *Placing Nature: Culture and Landscape Ecology*, Island Press, Washington DC.

Mellqvist, H., Gustavsson, R. and Gunnarsson, A. (2013) 'Using the connoisseur method during the introductory phase of landscape planning and management', *Urban Forestry & Urban Greening*, 12(2): 211–219.

Michelin, Y., Gustavsson, R., Pinto Correia, T., Briffaud, S., Geelmuyden, A. K., Konkoly-Gyuró, É. and Pirnat, J. (2008) 'The Landscape Ambassador Experience' *8th European IFSA Symposium. Satellite Session: Education in Landscape and Territory Agronomy*, Clermont-Ferrand, France, pp. 961–970.

Nielsen, B. S. and Nielsen, K. A. (2006) *En menneskelig natur. Aktionsforskning for baeredygtighed og politisk kultur*, Frydenlund Academic, Köbenhavn.

Orlowski, A. and Ruin, H. (1997) *Fenomenologiska perspektiv. Studier i Husserls och Heideggers filosofi*, Thales, Karlshamn.

Ostrom, E. (1990) *Governing the Commons: The Evolution of Institutions for Collective Action*, Cambridge University Press, Cambridge.

Pinto-Correia, T., Gustavsson, R. and Pirnat, J. (2006) 'Bridging the gap between centrally defined policies and local decisions—towards more sensitive and creative rural landscape management' *Landscape Ecology*, 21(3): 333–346.

Qviström, M. (2013) 'Searching for an open future: planning history as a means of peri-urban landscape analysis' *Journal of Environmental Planning and Management*, 56(10): 1549–1569.

Robbins, P. (2011) *Critical Introductions to Geography: Political Ecology* (2nd Edition), Wiley, Hoboken, NJ.

Sandström, C., Hovik, S. and Falleth, E. (eds) (2008) *Omstridd natur. Trender och utmaningar i nordisk naturförvaltning*. Umeå, Boréa bokförlag.

Schusterman, R. (2000) *Pragmatist Aesthetics. Living Beauty, Rethinking Art*, Rowman and Littlefield Publishers, Inc. Lanham, MA.

Selman, P. (2012) 'Landscape as integrating frameworks for human, environmental and policy processes', in T. Plieninger and C. Bieling (eds) *Resilience and the Cultural Landscape*, Cambridge University Press, Cambridge, pp. 27–48.

Swaffield, S. and Primdahl, J. (2006) 'Spatial concepts in landscape analysis and policy: some implications of globalisation', *Landscape Ecology*, 21: 315–331.

Swedish National Agency for Higher Education (1992) 'The Swedish Higher Education Act', Ministry of Education and Research, www.uhr.se/sv/Information-in-English/Laws-and-regulations/The-Swedish-Higher-Education-Act/ (accessed 25 November 2013).

Swyngedouw, E. and Heynen, N. C. (2003) 'Urban political ecology, justice and the politics of scale', *Antipode*, 35(5): 898–918.

Waldheim, C. (2006) *The Landscape Urbanism Reader*, Princeton Architectural Press, New York City, NY.

Widgren, M. (2012) 'Resilience thinking versus political ecology: understanding the dynamics of small-scale, labour intensive farming landscapes', in T. Plieninger and C. Bieling (eds) *Resilience and the Cultural Landscape*, Cambridge University Press, Cambridge, pp. 95–110.

Woodhill, J. and Röling, N. G. (2000) 'The second wing of the eagle: the human dimension in learning our way to more sustainable futures', in N. G. Röling and M. A. E. Wagemakers (eds) *Facilitating Sustainable Agriculture*, Cambridge University Press, Cambridge, pp. 46–69.

World Health Organization (2014) 'Ebola virus disease in Guinea', www.who.int/en/ (accessed 31 March 2014).

Vygotsky, L. S. (2012) *Thought and Language*, MIT Press, Cambridge, MA, and London, UK.

20 Art, Enchantment, and the Urban Forest

A Step, a Stitch, a Sense of Self

Kathleen Vaughan *

> I saw the world I had walked since my birth and I understood how fragile it was, that the reality I knew was a thin layer of icing on a great dark birthday cake writhing with grubs and nightmares and hunger. I saw the world from above and below. I saw that there were patterns and gates and paths beyond the real. I saw all these things and understood them and they filled me, just as the waters of the ocean filled me.
>
> (Gaiman, 2013: 143)

Just as the woods of my city fill me.

For as long as I can remember, I have been in love with the shifting patterns of light through the leaves of trees. Could this be a reflection of a happy early memory, when as a toddler I was lifted up towards the leafy bower of our backyard oak tree? I remember my father's hands on my ribs, my small hands rising to grasp the unexpectedly cool strength of the oak's horizontal branch, normally so far above my head, and my eyes up to the shifting cathedral of greens and blues, the branches and creatures even further above me. Perhaps more recent happy encounters reinforce this early touchstone: I have been nourished by years of habitual walks with my dogs in the urban forests[1] of the cities where I have lived. These experiences bring together the thematics that are core to my creative and research practices as an artist and academic: mobility, corporeality, relationality, identity, place, urbanity, the role of the more-than-human[2] in the city, and the place of the wildish in contemporary western life. Embedded within these are considerations both ethical and aesthetic.

Currently, I am developing a series of artworks about my walks in urban woods, each work a reflection of multiple engagements with a single place that has been officially designated or unofficially claimed as a wildish preserve. In my making, I am oriented to what Australian artist-theorist Paul Carter calls "the dapple of things," which he finds in "the pied beauty of clouds, foliage, and limestone walls [that] comes into view not as a background to important events but offering an alternative focus of its own" (2009: 1).

That is to say, I am interested in the shapes and shimmer of light through leaves and needles; the quiet thud of my footsteps forward on the earth, a steady pulse of

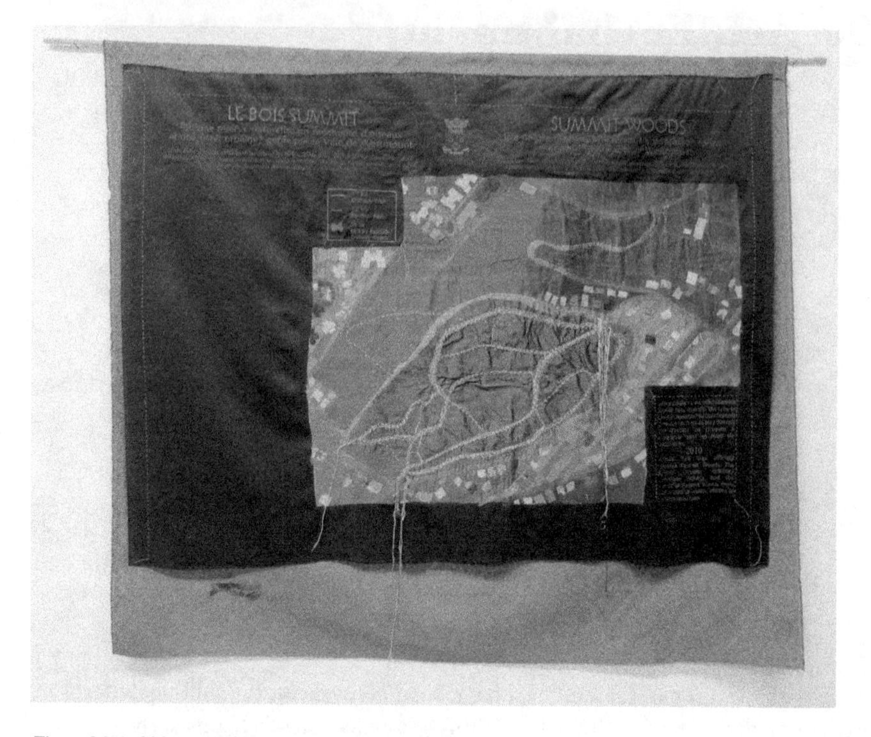

Figure 20.1 *Nel mezzo del cammin: Bois Summit/Summit Woods* (2013), digital and hand embroidery on textile assemblage, by Kathleen Vaughan.

Source: Kathleen Vaughan, 2013.

footfalls; the scarlet gift of a cardinal flitting among a stand of winterbare sugar maples (*Acer saccharum*), of a *trillium erectum* or 'wake-robin' in the first flush of the eastern woodlands' spring groundcover; the jubilant leaps of my dog, who will run ahead and then dash back, just for the pure joy of it, and who sometimes bounds off all four feet at once, like a bunny; the metallic scent of fresh snow that has settled a soft layer of quiet and yields to our footsteps; the suggestion of the untamed in the whispers of natural life around us—of Neil Gaiman's great darkness—that is accessible and an important reminder but not overwhelming; and the calming rhythms of my breath as I move through *this* space at *this* moment—and dozens and dozens of other specifically situated experiences in the woods that claim and delight me.

The alternative focus offered by these aspects of my regular experiences in the woods is that of "becoming with," as the phrase is used by cultural theorist Donna Haraway to describe the way that we humans as well as "species of all kinds, living or not, are consequent on a subject- and object-shaping dance of encounters" (2008: 3, 4). In the dapple of the woods, I paradoxically recognize myself more clearly and at the same time feel more keenly that my own identity is porous and changing, and entwined with those I encounter along the way. As much as I appreciate the opportunity to feel the contours of my own sense of self, I am also grateful for the chance to engage with what art theorist Miwon Kwon has termed the "locational identity" (2004: 8) of the woods. Scholars such as Kwon and geographer Doreen Massey (1991) suggest that one of our contemporary challenges with respect to place is to resist both the nostalgic vision of a static locality, bound by an objective [colonizers'] history, and the fashionable alternative, the belief that all is only flux, nomadism. To help us resist, Massey reminds us that, "If it is now recognized that people have multiple identities then the same point can be made in relation to places" (p. 28). Both provisional and in flux but situated and specific at the moments of our encounters, the woods and I are separate but connected as we 'become' together.

As an artist I strive to share the qualities of this "becoming with" through art making and related writing. Carter urges those of us aiming to represent our experience in language to "read—and write—our environment in a way that [does] not mummify its dynamic character" (p. 5), an exhortation that I would argue applies just as well to visual making. This is a challenge, since I am working in the static medium of textiles, which I aim to animate in ways I describe below. Also, I use language—specifically, the title of the series—to allude to the distinctly unmummified woods of myth and fairy tales, a potent and wildish place that laid claim to the imaginations of so many of us when we were young, and perhaps even still. And so the series title is *Nel mezzo del cammin* ["In the middle of the journey"], the opening words of Dante's *Inferno,* itself a story of a walk in the woods that leads to hell, purgatory, paradise and a consideration of questions of ethics, aesthetics, and human nature (Halpern, 1993).

While the full *Nel mezzo* series will reference woodswalks in several cities and nations, my first four works are oriented to greenspaces of urban Montreal, Canada: Summit Woods, the Morgan Arboretum, Angell Woods, and a segment of the linear park alongside the Lachine Canal. As it happens, these territories comprise the three sorts of spaces in which "ecological co-fabrication [occurs] between humans and other city inhabitants" (Whatmore and Hinchcliffe, 2010: 450): two are spaces oriented to conservation, one to cultivation, and the last to land restoration.

In the category of conservation, officially designated an urban forest, Summit Woods is a remnant woodlot on the lowest of Mount Royal's three peaks (201 meters above sea level), owned by the City of Westmount (a borough of Montreal) and technically a bird and wildlife preserve, just 23 hectares large (City of Westmount, 2011; Les amis de la montagne, undated). Here, the trees are the expected coniferous–deciduous mix of the Great Lakes/St Lawrence forest, the St Lawrence being the river that encircles the island of Montreal. Of the same orientation, Angell Woods is the name given to an informal patchwork of territories of hardwoods and marshlands about 100 hectares large, its aggregated pieces owned respectively by the Town of Beaconsfield, the City of Montreal, Ducks Unlimited, private developers, and the Association for the Protection of Angell Woods, with this latter non-profit, volunteer advocacy group mobilizing for conservation and working against residential development of the site (APAW/APBA, 2013). The Morgan Arboretum is 245 hectares of space owned and cultivated by McGill University as a teaching, research, and public education site, a forested reserve including examples of most of Quebec's native trees as well as 18 collections of trees and shrubs from around the world (Morgan Arboretum, 2012a).

A space of land restoration, the Lachine Canal and its adjacent parklands are property of the Canadian government and since 1978 have been managed by its agency, Parks Canada. Operating from 1825 to 1970 as the heartland of industrial Montreal's shipping, processing, and manufacturing, the Lachine Canal and its adjacent lands suffer from significant soil and water contamination, much of which has not been remediated on the advice of a joint provincial–federal review panel (Parks Canada, 2010). Even so, post-industrially, the terrain along the canal has been developed as a 14 km linear park whose notable amenity is a multipurpose path (biking, blading, walking) visited by humans more than a million times per April to November season. Unlike the previous three ecosystems described, here, the canal-side urban forest is sporadic and background to other local features, not even remarked among the 'natural wonders' mentioned on the Park's website (Parks Canada, 2013). However, the summer shade of the parkland's trees is important to the pleasure of its human users and the lives of other creatures.

Engaging with the sites in depth, my artwork takes the form of textile maps that integrate official visual-navigational tools of the woods in question with traces of my own progress through the space. That is, informational maps (often posted or distributed by the site itself) are the basis of imagery that I re-create with digital embroidery and hand piecing on cloth. Google Maps and other aerial photographs provide additional imagery of the neighbourhoods that surround the woods, contextualizing them as *urban* forests. I cut fabric into shapes of buildings, roadways and other structures both architectural and topographical, versioning what other cartographers have considered important.

Into these assembled surfaces—silks, wools, organzas, chintzes, all adhered to a blanket-weight wool—I hand-embroider the specific routes that my dog and I have taken over the course of a specific interval of time (a month, a fortnight, a week); each walk is a unique colour. As the stitched-in version of the walks accumulate on the cloth, the density of handwork reflects our repeated choices of, and preferences for, particular terrain or path-side amenities. In this way, the maps are a trail of breadcrumbs of my experience: they reflect my histories of walking in urban green-spaces, aiming to explore questions of knowledge and translation, but more profoundly, questions of love and belonging, and how these may be constituted in a postmodern ethical framework that considers political ecologies and ecologized politics (Hinchcliffe et al., 2005: 654).

Maps are wonderful things. Like photographs, they tend to present as authoritative information but in fact, as cultural geographer Brad Harley (1988) reminds us, maps are never value-free images. Maps are thus a perfect vehicle to explore questions of knowledge, its transmission and the power structures embedded therein. Also implicit within mapping are considerations of our relationship with

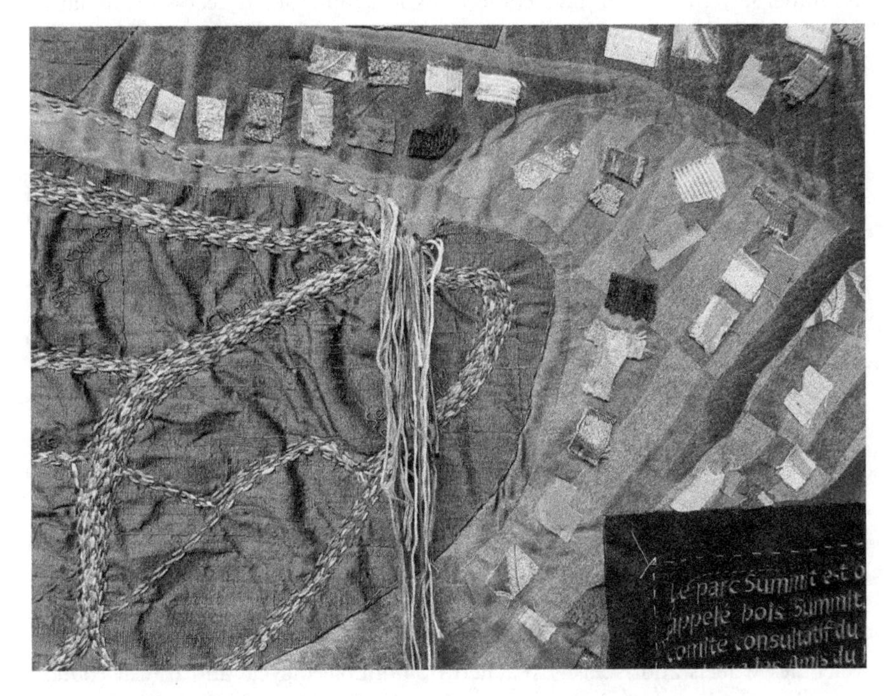

Figure 20.2 *Nel mezzo del cammin: Bois Summit/Summit Woods* (2013), detail, by Kathleen Vaughan.

Source: Kathleen Vaughan, 2013.

place, of human frailty and human mobility and issues of representation itself. Maps are thus much beloved of artists (Harmon, 2004; O'Rourke, 2013), myself included. Writing about artists' work with maps, theorist Stephen S. Hall has stated, "It is hard to look at a map without sensing, in our bones, private hopes and secret fears about change" (2004: 16). I return to the thematic of change, below, but for now will suggest that perhaps it is no accident that I started this mapping project not long after moving back to Montreal from two decades of life in another city 600 km away. This project took shape as I began the work of re-emplacing myself, of finding fresh woodlands in which to walk with my dog, and a new sense of self that comes with the ongoing habitual engagements with a new place: a vernacular ecology.[3]

Like any place, the four greenspaces of my maps are contentious and politicized, raising questions of power and access (for whom and by whom?) with respect to more-than-humans as well as human visitors. All sites aim to control use, limiting humans' access to certain hours or seasons or to fee-paying individuals (a day fee or annual membership being required at the Arboretum). Asserting their rights to the land's use, the private owners of segments of Angell Woods have gone so far as to occasionally post security guards to turn away hikers and dog walkers

(Kramberger, 2012) who have come to expect unhindered access to the land. Access to these spaces by dogs is even more structured and licensed, with special annual fees (and collar tags) required to walk a dog in Summit Woods and at the Arboretum, fees I have paid. With respect to the Summit, a complex series of negotiations is currently underway regarding the rights and responsibilities of various factions—birdwatchers and naturalists, dog owners, professional dog walkers, hikers and recreational walkers, and family users of urban space (Sweeney, 2013)—with associated discussions about what the woods are 'for'. Additionally, an academic research study is in progress, jointly sponsored by the City of Westmount, the Healthy Communities Research Network, and the University of Waterloo, to explore the more-than-human geographies of encounter at Summit Woods with emphasis on evolving connections of self, place and 'other' (Graham, 2013). The outcome of this study may well affect City of Westmount policy about Summit Woods and other parks.

Political implications exist with respect to physical access to these green spaces, at least to the three named 'woods' mapped. They are removed from the main flow of the city's activities and its areas of densest residency, on the western end of the island of Montreal or on the tip of one of its three peaks, in areas ill-served by public transit and so requiring vehicular access for those not immediately adjacent. None of these terrains is well suited to those in wheelchairs or other wheeled mobility supports. By contrast, the Lachine Canal parklands have a paved route that welcomes wheels and is easily reached by multiple bicycle and transit routes. Immediately adjacent to downtown, the Lachine Canal runs through the heart of what once were Montreal's industrial working-class neighbourhoods. A dynamic thoroughfare for 150 years until its closure in 1970, the Canal was the economic and social engine of adjacent workers' communities. However, given these neighbourhoods' recent decades of de-industrialization, decline, and subsequent post-industrial revitalization as sites of leisure, more subtle access issues are embedded in this greenspace, those of class: longtime local working-class residents can feel excluded by, and in opposition to, many of those along the Canal, that is, incoming residential gentrifiers or park visitors with attributes of privilege such as high-end condominiums, luxurious motor boats, expensive racing bicycles, even purebred dogs. I am always mindful, when I use these spaces, of the political aspects at play.

The maps that I am making hint at these issues, especially insofar as they are embedded in the source imagery I start from. For instance, the map of Summit Woods is posted within the centre of the woods, at a place of convergence of the pathways but away from any access point. The official map includes much ancillary text (not reproduced by me), detailing the history of the site as well as offering a definition of an urban forest (which I do include). While all the trails are named, no nearby roads are, meaning that the map becomes a prompt for conversation with walkers who want directions to the place where they have left their car. The Morgan Arboretum map handed out to visitors and available online (Morgan Arboretum/Arboretum Morgan, 2012b) describes topography, names trails, and marks parking sites, and is explicit about where and how dogs are permitted on the territory.

At Angell Woods, two maps posted together at an entrance describe the same territory, one showing the routes of the walking trails, the other the cadastral aspects of the terrain—an interesting distinction. With these two sets of information not connected, a walker will know how to navigate but not necessarily whose land she is traversing in the moment of each footfall. And Parks Canada's (2009) online maps of the Lachine Canal and its adjacent territories, current to its posting five years ago, indicate both the original corporate names of the neighbourhood's residual industrial buildings and their newer residential monikers, reflecting the locality's changed use from labour to lifestyle. (My own version updates the official maps with more recent condominium construction.) I am alert to these various political complexities and wish them to be part of the depths of meanings accessible in my artwork. At the same time, my maps do not explicitly promote a particular political agenda or propose a plan of action. Rather than identifying these pieces as 'political', I see them as 'critical' as articulated by art historian Abigail Solomon-Godeau (1990). She suggests that critical artwork is resistant to reproducing commonly accepted cultural norms and instead aims to provide a space for their contestation (pp. 75–76). The ways that my maps may do so can seem indirect or obscure, since as artwork, my works are non-propositional forms of communications. Non-propositional is the term used by philosopher Suzanne Langer (1953) to distinguish the orientation of visual art from that of discursive—that is, propositional—forms of communications, such as this text. With non-propositional forms, there is no explicit directive. If art is not, then, to promote a particular agenda, what *is* art for? This question precipitates a cascade of worthy answers including a personal favourite proposed by philosopher Maxine Greene (quoted in Csikszentmihalyi, 1997: 34): by engaging the viewer's body, spirit, and mind, a work of art enables an experience of radical happiness, a feeling that may orient our future engagements with the world.[4]

And so my maps aim to engage and delight by multiple aesthetic means: the lustre of colour, the tactile richness of soft wools, shimmer of slippery silks, and the gloss of organza. In their gift of beauty, the maps offer an invitation not only to engage deeply with the artworks themselves as material artifacts (their form) but also to consider the happiness and meanings of the woodswalks that the artworks represent (their content). I aim to create work that is aesthetically 'useful' in the way described in Suzi Gablik's seminal statement of credo, *The Reenchantment of Art* (1991: 141). I am hoping that the viewer may be moved by the pleasure of their experience of the artwork, to consider re-enacting the pleasure that prompted the works—the woodswalking itself—and hence to become more deeply engaged with notions of the value and significance of the urban forest itself. Thus, a viewer may be moved from personal aesthetic pleasure to greater awareness and even advocacy, a change of stance, mind, and behaviour.

I see as a primary audience for my work people who may not walk in the woods themselves, especially those who simply are not aware of their existence.[5] Speaking to neighbours, colleagues, and friends—individuals who like me tend to live in city neighbourhoods distant from urban woods—I am always amazed by how many people have no idea of the multiple forests available to them nor a strong wish to investigate. Perhaps their own urban experience makes them less tree-oriented: that is, their lack of engagement may reflect a relatively low level of treed green in their own stomping grounds. In Montreal's post-industrial South West borough where I live, definitely a "neighbourhood in transition," we have an estimated 2.7 hectares of greenspace per 1000 individuals, just over the minimum of 2.0 hectares per thousand inhabitants recommended by urban environmentalists (Rocha, 2011). Here, most residential buildings have no set-back from the sidewalk—a reflection of their architectural roots in nineteenth-century workers' housing. Our neighbourhood has an increasing number of lollipop sidewalk trees that provide some green, some shade, and some mitigation from the worst of the summer heat. Many of these are relatively new, perhaps thanks to local gentrifiers, whose property values increase with neighbourhood tree planting and so may push for more green (Perkins et al., 2004; Conway et al., 2011).

My acquaintances' lack of knowledge about, and intense engagement with, the urban forests I love may also be linked to their distance from the sites in question: research shows that residents who live near large amounts of green space are the ones who tend to have strong, deeply felt feelings about them (Lober, 1995; Lober and Green, 1994). Interestingly, literature in the field of ethics also suggests that proximity can make a difference to a person's feeling of responsibility for another, enhancing its likelihood (see Bauman, 1989 and Vetlesen, 1993). Of course locality matters; but perhaps the repeated engagement with more-distant sites and their ecosystems can build a bond that becomes a link of care. And perhaps art can help to promote that engagement and ethical implication.

In fact, theorists suggest that this is exactly what beauty—enchantment—can do. Dave Hickey, for one, reminds us that by offering visual pleasure without restraint or discrimination, the beautiful "remains a potent instrument for change in this civilization" (1993: 24)—change of the individual as much as social change. Or, adding on to philosopher Jane Bennett's observation, "the mood of enchantment may be valuable for [the kind of] ethical life" (2002: 3) that would have us more engaged with the human and more-than-human world around us.

Certainly a quality of enchantment is present in my ongoing walks in woods, as I repeatedly walk, observe, register, and I am prompted to research, learn and respond, creatively to the woods and their surprises. I am learning to be affected by the complex network of life that is the woods and also by the shifts, developments, and consistencies of Baloo, my animal companion in these walks. My dog, the woods, and I are all companion species, in the way that cultural theorist Donna Haraway (2008) has constituted the notion to include the whole range of flora, fauna, and organisms that are part of the world that we also inhabit, and that are even part of ourselves. She writes:

> I love the fact that human genomes can be found in only about 10 percent of all the cells that occupy the mundane space I call my body; the other 90 percent of the cells are filled with the genomes of bacteria, fungi, protists, and such, some of which play in a symphony necessary to my being alive at all, and some of which are hitching a ride and doing the rest of me, of us, no harm. I am

> vastly out-numbered by my tiny companions; better put, I become an adult human being in company with these tiny messmates. To be one is always to become with many.
>
> (2008: 3–4)

I am also "becoming with" my maps in the process of creating them. I work back and forth between the walks themselves and the stitching that traces the pathways of the walk took, remembering the route I picked as I render it in coloured floss. Out in the woods, I choose trails based on my aesthetic response to the map-in-progress: where, compositionally, does my map need some stitching? Or, the walking corollary, which terrains have I insufficiently explored? The artistic process influences my lived reality just as the actual walks inspire the making, integrating the stitches and the steps more and more closely. My mapping seems a necessary extension of my experience of the woods.

Their capacity to imply a sense of co-implication, relationality, and co-evolution is a reason that I am working through textiles and not a form of pictorial representation such as painting or photography. I am not interested in evoking a kind of "scenic aesthetic" (Gobster, 1999: 55) that portrays the woods as an arm's-length object, a reified other, even if that is the familiar way that many urbanites relate to

Figure 20.3 Nel mezzo del cammin: Bois Summit/Summit Woods (2013), detail, by Kathleen Vaughan.

Source: Kathleen Vaughan, 2013.

the distant forest. I also appreciate textiles' material complexity, cloth being a familiar, everyday artifact and fibres a creative practice with a disciplinary history. With this dense cultural history also comes multiple embedded politics, summarized by American artists and theorists Joan Livingstone and John Ploof:

> Originating with the history of survival, cloth manufacture, and its accompanying division of labour, expands to impact all spheres of culture and power. Additionally, the physical and intimate qualities of fabric allow it to embody memory and sensation and become a quintessential metaphor for the human condition. Crossing between arenas of function, craft, art, and ritual, the meaning of cloth from its most banal to its most splendid form affects our daily lives and welfare in terms of gender, race, ethnicity, sexuality, class, invention and technology, commerce and work.
>
> (2007: vii)

Before the industrial period, when it was produced close to the family and invested with ritual and status, cloth tended to be viewed as regenerative, protective, somewhat magical (Weiner and Schneider, 1989)—and was highly costly: in medieval times, each person had to dedicate more than a quarter of the day's work to cover personal textile needs. By contrast, today's consumer in an industrialized nation puts in just minutes per day for the same effect (Tobler-Rohr, 2011: iv). Through mass manufacture under industrialized capitalism, cloth became distanced from familial and ritual sites of power and so—it is proposed—"disenchanted" (Weiner and Schneider, 1989: 13). Contemporary textile manufacture is further tainted by recent workers' deaths in horrific conditions in Bangladeshi garment factories (Calamur, 2013). However, with art theorist Mary Jane Jacobs, I would like to believe that through art, cloth may become "re-enchanted" (2007: 309), in part via artists' [drawing] attention to the political ecologies of its materiality and its production. This means not just the specifics of the ethical, human issues, but also the sustainable, environmental ones.

And so as I walk and work, stitch and think, I am increasingly concerned about the environmental impact of the materials that I am using: my silks, wools, organzas, and chintzes are all beautiful—and all commercially produced. I am beginning to understand more about what this means—in terms of fibre production, for instance, with respect to the treatment of the land that grows the crops or the sheep that provide the wool (Henry, 2012); and in terms of the textile processing and technologies in yarn and fabric production, with one 'for instance' being the use of/environmental contamination by dangerous chemicals (Greenpeace, 2012). I will take this research deeper and probably adjust my practice as I produce future maps.

Such research and adaptations are a natural part of my artistic practice, which—given that I am situated within a university—falls into the category the academy calls *practice-based research*, a term that developed in the United Kingdom to describe "research initiated in practice and carried out through practice" (Gray, 1998: 1), meaning artistic practice. In Canada, we have a particular term for this process, recognized by the major academic funding bodies[6] and used for certain streams of academic art production: "research-creation." Research-creation is oriented to professional levels of practice as determined by a peer review, deliberate contributions to knowledge, and some orientation to method (Government of Quebec, 2008; SSHRC, 2013). For me and the graduate artist-researchers with whom I work, research-creation is distinguished from the kind of research that any artist might do to prepare a body of work by the self-consciousness about method and articulation of a context of practice, an elaboration of one's alignment with, and differences from, other relevant creative and intellectual work. My own method draws on the collage practices of arts traditions (Vaughan, 2005, 2007, 2009) and, like research in the social sciences, requires that my work address one or more research questions. In the case of the *Nel mezzo* maps, the entwined questions that guide their making are the following:

- What can urban woods offer the walker? Reciprocally, what can the walker offer the urban woods?
- What is it we come to know through regular, devotional walking in a particular place?
- What kinds of ethical questions and responsibilities are foregrounded through animal companionship? What can close living with a dependent other of another species help us understand about relationality and identity?
- How does artistic practice, specifically the slow work of creating digital and analogue textiles, enable my deep reflection on these issues? How might artistic engagement enable others to feel into some of these same themes and questions?

I have begun to articulate preliminary answers to some of those questions here in this text but of course I have more learning and creating to do. I hope and like to believe that I am engaged with questions that, in the words of cultural theorist David Scott are "*worth having answers to*" [emphasis in original] (1999: 7) at this moment in our earth's and our own histories. I am perpetually mindful that we are now working in what Haraway (2012) has described as a time of great extinctions—which is extraordinary motivation to consider critically the relational implications and ethics of our practices.

It is at this heartfelt level that the urban woods, my walks there, and my maps of my walks there matter to me. If indeed it is true that "[t]hings do not look too rosy for urban wilds. Not pure enough to be true and not human enough to be political, urban wilds have no constituency" (Hinchcliffe et al., 2005: 645), then I hope that my maps may help to build a constituency. I hope for a constituency that is re-enchanted and moved to mindful stewardship—and also to delight in the dapple of our urban forests.

Notes

* I dedicate this text to Rishma Dunlop, poet, friend, and mentor, whose own "tawny grammar" (2002) about place, memory, and belonging speaks of the Beaconsfield of her youth (and other sites) and often comes to mind when I walk the trails of Angell Woods.

1 The definition of urban forest that I use is printed on the visitor's map posted within Summit Woods and attributed to the Canadian Urban Forest Network: "An urban forest is defined as trees, forests, greenspaces and related non-living, living and cultural components in and around cities and communities." (See also the Canadian Urban Forest Network's document, *Canadian Urban Forest Strategy 2013–2018* (2012).) Because my map of Summit Woods includes text from the posted map on-site, this definition is also versioned in digital stitching onto my artwork, offering the casual viewer the opportunity to consider that an 'urban forest' is a specific, recognized and complex entity rather than a haphazard agglomeration of trees.

2 I adopt the term 'more-than-human' in part inspired by David Abram's *The Spell of the Sensuous: Perception and Language in a More-Than-Human World*, whose extended philosophical and poetic discussion of our human orientation to sense-based relationality and thinking has been important to me since it first appeared in 1996. I appreciate that the formulation 'more-than-human' can suggest grandeurs that we don't yet understand and yet possible connections to the human. While I agree with Hinchcliffe et al. (2005) that "The term 'nonhuman' signals a worldliness of worlds, suggesting that cultures and societies are shaped by more than human geographies" (p. 644), to me 'non-human' also seems to suggest realms of others too separate and far away, at least from the perspective of my own intersubjective orientation and experience.

3 This concept of vernacular ecology has been variously discussed, including by geographers Sarah Whatmore and Steve Hinchcliffe (2010), who emphasize the ways in which the "intimate interweavings of the life patterns and rhythms of people and other city dwellers refigure the landscapes of everyday life" (p. 450). Vernacular ecology adapts the term vernacular landscape popularized by philosopher Michel de Certeau (1984) and J.B. Jackson (1984), to describe a place that is constituted through the everyday practices and routines of daily life.

4 This is not of course to say that art cannot be oriented to outrage, to difficult content, or to social justice—even at the same time as a work invokes a sense of happiness: in fact, another of the purposes of art proposed by Maxine Greene in the same article is to hold the world at bay while not denying it (Csikszentmihalyi, 1997: 34). Given its depths and densities—and the wide variety of viewers—art is not easily instrumentalized, that is, oriented to a particular outcome. Art can be very disobedient to the artist's intentions or political need.

5 This is a reason that I am not working in site-specific practices, that is, locating my creations within the urban forest itself.

6 The two major Canadian funding bodies of art making as a research practice within the academy are the Social Sciences and Humanities Research Council (SSHRC) and the Fonds Québec de Recherche en Société et Culture (FQRSC), with funding from the latter being available only to researcher-creators working within the province of Quebec.

References

Abram, D. (1996) *The Spell of the Sensuous: Perception and Language in a More-Than-Human-World*, Pantheon Books, New York City.

Les amis de la montagne (undated) 'Mount Royal, a territory to discover: Westmount Summit, 201m above sea level', www.lemontroyal.qc.ca/carte/en/html/Summit-Park-30.html (accessed 29 July 2013).

APAW/APBA Association for the Protection of Angell Woods/L'association pour la protection du bois Angell (2013) 'Angell Woods', www.apaw.ca/?page_id=27 (accessed 29 July 2013).

Bauman, Z. (1989) *Modernity and the Holocaust*, Polity Press, Cambridge, UK.

Bennett, J. (2001) *The Enchantment of Modern Life: Attachments, Crossings, Ethics*, Princeton University Press, Princeton, NJ.

Calamur, K. (2013) 'Bangladesh textile exports surge; Another factory worker dies', *Capital Public Radio/NPR*, 12 August, www.capradio.org/news/npr/story?storyid=211367798 (accessed 14 August 2013).

Canadian Urban Forest Network/Réseau canadien de la forêt urbaine (2012) *Canadian Urban Forest Strategy 2013–2018*, www.tcf-fca.ca/programs/urbanforestry/cufn/resources/pages/files/CUFS_July20_2012.pdf (accessed 30 July 2013).

Carter, P. (2009) *Dark Writing: Geography, Performance, Design*, University of Hawaii Press, Honolulu.

Certeau, M. de (1984) *The Practice of Everyday Life* (trans. S. Rendall) University of California Press, Berkeley, CA.

City of Westmount (2011) 'About Westmount: Summit Woods and Summit Lookout' [date modified 4 November 2011], www.westmount.org/page.cfm?Section_ID=2&Menu_Item_ID=41 (accessed 29 July 2013).

Conway, T.M., Shakeel, T., and Atallah, J. (2011) 'Community groups and urban forestry activity: Drivers of uneven canopy cover?', *Landscape and Urban Planning*, 101(4): 321–329.

Csikszentmihalyi, M. (1997) 'Assessing aesthetic education: Measuring the ability to ward off chaos', *Arts Education Policy Review*, 99(1): 33–38.

Dunlop, R. (2002) 'In search of tawny grammar: Poetics, landscape and embodied ways of knowing', *Canadian Journal of Environmental Education*, 7(2): 23–37.

Gablik, S. (1991) *The Reenchantment of Art*, Thames and Hudson, New York City.

Gaiman, N. (2013) *The Ocean at the End of the Lane*, HarperCollins, New York City.

Gobster, P. (1999). 'An ecological aesthetic for forest landscape management', *Landscape Journal*, 18(1): 54–64.

Government of Quebec (2008) 'Fonds Québec de recherche société et culture (FQRSC): Post-doctoral Research/Creation Scholarships', www.fqrsc.gouv.qc.ca/en/bourses/id-21.php (accessed 17 August 2013).

Graham, T. (2013) 'Further inquiry about your Summit Woods research' [e-mail], personal communication, 22 June 2013.

Gray, C. (1998) 'Inquiry through practice: Developing appropriate research strategies', Keynote in *No Guru, No Method? Discussions on Art and Design Research* [conference proceedings], University of Art & Design, UIAH, Helsinki, Finland.

Greenpeace (2012) *Toxic Threads: The Big Fashion Stitch-Up*, www.greenpeace.org/international/Global/international/publications/toxics/Water%202012/ToxicThreads01.pdf (accessed 10 August 2013).

Hall, S. (2004) 'I, Mercator' in K. Harmon (ed) *You Are Here: Personal Geographies and Other Maps of the Imagination*, Princeton Architectural Press, New York City.

Halpern, D. (1993). *Dante's Inferno: Translations by 20 Contemporary Poets*, Ecco Press, New York City.

Haraway, D. (2008) *When Species Meet*, University of Minnesota Press, Minneapolis, MN.

Haraway, D. (2012) 'Cosmopolitical critters: Companion species, SF, and staying with the trouble', John Coffin Memorial Lecture at the Cosmopolitan Animals conference, 26–27 October, London.

Harley, J.B. [Brad] (1988) 'Maps, knowledge, and power,' in D. Cosgrove and S. Daniels (eds) *The Iconography of Landscape: Essays on the Symbolic Representation, Design and Use of Past Environments*, Cambridge University Press, Cambridge, UK.

Harmon, K. (ed) (2004) *You are Here: Personal Geographies and Other Maps of the Imagination*, Princeton Architectural Press, New York City.

Henry, B. (2012) 'Understanding the environmental impacts of wool: A review of life cycle assessment studies', A report prepared for Australian Wool Innovation and International Wool Textile Organization, www.iwto.org/uploaded/publications/Understanding_Wool_LCA2_20120513.pdf (accessed 10 August 2013).

Hickey, D. (1993) *The Invisible Dragon: Four Essays on Beauty*, Art Issues Press, Los Angeles, CA.

Hinchliffe, S., Kearnes, M.B., Degen, M., and Whatmore, S. (2005). 'Urban wild things: A cosmopolitical experiment', *Environment and Planning D: Society and Space*, 23: 643-658.

Jackson, J.B. (1984) *Discovering the Vernacular Landscape*, Yale University Press, New Haven, CT.

Jacobs, M.J. (2007) 'Material with a memory: J. Morgan Puett's *Cottage Industry*', in J. Livingstone and J. Ploof (eds) *The Object of Labour: Art, Cloth and Cultural Production*, School of the Art Institute of Chicago Press and the MIT Press, Chicago, IL, and Cambridge, MA.

Kramberger, A. (2012) 'Guards keep walkers out of Angell Woods', *The Gazette*, 1 February, www2.canada.com/montrealgazette/news/westisland/story.html?id=c0180691-1def-4dd2-8990-13c84d2bb6d6&p=1 (accessed 26 July 2013).

Kwon, M. (2004) *One Place after Another: Site-Specific Art and Locational Identity*, MIT Press, Cambridge, MA.

Langer, S. (1953) *Feeling and Form*, Charles Scribner's Sons, New York City.

Livingston, J. and Ploof, J. (eds) (2007) *The Object of Labour: Art, Cloth, and Cultural Production*, School of the Art Institute of Chicago Press and the MIT Press, Chicago, IL, and Cambridge, MA.

Lober, D.J. (1995) 'Why protest? Public behavioral and attitudinal response to siting a waste disposal facility', *Policy Studies Journal*, 23(3): 499–518.

Lober, D.J. and Green, D.P. (1994) 'NIMBY or NIABY? A logit model of opposition to solid-waste-disposal facility siting', *Journal of Environmental Management*, 40(1): 33–50.

Massey, D. (1991) 'A global sense of place', *Marxism Today*, June: 24–29.

Morgan Arboretum/Arboretum Morgan (2012a) 'Welcome to the Arboretum' (date modified 19 November 2012), www.morganarboretum.org/fma/ (accessed 29 July 2013).

Morgan Arboretum/Arboretum Morgan (2012b) 'Trail map', www.morganarboretum.org/Downloads/arbomap_trails.pdf (accessed 27 July 2013).

O' Rourke, K. (2013) *Walking and Mapping: Artists as Cartographers*, MIT Press, Cambridge, MA.

Parks Canada (2009) 'Lachine Canal national historic site of Canada: geographical maps: boundaries of the site and cultural resources' (date modified 13 July 2009), www.pc.gc.ca/lhn-nhs/qc/canallachine/plan/plan2.aspx (accessed 26 July 2013).

Parks Canada (2010) 'Background to the Lachine Canal Decontamination Project' (date modified 14 July 2010), www.pc.gc.ca/APPS/CP-NR/release_e.asp?bgid=261&andor1=bg (accessed 29 July 2013).

Parks Canada (2013) 'Lachine Canal National Historic Site: Natural Wonders & Cultural Treasures' (date modified 29 January 2013), www.pc.gc.ca/eng/lhn-nhs/qc/canallachine/natcul.aspx (accessed 29 July 2013).

Perkins, H.A., Heynen, N., and Wilson, J. (2004) 'Inequitable access to urban reforestation: The impact of urban political economy on housing tenure and urban forests', *Cities*, 21(4): 291–299.

Rocha, R. (2011) 'Mapping green spaces in Montreal', *Montreal Gazette Blog*, September 26, http://blogs.montrealgazette.com/2011/09/26/mapping-green-spaces-in-montreal/ (accessed 12 April 2013).

Scott, D. (1999) *Refashioning Futures: Criticism after Postcoloniality*, Princeton University Press, Princeton, NJ.

Solomon-Godeau, A. (1990). 'Living with contradictions: Critical practices in the age of supply-side aesthetics', in C. Squiers (ed) *The Critical Image: Essays on Contemporary Photography*, Bay Press, Seattle, WA.

SSHRC (Social Sciences and Humanities Research Council) (2013) 'Definitions of terms' (date modified 6 June 2013), www.sshrc-crsh.gc.ca/funding-financement/programs-programmes/definitions-eng.aspx#a22 (accessed 17 August 2013).

Sweeney, L. (2013) 'Dogs at Summit Woods capped at four: Max was almost three', *Westmount Independent*, 9 July, p. 8.

Tobler-Rohr, M. (2011) *Handbook of Sustainable Textile Production*, Textile Institute and Woodhead Publishing, Manchester and Cambridge, UK.

Vaughan, K. (2005) 'Pieced together: Collage as an artist's method of interdisciplinary research', *International Journal of Qualitative Methods*, vol 4, no 1, http://ejournals.library.ualberta.ca/index.php/IJQM/article/view/4452/3557 (accessed 18 August 2013).

Vaughan, K. (2007) 'Finding home: Knowledge, collage and the local environments', PhD thesis, York University, Toronto, ON.

Vaughan, K. (2009) 'Collage as form, collage as method: An approach to interdisciplinary qualitative research using visual art practices and theory', Paper presented at the annual conference of the American Educational Research Association, April 13–17, San Diego, CA.

Vetlesen, A.J. (1993) 'Why does proximity make a moral difference? Coming to terms with a lesson learned from the Holocaust', *Praxis International*, 12(4): 371–386.

Weiner, A.B. and Schneider, J. (eds) (1989) *Cloth and Human Experience*, Smithsonian Institution Press, Washington, DC.

Whatmore, S. and Hinchcliffe, S. (2010) 'Ecological landscapes', in D. Hicks and M.C. Beaudry (eds) *The Oxford Handbook of Material Culture Studies*, Oxford University Press, Oxford, UK.

Index

Abrams, David 318n2
acacia 270–2
ACC *see* African Centre for Cities
activism: Berlin 213; Cape Town 270, 272; Gezi Park protests 11–12, 227–8, 231–9; students 285
Actor Network Theory (ANT) 5, 9–10, 133, 143, 213
actors 37, 38–40, 44
aesthetics 21, 307
affect 9, 112, 115, 116–22, 125, 126–7
affordances 125
afforestation: Flanders 51–2, 57, 59; MillionTreesNYC Campaign 253; population growth 103; Tokyo 151, 153–4, 156; Wallonia 48; *see also* planting of trees; reforestation
African Centre for Cities (ACC) 261, 266, 267–8
agency 6, 207–8, 210; nature 118; places 120; trees 9–11, 112, 113, 115, 122, 126, 143
agriculture: Bräkne River Valley 297, 298, 303; Flanders 57; forest transition theory 94; Tokyo 154, 156–7
agroforestry 98, 101
Aird, Paul 164
Akbulut, Bengi 11–12, 13, 227–41
Alabama 67, 68, 69, 73
Allee, Warder C. 199
allelopathy 194, 199, 204n6
Almey crab apple 166, 167–9
American elm 25, 164–5
Amin, A. 118, 122
Amsterdam 165
Anderson, B. 117, 127
Anderson, Pippin 12, 261–76
Angell Woods 309–10, 313, 317
ANT *see* Actor Network Theory
arboricultures 10, 147–61
Aronczyk, M. 89
art 13, 125–6, 220, 307–21; affective realm 119; 'Ghetto Palm' 11, 192, 201–2, 203; Tree of Heaven Woodshop 199–200, 203
Arts, B. 35
Atlanta 29

Baltimore 24, 25, 29
Banham, R. 148
Bardekjian, Adrina 1–16
Barnett, C. 118
Barthes, Roland 149, 151
Bass, J.O. Joby 8–9, 93–107
beautification 27, 77, 82, 164, 169
beauty 307, 313, 314
Becher, Drew 249
'becoming with' 308–9, 315
Belgium 7–8, 47–60
Benepe, Adrian 246, 249
Bennett, Jane 314
Berkeley 155
Berlin 1, 11, 207–23
biodiversity 3, 20, 43, 127, 164; Berlin 211, 212–13; Bräkne River Valley 293, 297, 299; Cape Town 263–4, 267, 270, 271, 273; habitats 49; Honduras 101–2; poverty alleviation 267; Singapore 87, 88, 89; threats to 194, 195, 201, 211; urban 114
biotopes 212, 213, 220
Birkenholtz, T. 5
Bloomberg, Michael 12, 242, 245, 246, 247, 249, 250
Blume, Hans-Peter 210
Boboli Gardens, Florence 39
bodies 116, 117, 120, 124–5
Bolivia 3
Bolthouse, Jay 10, 147–61
Bolton Primary School 288, 289

bonsai 151–2
Boones Ferry Primary School 286–8
Boston 134, 138, 139, 141, 282–5
Boucher, D. 199
Boudreau, Julie-Anne 243, 255
Bradley, Jyll 125
Bräkne River Valley (BRV) 13, 293, 296–304
Brand, P. 244
Braverman, Irus 9–10, 116, 128, 132–46, 171–2
Brazil 3
Brinkmann, F. 207, 209–10, 212, 216, 220
Bristol 111–12, 120, 126
Brookline 136, 139, 140, 142
Brown, Bill 134
BRV *see* Bräkne River Valley
Buijs, A.E. 184–5
Bürgerinitiativen 212
Butt, Sadia 1–16

Callon, M. 139
Campanella, Thomas 164–5
Campbell, Lindsay K. 12, 13, 242–60
Canada: Central Experimental Farm 162; deforestation 69; federal government 8, 66–7, 70–2, 73, 74; 'leapfrog development' 75n2; legal institutions 61; Norway maple 10, 176–90; private-property owners 62; 'research-creation' 316–17, 318n6; textile maps 309–15, 317; tree narratives 162–3, 164, 166–71; urban forestry in 4–5; urban sprawl 64
Canadian embassy, Berlin 11, 207, 208, 215–21
canals 152, 310
Canberra 125
Cape Town 1, 12, 261–76
capitalism 1, 8, 22, 80; neoliberalism 7; Singapore 82, 88; Turkey 228, 236
carbon emissions 3, 66, 69
carbon sequestration 3, 19, 31, 68, 74, 94, 103, 163, 185
carbon sinks 3, 66, 68, 166
Carter, Paul 307, 309
Cashore, B. 71, 73
Central Park, New York City 1, 40, 41, 135
Certeau, Michel de 318n3
Chambers, D. 120
Chang, T.C. 87
citizen participation 4, 244, 285–6; Berlin 212; governance 39–40; Singapore 84, 86–7; Tokyo 157–8; volunteerism 29–30
citizenship 181–2, 288
'city gardens' 297
City in a Garden brand 84–8, 89
CityLab programme, Cape Town 12, 266–73
civil society 243, 250–1, 254, 255; *see also* citizen participation
class 7, 20, 22, 31, 116; Cape Town 270, 271; Lachine Canal 312; tree haters 171; United States 25–7
climate change 3, 20, 31, 58, 163, 256; impact on sugar maple 187; legal constraints 66; political ecology 295; *see also* carbon sequestration
Cloke, Paul 40, 148, 195
'close-to-nature forestry' 49–50
co-management 43
coalitions 37–8, 40
Coates, Peter 181
Cohen, S. 272
colonialism 195, 201; Africa 261–2; Latin America 95–6; Singapore 81
commodification 22, 84, 85, 89
common pool resources (CPR) 293
commons 38, 40, 157; Gezi Park 235–7; 'new commons' 43, 44
communalism 235–7
community-based education 277–91
community forestry 243, 255
community gardening 40, 86
conservation: Bräkne River Valley 296; Cape Town 263–6, 267, 270, 272, 273; Flanders 49; Honduras 93, 94, 97, 101–2; Singapore 83, 84; Tokyo 155; western NGOs 93–4
consumption 231, 232
Cope, Mitchell 199
Coppice Club 157–8
Corbyn, Z. 267
Corlett, R.T. 87
Cosgrove, D. 95
Coventry 125–6
CPR *see* common pool resources
crab apple 162, 166, 167–9, 171, 172
crime 264, 279, 281
Cronon, William 185, 214, 220, 297
Cronquist, A. 182
Crosby, A.W. 261, 262
cultural diversity 89
cultural ecosystem services 270, 271

cultural landscapes 54, 56–7, 103
culture 20, 111, 119; Honduras 93; nature-culture divide 114, 118, 127, 128, 186, 212, 214; queer theory 192

Damasio, A. 119
Darwin, Charles 119
David, Joshua 197
Davies, R.J. 262
Dean, Joanna 10, 162–75
decentralization 62, 64, 74
decolonization 278, 285
deer 14n1
deforestation 8, 69, 103; assumptions about 93; Flanders 48; Honduras 97; impact on climate change 66
Deleuze, Gilles 194
democracy 236, 238, 293, 296
Denen Chofu 154–5
Denmark 39–40, 294
Department of Parks and Recreation (DPR), New York City 242, 245–6, 248–50, 252, 253, 255
Detroit: bankruptcy 29; inequalities 25, 26; Tree of Heaven Woodshop 10–11, 191–2, 199–200, 203
developing countries 2–3, 8
Dewey, John 300
Dewsbury, J.-D. 121–2
Dig-Safe procedure 132–3, 138, 140–2, 143
Dimopoulos, Konstantin 126
direct democracy 236, 238
disability 134, 312
discourses 7–8, 37–8, 42–3, 244; city-centred 48, 53; Flanders 47–8, 52–3, 58–9; forest-centred 47, 53, 58; Honduras 97, 101, 102; people-centred 58; perception shaped by 102, 103–4; 'tree-centric' 104
diseases 21, 186; *see also* Dutch elm disease
distance rules 134
distributional justice 250, 255, 256
diversity 89, 183–4
Douglas fir 172
DPR *see* Department of Parks and Recreation
Dunlop, Rishma 317
Dutch elm disease (DED) 21, 25, 26, 39, 167, 180

Earth Summit (Rio, 1992) 83
ecofeminism 116–17
ecological modernization 79
economic development: North America 69, 73; pathways of 101; Singapore 79, 81, 88; Turkey 228–9, 230
economic value 3, 256; Honduras 102; i-Tree software program 163; neoliberal discourse 42–3; New York City 246; Singapore 88
ecosystem services 3, 19, 44, 68–9, 73, 264; Cape Town 262, 269–70, 271; Gezi Park 232; narrative of service 162–3, 165–6; neoliberal discourse 42–3; Norway maple 185
education 12–13, 277–91; Honduras 102; Landscape Ambassador course 299–304; MillionTreesNYC Campaign 252–3
ELC *see* European Landscape Convention
elites 8, 38, 40, 43
elm 21, 25, 162–3, 164–5, 167
embodiment 121
emotions 9, 112, 115, 117, 120, 122, 125
employment 271
enchantment 314
energy sector 230
engineers 139, 143
environmental justice 20, 31, 115, 242, 250, 256
environmentalism: North America 8; personal responsibility 244; 'roll out' 243; Singapore 83, 84; Turkey 237
epistemic communities 4
Epping Forest 157
ethics 195, 314
ethnicity 20, 22–5, 30, 31, 165–6, 172, 184; *see also* race
Europe 2, 36, 38, 47, 292
European Forum on Urban Forestry 4
European Landscape Convention (ELC) 292, 294
exclusion 47, 50, 53, 59
exotics 2, 10, 184; *see also* invasive species

federal government 8, 48, 61–7, 70–4
feminist groups 237
fence-line forest 24–5
Fernow, Bernhard 4, 163, 172n2
field trips 281
Finland 40, 43
fir 172
fires 263–4, 271
firewood 99, 271
Fisher, D.R. 255
Flanders 7–8, 47–60

flooding 264
Florence 39
Flyvbjerg, B. 295
Ford, Richard 111
forest transition theory 94, 101, 103
foresters 2, 7, 48–50, 51, 53, 139, 142, 243
forestry 1, 243; 'close-to-nature' 49–50; gender and 27–8; Honduras 101; multi-functional 48, 50–1, 53; struggles for political attention in Flanders 48–9; *see also* trees; urban forestry
Forgan, Anne 125–6
Foucault, Michel 47, 59, 133, 203
Frank, Anne 165, 166
Freeman, Alan 216
Freund, Alexander 209
fruit trees 100
funding: Canadian art funding 318n6; Flanders 58; MillionTreesNYC Campaign 248, 253, 254–5; quantification of benefits 165; Tokyo 156
Fynbos 12, 263, 265, 270

Gablik, Suzi 313
Gaiman, Neil 307, 308
Galloway, Matt 183, 185
Gandy, Matthew 192–3
Garden City campaign, Singapore 8, 77, 79–80, 81–3, 89
garden city concept 154, 209
Gardens by the Bay, Singapore 1, 77–8, 84, 85, 87–8, 89
gender 20, 22, 27–8, 31, 192, 209
gentrification: High Line redevelopment 196–7, 198; Lachine Canal 312; queering 202; Turkey 229, 230, 231
Geographic Information Systems 3
Germany 207, 209, 212, 220
Gezi Park, Istanbul 11–12, 227–9, 231–9
'Ghetto Palm' 11, 192, 201–2, 203
Ghomeshi, Jian 183, 185
Gleason, H.A. 182
Gleisdreieck 212, 213, 220
Goebbels, Paul Joseph 216
Goh Chok Tong 83
governance 4, 7, 35–46, 243–4, 256; definition of 35; landscape 293, 294; legal institutions 61; MillionTreesNYC Campaign 242, 250–2; networked 40, 242, 243, 250–2, 255; policy arrangement approach 37–8; regulatory jurisdiction 65–6; of trees 132, 136–8, 143; underground 138–42; *see also* law; policy; regulation
government 7, 35, 44; Bräkne River Valley 304; Cape Town 264; federal 61–7, 70–4
governmentality 133, 143
Grady, Wayne 172
grate, the 132–3, 136–8, 143
Gray, C. 316
Greater Egleston Community High School 282–5
green city visioning 77, 78, 79–80, 81–9
Green Point Urban Park, Cape Town 268, 269
green roofs 217, 218, 219, 220
'green wedge' 207, 209–10, 212, 213, 216, 220
Greene, Maxine 313, 318n4
Greenhough, B. 120
greening: Greening Milwaukee 29, 30, 243; Singapore 8, 77, 78, 81–9; Tokyo 149, 151, 152–3
grid, the 132–3, 135–6, 143
Grosz, E. 220, 221
growth boundaries 73, 75n2
Gruenewald, David 278
Guattari, Félix 127, 194
Gulsrud, Natalie Marie 8, 77–92
Gurin, D. 69
Gustavsson, Roland 13, 292–306

Haas, E. 4
habitats 49, 84, 87, 212
Hahn Oberlander, Cornelia 11, 207, 208, 217–18, 220
Hajer, M. 37
Häkli, Jouni 207, 212–14, 218–20
half-second delay 118
Hall, Stephen S. 311
Hammond, Cynthia Imogen 11, 207–23
Hammond, Robert 197
Haraway, Donna 308, 314–15, 317
Harley, Brad 310
Harris, Patricia 247, 249
Harrison, F. 122–3
Harrison, Robert Pogue 114, 125
Harvey, David 21, 79
Haussmann, Georges-Eugène 135, 163–4
health issues 283–5
heat island effect 3, 10, 19, 31, 147, 151, 153, 253
heatwaves 31
hedonic pricing 43

Heidegger, M. 300
hemerochores 211, 212
Hemmings, C. 121
HEPPs *see* hydropower plants
heritage, narrative of 164–5
Heynen, Nik 12, 24, 80, 147, 163, 172, 243, 244, 254, 282
Hickey, Dave 314
High Line redevelopment, New York City 196–8, 203
Hinchcliffe, Steve 309, 317, 318n2, 318n3
Hoggett, P. 118
Holm, A. 214
homeownership 26
Honduras 8–9, 93–107
horse chestnut 165, 166
housing: Bräkne River Valley 297, 303; Cape Town 262–3; class inequalities 26; High Line redevelopment 198; household tree cover in Honduras 93, 95, 97–100, 103; racial inequalities 23; Singapore 83; trees sacrificed for 1
Howard, Ebenezer 154, 209, 220
Huang, S. 87
Hudson, Blake 8, 61–76
Hughes-Gibb, E. 124
Husserl, Edmund 300
Huyssen, Andreas 215, 216
hydro-electric plants (HEPPs) 230

i-Tree software program 3, 6, 43, 44, 163, 165, 173n4, 244
identity 117, 122, 125; German 209, 212; locational 308–9; professional 47, 49–50; Singapore 78–9, 80, 83, 89
image-making 49, 53–6, 79
immigrants 171, 183, 184–5, 232
Imperial Palace, Tokyo 149, 150, 151
India 3
indigenous peoples 180, 185, 262
inequalities 7, 31–2, 165–6, 228, 244, 282; Africa 266; Cape Town 273; class 25–7; gender 27–8; MillionTreesNYC Campaign 250; race 22–5
infrastructure 256, 293
Ingold, T. 114, 120
Ingram, Gordon Brent 192
insects 14n1
Institute of Ecology, Berlin 210–12, 221
institutions 59, 61–2, 65, 70–3, 74, 290
integration 36
interdisciplinary work 4, 272–3, 299
interests 53, 93
invasive species 184, 185, 187n3, 195, 201; Cape Town 263, 269, 270–2; Norway maple 10, 180–1; Tree of Heaven 10–11, 191
Istanbul 11–12, 227–41
Ivy, M. 155

Jackson, J.B. 318n3
Jacobs, Mary Jane 316
Jones, Owain 9, 11, 40, 111–31, 148, 195
Jorgenson, Erik 164, 173n5

Keil, Roger 243, 255
Kentucky bluegrass 181
Killerman, Sam 204n7
Kjær, A.M. 35
Konijnendijk van den Bosch, Cecil C. 7, 35–46, 114–15, 128, 248, 255
Kowarik, I. 191
Kropotkin, Peter 199
Kuhn, A. 214
Kuhns, M. 28
Kunick, Wolfram 210
Kunikida Doppo 154
Kwon, Miwon 309

Lachine Canal 310, 312, 313
Lachmund, Jens 210, 211–12, 213, 221n2
Lamarque, L.J. 181
LAMB *see* Landscape Ambassador course
land reclamation 153
land-use planning: Cape Town 268; Flanders 52, 57; Singapore 80; United States 65, 70; *see also* planning
Landscape Ambassador (LAMB) course 299–304
landscapes 54–8, 93–4, 103, 104, 120, 125; affective processes 122; cultural 103; landscape planning 13, 292–306; neoliberal discourse 42–3; ruderal 207, 211, 212, 213, 220
Langer, Suzanne 313
Larson, Brendon M.H. 10, 176–90
Latin America 94, 95–6
Latour, Bruno 5, 118, 133, 138, 139, 140, 143, 295, 296
Laurie, Ian 214
law 61, 62, 70–1, 133, 134; legal changes in Turkey 229; legal constraints 66–7; legal norms 142; legislation 38–9, 229, 230; *see also* policy; regulation
Law, John 5
LEADER Program 294, 295

LEAF *see* Local Enhancement & Appreciation of Forests
'leapfrog development' 75n2
Lee Hsien Loong 84
Lee Kuan Yew 77, 81, 82, 83
Leipziger Platz 207, 214–16, 220
Lekan, Thomas 210
Lenné, Josef 216
Leroy, P. 35
Ley, David 113
livelihoods 13, 271–2, 296
Livingstone, Joan 316
lobbying 54
Local Agenda 21 42
Local Enhancement & Appreciation of Forests (LEAF) 183
local government 39, 62, 65, 74, 254, 264
local residents 242, 265; Flanders 53, 56–7, 59; involvement in governance 35, 39–40, 42; landscape planning 292, 295–6, 297, 299, 303, 304; opposition to trees 25, 171–2; power relations 42; researcher engagement with 295; *see also* stakeholders
Lombardy poplar 162, 166, 169–71, 172
London 114, 126, 157, 192–3
Los Angeles 148
Luttik, J. 116

Mackenzie River 217–18
Macnaghten, P. 125
Manitoba maple 162, 163, 166–7, 171, 172, 177
maple: Manitoba maple 162, 163, 166–7, 171, 172, 177; Norway maple 10, 163, 167, 176–90; persistent narratives 162–3; sugar maple 163, 164, 176–80, 182–3, 185, 186–7
maps, textile 13, 308, 309–15, 317
Marder, Michael 172, 194
marketization 37, 42, 89
markets 20, 22, 31–2
Massey, Doreen 309
materiality 120–1, 122, 137, 166
Matheson, J.R. 178–9, 182
Matthews, William Ezra 169, 173n13, 173n15
McClintock, N. 290
McDermott, C.L. 71, 73
McKibbon, Mabel 169–71, 172, 173n16, 173n17
Meiji Shrine, Tokyo 1, 147, 149–50, 153–4
Mellqvist, Helena 13, 292–306
memory 111, 117, 119
meshwork 120
Midler, Bette 12, 242, 246–7, 249, 250
MillionTreesNYC Campaign 12, 242–60
Millward, A.A. 185
Milwaukee: Greening Milwaukee 29, 30, 243; inequalities 23–7; place-based education 279–82; weed trees 172
minority groups 22–5, 30, 31, 171, 229–30; *see also* ethnicity; race
modernism 117, 164
modernization 35, 79, 157, 228–9, 230, 239n4
Montpellier, Paul 134, 171
Montreal 13, 179, 187, 309–15
Morgan Arboretum 309, 310, 312
multi-functional forestry 48, 50–1, 53
multiculturalism 183, 184
Munich 123, 124
Musashino 154–6
music 119
mutualism 192, 193, 199–200
My Neighbor Totoro (film) 155–6

Nakashima, K. 149
narratives 162–75; counter-narratives 10, 165–6; of exchange 112; persistent 10, 162–5; unruly tree 10, 162, 166–72
national symbolism 176–7, 182, 186
native species 2, 103, 195; Canada 10, 177, 178, 179–81; Cape Town 268, 272; Singapore 81–2, 83–4
'naturalization' 203
nature 9, 49, 80, 133; Canada 180; cultural symbols 221; discursive construction of 244; first, second and third 214; fourth 218–20; German love for 209; governance of 132; nature-culture divide 114, 118, 127, 128, 186, 212, 214; perceived as rural endeavour 264; queer theory 192; understandings of 49, 244
nature reserves 296, 298, 304; *see also* protected areas
Nature Society of Singapore 77–8, 84
Neighborhood Environmental Education Project (NEEP) 281–2
Neighborwoods program 29–30
Nel Mezzo maps 308, 309–15, 317
neoliberalism 6–7, 8, 11–12, 147–8, 256; discourse of 42–3; economic insecurity under 286; governance 35–6, 37, 40, 44, 88, 243; Honduras 102; i-Tree software program 43;

MillionTreesNYC Campaign 242; place-based education 289, 290; restructuring of space 237; southeastern US states 72, 73; state-civil society relations 254; Turkey 228–9; volunteerism as a result of 243–4
networked governance 40, 242, 243, 250–2, 255
networks, biotope 212
New Haven's Urban Resources Initiative 255
New Orleans 23
New York City: carbon sinks 166; Central Park 1, 40, 41, 135; grid design 135; MillionTreesNYC Campaign 12, 242–60; opposition to trees 25; Tree of Heaven 10, 191, 196–8, 203
New York Restoration Project (NYRP) 242, 246–7, 248–50, 252, 254, 255
Ng, P.K.L. 87
non-governmental organizations (NGOs) 58, 93–4; Honduras 102; i-Tree data 43; Taksim Solidarity 232, 233; Tokyo 153
nonprofit organizations 29, 30, 255
non-representational theory 116, 122
North America 2, 36, 39; Dig-Safe procedure 140; narrative of service 162–3; street trees 164; Tree of Heaven 191; volunteerism 40; *see also* Canada; United States
Norway maple 10, 163, 167, 176–90
Nowak, D.J. 180
NYRP *see* New York Restoration Project

O'Farrell, P.J. 268–9
Olmstead, Frederick 279
Olwig, K.R. 38
Onder, Sirri Sureyya 233
Ontario 164, 179–80, 181, 183, 184, 186, 187
Ooi, Can-Seng 8, 77–92
oppression 278, 285, 286
Oregon 73, 75n2, 285–9
ornamental crab apple 162, 167–9, 171, 172
Ostrom, Elinor 43, 293
'other space' 125
Ottawa 162, 163, 164, 166–71, 179, 217
over-decentralization 62, 74
oxygen 163

Paris 135, 163–4
Park Forest Ghent 55–8, 59
Parnell, S. 266
Patrick, Darren 10–11, 191–206
Pearson, Lester 178–9
Peckham, S.C. 115
Pels, P. 139
peri-urban forests 8, 9, 52, 58
Periscape group 299, 300, 303
Perkins, Harold 7, 12, 19–34, 37, 40, 147, 243
Perlman, M. 127
Philippines 3
photosynthesis 21
pipes 139, 140–2, 143
place-based education 277–91, 303
place identity-making 78, 79, 80, 83
places 120–2
planning 13, 245, 292–306; Cape Town 268, 269; Flanders 7, 51–2, 53, 54–5, 56, 57; green 114; 'green wedge' in Berlin 207, 209–10; Tokyo 152
plantation forestry 270
planting of trees 19, 21, 22, 26, 127; Cape Town 272; Flanders 58–9; MillionTreesNYC Campaign 242, 245–56; Ottawa 168; risk mitigation 94; Singapore 81, 82, 83–4; Tokyo 153; volunteer stewardship 244; *see also* afforestation; reforestation
plazas 8, 93, 95–7, 100
Ploof, John 316
Plumwood, Val 116–17, 195
pluralism 4
policy 2, 5, 7; European Landscape Convention 294; federal government 8, 61–2, 64–5, 66–7, 70–1, 74; Flanders 48; Honduras 94, 97, 101; landscape planning 292; perception shaped by 102, 103; political modernization 35, 37; Singapore 79–80; *see also* law; planning; regulation
policy arrangement approach 37–8
political ecology 2, 5–6, 13, 21, 93, 115, 147–8, 242–4; arboriculture 159; Cape Town 268, 273; Gezi Park protests 228, 238; governance 35–6, 44, 243–4; institutional and legal context 62, 70–3, 74; landscape planning 295–6; narrative of service 163; Norway maple 178; place-based education 278; queer 193; risk 94; Singapore 78, 79–80
political modernization 35
politics: affective 116, 118; Bräkne River Valley 303–4; Flanders 48–9, 54–5; Gezi Park protests 238, 239; 'green wedge' in Berlin 207; institutional and

legal context 70–3; MillionTreesNYC Campaign 242; neoliberal agenda in Turkey 228–30; Singapore 79, 80–1; Tree of Heaven 194; trees and 116
Pollan, Michael 181
pollution 279, 283–5
poplar 162, 166, 169–71, 172
population growth 103, 266; *see also* urbanization
Portland, Oregon 285–9
Potsdamer Platz 211, 214–16, 218
potted plants 151–2
poverty 266, 267
power 1, 38–9; affective turn 116; governance 4, 36, 37–8, 42; narrative of 163–4, 166; political ecology 2; urbanization 80
Powers, D. 89
practice-based research 316–17
preservation 65, 73
private ownership 8, 22, 44, 62, 71–3, 74
private sector 86, 229
privatization 229, 231
professional forest management 47, 49–50, 101
professional organizations 4–5
profit accumulation 88
property 22
protected areas 102, 213, 264; *see also* nature reserves
psychogeography 125
public access 50, 311–12
public ownership 8, 22, 71, 72
public-private partnerships 247–50, 255
public spaces 111–12; activism in Turkey 227–8, 231–9; Cape Town 270; Honduras 93, 95–7, 104; regulation of trees in 134–5

quantification 43, 165–6, 244, 256; *see also* economic value
queering 11, 192–5, 202, 203

race 7, 20, 22–5, 30, 31, 116; apartheid in South Africa 262–3, 268–9; Singapore 88
racism 184
Rademacher, Anne 198
rationality 118
'real world studios' 300
recalcitrance 140
recreation 3, 69; Bräkne River Valley 298; Cape Town 270; Flanders 51, 54, 56; Singapore 77, 83
recycling 153
redevelopment projects: Gezi Park 227, 231–9; High Line 196–8, 203; Tokyo 153; Turkish policy 229–30
reforestation 3, 20, 63; class inequalities 26; Honduras 8–9, 102; MillionTreesNYC Campaign 253; Singapore 83–4; *see also* afforestation; planting of trees
regulation: European Landscape Convention 294; the grate 138; regulatory jurisdiction 65, 70; southeastern US states 8, 71–2; trees and humans 134–5; underground governance 138–42; *see also* law; policy
reinhabitation 278, 285
relationality 118, 121
'research-creation' 316–17, 318n6
Respighi, Ottorino 119
Richardson, Miles 95
risk 94, 296
roads 134–5
Rocheleau, D. 1, 5
Roe, E. 120
Ronneby 293, 295, 301, 302, 304
roots 9, 133, 138–42, 166
Rosemont Ridge Middle School 288–9
Roth, J.H. 151
Roth, R. 1, 5
Rowntree, R.A. 180
ruderal ecologies 192, 204n5
ruderal landscapes 207, 211, 212, 213, 220
rules of the game 37–8, 40, 248
Rural School and Community Trust 277–8
rural-to-urban migration 1, 20, 95, 101, 103

Sabir, S. 185
San Francisco 148, 184
Sandberg, L. Anders 1–16
satoyama (village-forest) 155–6
Säumel, I. 191
Saunders, William 4, 162–3, 164, 166, 167, 172n2
Schmaling, Sebastian 214–15
Science and Technology Studies (STS) 133
scientists 139
Scott, David 317
Scott, J. 244
Sea Forest, Tokyo 153–4
self 116, 117, 122
self-regulation 72
Selman, P. 292, 293

Sennett, Richard 135
service, narrative of 162–3, 165–6
settlement forests 9, 99–100, 103
sexuality 192, 203
shade 2, 3, 28, 31, 119; Cape Town 270; narrative of service 162–3; Norway maple 180
Shimousa Plain 156–7
silver maple 163, 164, 186
Singapore 1, 8, 77–92
Sky Tree, Tokyo 147
Slaby, J. 111, 119, 122
slums 88
smallholders 94, 100, 101, 103
Smith, Gregory 12–13, 277–91
Smith, Neil 115, 196
Sobel, David 277
social change 290
social constructionism 118
social forestry 50
social inclusion 36
social movements 236–7, 239
'social nature' 21
Solomon-Godeau, Abigail 313
Soros, George 247
South Africa 261–76
spatial development 52
spatial form 268
spatial order 132, 134–5, 163–4, 166
Spatial Structure Plan 52–5
spatial technologies 132–3
Speer, Albert 216
stakeholders: governance 4, 36, 37, 42; MillionTreesNYC Campaign 250–2; researcher engagement with 295, 301, 303; *see also* local residents
Stephens, John B. 96–7
stewardship 242–4, 250–3, 255, 288, 317
Stewart, K. 121
Stimmann, Hans 214
Stockholm 40, 41, 42
storyline 47–8, 51–3, 58–9
STRATUM model 244, 246
street trees: grates 136–8, 143; grid design 135–6, 143; MillionTreesNYC Campaign 12, 246, 247, 248, 250, 253; Montreal 314; narrative of power 164, 166; Norway maple 180; problems with 166–7; regulation of 134–5; Tokyo 151, 152; underground governance 138–42, 143
STS *see* Science and Technology Studies
subconscious 118–19
subnational government 61–2, 64–7, 70–3
Südgelände 212–13, 220
sugar maple 163, 164, 176–80, 182–3, 185, 186–7
Sukopp, Herbert 192, 207, 210–12
Summit Woods 309, 310, 311, 312, 318n1
sustainability 51, 79, 114; MillionTreesNYC Campaign 242, 245; place-based education 286; Singapore 87, 88, 89
Suzuki, David 172
Sweden 13, 40, 293, 294, 296–304
Swyngedouw, E. 40, 42, 80, 254

Table Mountain National Park, Cape Town 1, 264, 270
Taksim 227–8, 231–9
Tampa 23, 26
technology 3–4
textile maps 13, 308, 309–15, 317
textile production 316
Thing theory 133
thingness 134, 138, 143
Thrift, N. 116, 117, 118, 120, 122
Tiergarten, Berlin 1
Tiganas, Vadim 192, 201, 203
Tiron, Stefan 192, 201, 203
Tokyo 1, 10, 147–61
Toronto 140, 142, 179, 181, 183, 185
tourism 292; agro-tourism 293; Bräkne River Valley 293, 296, 297; Singapore 8, 84, 85, 87; Taksim Square 232
'translation' 139
transport, pollution from 283–5
Tree of Heaven 10–11, 112, 191–2, 194, 195–203, 211
Tree US Aid 3, 4
trees 6, 111–31; affective processes 9, 112, 115, 116–22, 126–7; awareness of the benefits of 19; beauty and enchantment of 13, 307–8, 314; biological processes 19–21; Cape Town 263, 270–2; class inequalities 25–7; control over 9–10, 132–46, 163–4, 166; counter-narratives 165–6; heatwave mitigation 31; Honduras 93–107; local residents' opposition to 25, 171–2; persistent narratives 10, 162–5; as property 22; queering the urban forest 193–5; racial inequalities 22–5; 'social nature' 21; Tokyo arboricultures 10, 147–61; 'tree-centric' discourse 49, 104; unruly tree narrative 10, 162, 166–72; *see also* forestry; maple; street trees; urban forestry

Tropical City of Excellence campaign, Singapore 83–4
Turkey 11–12, 227–41

United Kingdom 40, 316
United Nations (UN) 128
United States: American elm 25, 164–5; class inequalities 25–7; critical view of trees 116; ecosystem services 43; federal government 8, 65, 66–7, 70–4; forest loss 69, 73; i-Tree software program 3, 163, 173n4, 244; legal institutions 61; private-property owners 62; professional forest management 101; racial inequalities 7, 22–5; sugar maple 179; Tree US Aid 3, 4; urban sprawl 64; voluntarism and volunteerism 29–30; women in forestry 27–8
unruly tree narrative 10, 162, 166–72
Urban Ecology Center, Milwaukee 279–82
urban forestry 114–15; birth of 39; definitions of an urban forest 52, 54, 58, 63, 318n1; dominant themes 2–5; gender and 27–8; governance 36–7, 43–4, 243–4; grid design 135; hybrid models 40, 248, 255; institutional and legal context 61; MillionTreesNYC Campaign 242, 248; political ecology 5–6; power 42; quantification of benefits 43, 244; as 'regenerative tool' 128; *see also* forestry; trees
urban sprawl 64, 66, 71, 73
urbanization 64, 73, 80; Africa 266; Canada 69; Cape Town 263; Flanders 48, 53; forest transition theory 94; Honduras 95; landscape planning 292; Latin America 94; Singapore 81; smallholders influenced by 100; Tokyo 155
Urry, J. 125

value of trees 103, 185, 252–3; *see also* economic value
values 36, 68–9, 73, 74, 244; Honduras 102; landscapes 94; planning 303; social 115
Van Herzele, Ann 7–8, 47–60
Van Tatenhove, J.P.M. 7, 35
Vancouver 113, 134, 135–6, 140, 171, 192, 217, 218
Vannini, P. 120–1
Vaughan, Kathleen 13, 307–21
vernacular ecology 311, 318n3
Versteeg, W. 37
Vetter, Ingo 199
Victoria Square, Bristol 111–12, 114, 125, 126
visibility 142–3
voluntarism 29, 30
volunteerism 7, 29–30, 148, 243–4; Denmark 39–40; environmental mitigation projects 290; MillionTreesNYC Campaign 253–4, 255, 257n5; Milwaukee Urban Ecology Center 281–2; Tokyo 10, 153, 155, 157–8, 159; *see also* citizen participation
Vygotsky, L.S. 300

Waley, P. 151
Walker, R. 148, 155
Wallonia 48
Washington 73
wastelands 210–11, 212, 213
water 3, 218, 230, 264, 271, 288
Waugh, Richard 166–7
Weaver, C. 177
weeds 194, 210, 212
Wegner, Phillip E. 156
Weisser, Annette 199
well-being 3, 6, 13, 115, 128, 266–7
Wertheim, Georg 216
West Linn 288–9
West, R.C. 96–7
Westoby, Jack 1
Whatmore, Sarah 195, 309, 318n3
'wicked problems' 37
wilderness 178, 184–5, 214; Berlin 213; Bräkne River Valley 297; Cape Town 264
wildlife 3, 14n1, 213, 262, 268; *see also* habitats
Wilsonville 286–8
Wittgenstein, Ludwig 300
Wolfe, Cary 172
women 27–8, 209
woodswalking 13, 307–9, 310, 311–13, 314–15, 317
Wright, Erik Olin 290

Yamanote 149–51

CPSIA information can be obtained
at www.ICGtesting.com
Printed in the USA
LVHW05s2241300818
588576LV00003B/3/P

9 781138 282575